Environmental Issues and Waste Management Technologies in the Ceramic and Nuclear Industries X

T0329150

Technical Resources

Journal of the American Ceramic Society

www.ceramicjournal.org

With the highest impact factor of any ceramics-specific journal, the *Journal of the American Ceramic Society* is the world's leading source of published research in ceramics and related materials sciences.

Contents include ceramic processing science; electric and dielectric properties; mechanical, thermal and chemical properties; microstructure and phase equilibria; and much more.

Journal of the American Ceramic Society is abstracted/indexed in Chemical Abstracts, Ceramic Abstracts, Cambridge Scientific, ISI's Web of Science, Science Citation Index, Chemistry Citation Index, Materials Science Citation Index, Reaction Citation Index, Current Contents/ Physical, Chemical and Earth Sciences, Current Contents/Engineering, Computing and Technology, plus more.

View abstracts of all content from 1997 through the current issue at no charge at www.ceramicjournal.org. Subscribers receive full-text access to online content.

Published monthly in print and online. Annual subscription runs from January through December. ISSN 0002-7820

International Journal of Applied Ceramic Technology

www.ceramics.org/act

Launched in January 2004, *International Journal of Applied Ceramic Technology* is a must read for engineers, scientists,and companies using or exploring the use of engineered ceramics in product and commercial applications.

Led by an editorial board of experts from industry, government and universities, *International Journal of Applied Ceramic Technology* is a peer-reviewed publication that provides the latest information on fuel cells, nanotechnology, ceramic armor, thermal and environmental barrier coatings, functional materials, ceramic matrix composites, biomaterials, and other cutting-edge topics.

Go to www.ceramics.org/act to see the current issue's table of contents listing state-of-the-art coverage of important topics by internationally recognized leaders.

Published quarterly. Annual subscription runs from January through December. ISSN 1546-542X

American Ceramic Society Bulletin

www.ceramicbulletin.org

The *American Ceramic Society Bulletin*, is a must-read publication devoted to current and emerging developments in materials, manufacturing processes, instrumentation, equipment, and systems impacting the global ceramics and glass industries.

The *Bulletin* is written primarily for key specifiers of products and services: researchers, engineers, other technical personnel and corporate managers involved in the research, development and manufacture of ceramic and glass products. Membership in The American Ceramic Society includes a subscription to the *Bulletin*, including online access.

Published monthly in print and online, the December issue includes the annual *ceramicSOURCE* company directory and buyer's guide. ISSN 0002-7812

Ceramic Engineering and Science Proceedings (CESP)

www.ceramics.org/cesp

Practical and effective solutions for manufacturing and processing issues are offered by industry experts. CESP includes five issues per year: Glass Problems, Whitewares & Materials, Advanced Ceramics and Composites, Porcelain Enamel. Annual subscription runs from January to December. ISSN 0196-6219

ACerS-NIST Phase Equilibria Diagrams CD-ROM Database Version 3.0

www.ceramics.org/phasecd

The ACerS-NIST Phase Equilibria Diagrams CD-ROM Database Version 3.0 contains more than 19,000 diagrams previously published in 20 phase volumes produced as part of the ACerS-NIST Phase Equilibria Diagrams Program: Volumes I through XIII; Annuals 91, 92 and 93; High Tc Superconductors I & II; Zirconium & Zirconia Systems; and Electronic Ceramics I. The CD-ROM includes full commentaries and interactive capabilities.

Environmental Issues and Waste Management Technologies in the Ceramic and Nuclear Industries X

Ceramic Transactions Volume 168

Proceedings of the 106th Annual Meeting of The American Ceramic Society, Indianapolis, Indiana, USA (2004)

Editors

John Vienna

Connie Herman

Sharon Marra

Published by
The American Ceramic Society
PO Box 6136
Westerville, Ohio 43086-6136
www.ceramics.org

Environmental Issues and Waste Management
Technologies in the Ceramic and Nuclear Industries X

For information on ordering titles published by The American Ceramic Society, or to request a publications catalog, please call 614-794-5890, or visit our website at www.ceramics.org

ISBN 1-57498-189-7

Contents

Nuclear and Hazardous Waste Forms and Fuels— Processing and Technology

Glass Waste Forms—Modelling, Properties, and Testing

Ceramic Waste Forms—Formulation and Testing

Preface

A symposium on Environmental Issues and Waste Management Technologies in the ceramic and nuclear industry took place in Indianapolis, IN, April 18–21, 2004. This was the tenth in this series. A separate session on Environmental Issues in the Ceramic Industry was also organized and is included in this proceedings. The symposium was held in conjunction with the 106th Annual Meeting of The American Ceramic Society, and was sponsored by the Nuclear and Environmental Technology, Glass and Optical Materials, Cements, Basis Science, and Whitewares and Materials Divisions. This volume documents a number of papers presented at the symposium.

The success of the symposium and the issuance of the proceedings could not have been possible without the support of Greg Geiger at The American Ceramic Society Headquarters and the other organizers of the program. Greg's assistance with the technical review and submittal of final abstracts was invaluable. The program organizers included Lou Vance, Michael Hu, Alex Cozzi, and James Marra. Their assistance with organizing the sessions and reviewing the manuscripts was invaluable in ensuring the creation of quality proceedings.

John Vienna
Connie Herman
Sharon Marra

Nuclear and Hazardous Waste Forms and Fuels—Processing and Technology

VITRIFICATION TESTING AND DEMONSTRATION FOR THE HANFORD WASTE TREATMENT AND IMMOBILIZATION PLANT

J.M. Perez, Jr., S.M. Barnes,
S. Kelly, Jr. and L. Petkus
2435 Stevens Center Place
Waste Treatment Plant Project
Richland, Washington 99352

E.V. Morrey
Battelle, Pacific Northwest Division
902 Battelle Blvd.
PO Box 999
Richland, Washington 99352

ABSTRACT

The U.S. Department of Energy's Office of River Protection has commissioned the design, construction and demonstration of the world's largest waste immobilization plant at the Hanford site -- the waste treatment and immobilization plant (WTP). At this plant, high-level tank waste (HLW) will be separated to generate a low-activity waste (LAW) and a HLW fraction that will be separately vitrified. The HLW and LAW melters are required to be larger and have significantly higher throughput than any previously used U.S. HLW glass melter. Therefore, significant design improvements must be developed and tested. The ongoing testing and demonstration programs for WTP melters are presented in this paper.

INTRODUCTION

The Hanford Reservation near Richland, Washington had been the site of almost fifty years of nuclear weapons material production. The nuclear and chemical processes that were employed to produce nuclear materials generated millions of gallons of hazardous and radioactive wastes. The U.S. Department of Energy's Office of River Projection is responsible for the retrieval and treatment of approximately 204 million liters of highly radioactive waste. The purpose of the WTP is to prepare and process Hanford HLW and LAW streams into glass waste forms that meet requirements established in the WTP Project Statement of Work (SOW). The WTP is being designed and built by Bechtel National, Inc. and its prime subcontractor Washington Group International. The WTP has made considerable progress since construction activities began in October 2001 (see Figure 1). The Pretreatment, High-Level Waste and Low-Activity Waste vitrification facilities along with the Balance of Facilities teams have safely completed over 900 000 m^3 of earthwork and placed more than 68 351 m^3 of concrete through December 2003. They have also installed over 31 000 m of piping, 54 000 m of electrical raceway and in excess of 32 000 kg of heating, ventilation and air conditioning ductwork. To date, over 75% of the research and technology activities, approximately 60% of the design and engineering effort and greater than 22% of the construction work has been completed. The Project is working towards its milestones of beginning nonradioactive commissioning of the major facilities by February 2009; followed by radioactive commissioning in December 2009. Following a successful commissioning period, Bechtel will formally turn the facility over to the operating contractor in July 2011. One role of the Project's Research and Technology (R&T) function is to demonstrate the WTP design concept by conducting testing of key process steps and performing activities that define acceptable glass formulations for waste immobilization. The melter design and supporting R&T testing are being provided by Duratek, Inc. and their subcontractor, Catholic University of America's Vitreous State Laboratory. A review of the WTP design and supporting R&T activities are presented in this paper.

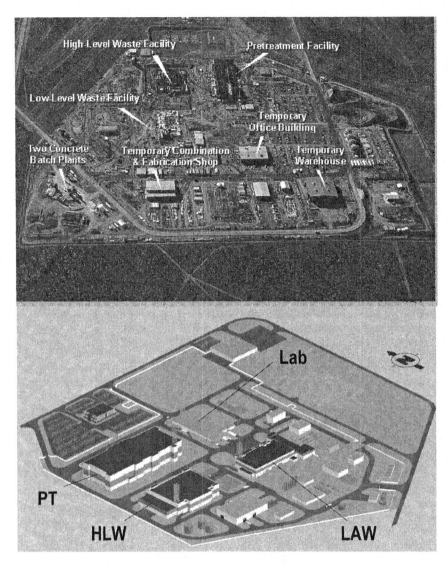

Figure 1. Top - WTP Site Construction Progress - September 2003,
Bottom - Conceptual Drawing of WTP at Completion

WTP FLOWSHEET OVERVIEW

Processing will be accomplished in three primary facilities. The Pretreatment facility will receive transfers of LAW and HLW from the Hanford tank farms. Schematics of the WTP flowsheet unit operations for Pretreatment, LAW Vitrification and HLW Vitrification are presented in Figure 2. The LAW is primarily a liquid supernatant comprised of sodium hydroxide and nitrate. It also contains trace metals, sulfate, aluminum and potassium and up to 3.8 weight percent entrained solids. LAW will be transferred to the Pretreatment facility in batches up to 5 700 000 liters. If required, the LAW stream will be concentrated up to a 5-molar sodium concentration and filtered to remove any entrained solids. As necessary, strontium and transuranics (TRU) will be removed using a permanganate precipitation process and blended with the HLW feed. Cesium will be separated next using an ion exchange resin for ultimate blending with the HLW feed. The treated LAW feed will then be concentrated along with any LAW vitrification recycle solutions to a concentration of between 1 and 8 molar sodium. The target concentration of the treated LAW will be defined by the relative concentrations of sodium and sulfate. These two constituents primarily define the ratio of wastes and glass formers in the LAW waste form to be produced in the LAW Vitrification facility. The Pretreatment facility is designed to provide sufficient pretreated LAW feed to produce 90 000 kg of LAW glass per day.

The HLW waste feed will contain 2 to 16 weight percent solids and will be transferred to the Pretreatment facility in batches up to 600 000 liters. If required, the HLW stream will be concentrated up to a 5 molar sodium concentration prior to washing and filtration. The permeate will be treated as an LAW stream. The solids will be filtered to between 17 and 20 weight percent undissolved solids, blended with any cesium and strontium/TRU products and transferred to the HLW Vitrification facility. The Pretreatment facility is designed to provide sufficient HLW solids to produce 6000 kg of HLW glass per day.

The LAW facility is unique among the WTP facilities in that it is designed to allow contact maintenance of the process equipment. The facility is designed to house three vitrification lines; each designed for a daily glass production rate of 15 000 kg. Factoring in expected operating availability, the facility will be capable of producing 30 000 kg of glass per day. However, under the requirements of the WTP contract Bechtel National, Inc. will install and commission two vitrification lines.

For LAW vitrification, 34 000 liter batches are received from the Pretreatment facility into Concentrate Receipt Vessels (CRV). Batches are transferred from the CRV to the Melter Feed Preparation Vessel (MFP) where the waste is combined with glass former chemicals (GFC). Based on composition and radionuclide analyses, glass formulation calculations will specify the GFC batch to be added. The GFC quantities will be delivered pneumatically to each vitrification facility. The glass former batch plant will provide a mixture of the required GFC from the following available chemicals; silica, zinc oxide, titanium oxide, ferric oxide, zirconium silicate, lithium carbonate, boric acid, hydrated sodium borate, aluminum silicate, magnesium silicate, calcium silicate. Sucrose is also included with the GFC as a reductant. Following the addition and mixing of the GFC with the pretreated waste the melter feed is transferred into the Melter Feed Tank; from where it is continuously pumped to the melter. Typical properties of the LAW and HLW melter feeds are presented in Table I.

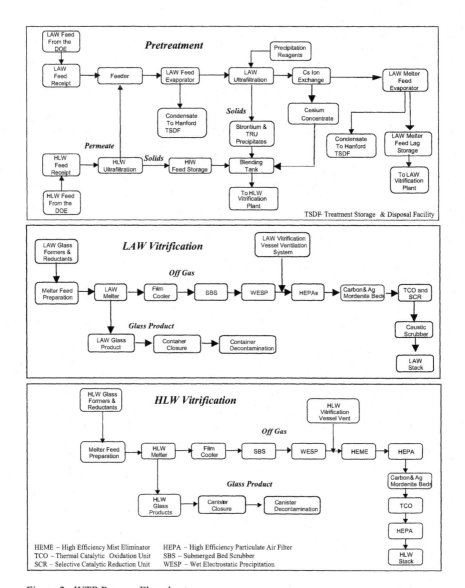

Figure 2. WTP Process Flowsheets

Table I. Characteristic Properties of LAW and HLW Melter Feeds

Property	LAW Melter Feed	HLW Melter Feed
Specific Gravity	1.6 to 1.8	1.2 to 1.5
pH	7.4 to 12.4	10.1 to 10.9
Weight Percent Solids	50 to 60	27 to 46
Equivalent Glass Content grams oxide per liter	850 to 950	280 to 570

Steam, decomposition products, particulate entrainment and inleakage and process air are drawn from the melter and treated by multiple off-gas treatment equipment. Air is injected into the off-gas steam through the film cooler device located at the melter entrance to the off-gas line. This minimizes the potential for solids to deposit in the line. Initial quenching and particulate removal are achieved by the primary off-gas system comprised of the submerged-bed scrubber (SBS), wet-electrostatic precipitator (WESP), and high-efficiency particulate air (HEPA) filters. The vitrification lines will share a common secondary off-gas treatment system. It is designed to reduce organic and NOx emissions to meet permit conditions and to capture mercury (sulfur-impregnated activitated carbon beds) and iodine (caustic scrubbing).

LAW glass product will be poured into 2.1 m-tall by 1.2 m-diameter right circular cylinder containers. Each container will hold approximately 6000 kg of glass product. Following initial cooling the containers will be sealed and decontaminated using a CO_2 pellet cleaning process. The containers are then ready for transporting to the permitted disposal facility on the Hanford Site.

The processing flowsheet within the HLW vitrification facility is very similar to LAW. The facility is designed to house two vitrification lines; each designed for a daily glass production rate of 3000 kg. Factoring in expected operating availability, the facility will be capable of producing 4000 kg of glass per day. Feed preparation is virtually identical to LAW. The off-gas treatment system is also very similar. Differences include the addition of a high-efficiency mist eliminator (HEME) after the WESP; and the use of silver mordenite for iodine capture. In addition, each vitrification line will have individual primary and secondary treatment trains.

The molten HLW glass product will be poured into 4.6 m-tall by 0.6 m-diameter right circular cylinder canisters. Each canister will hold approximately 3000 kg of glass product. Following initial cooling, canister lids will be welded and the canisters will be decontaminated using a cerium-IV, nitric acid chemical milling technique. The canisters will be transported to the on-site canister storage facility for interim storage prior to shipment to the federal HLW disposal repository.

HLW MELTER THROUGHPUT
Melter Description

The HLW melter, shown conceptually in Figure 3, is designed to have a five-year operating life. The glass tank dimensions are 2.44 m by 1.52 m (on the electrode wall) and 1.14 m deep. This results in a glass surface of 3.72 m^2 and a glass inventory of approximately 10 000 kg. The glass contact refractory is a high chromic oxide-alumina refractory backed up by an alumina-zirconia-silica refractory layer. The lid is constructed using courses of various alumina-based refractories. Three nickel-based alloy electrodes provide up to 600 kW of power to convert the melter feed slurry to glass and maintain the tank inventory at an average bulk glass temperature of 1150°C. Normal operation assumes conduction between the two side electrodes only. Five bubblers will be extended through the roof of the melter to within a few inches of the tank floor to provide agitation. Melter feed slurry will be brought into the melter through two feed nozzles at a combined rate of 250 to 500 L·h⁻¹. Glass will be discharged from the tank through two discharges. A bubbler placed in the riser will be used to lift glass into the pour trough during the periodic glass discharges. Silicon carbide U-bar heaters will provide supplemental heat in each discharge.

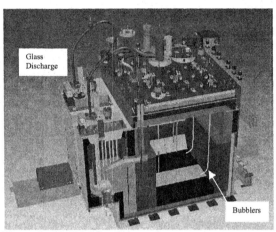

Figure 3. HLW Melter Schematic

High-Level Waste Melter Testing
The HLW melter technology has benefited from decades of development, both in this country and internationally. Therefore, the key objectives of HLW melter research and testing are specific to verification of the design and flowsheet assumptions, determination of production rates and product quality for the four HLW waste compositions defined in the SOW, and to generate data in support of regulatory permitting activities. To achieve these objectives the project is conducting testing in a prototypic pilot plant. Testing is being performed at the Catholic University of America's Vitreous State Laboratory (VSL) under contract to Duratek, Inc. Compared to the plant melter, the pilot plant melter has 32% of the glass surface area and 57% of the depth. Prototypic off-gas and feed equipment are also provided in the pilot plant. Prior to 2003, the SOW required the HLW facility to include two melters, each having a baseplate rate of 1500 kg/d; with the ability to achieve 3000 kg/d with "enhancements" to the melter. These enhancements were understood to mean that a sufficient number of bubblers could be inserted into the melter to double its baseplate rate. Testing results to determine production throughput capabilities both with and without bubblers is presented in the following paragraphs.

Initial Testing to Determine Throughput Rate: Testing performed in non-bubbled DM1200 testing during the last half of 2001 assessed; 1) the addition of sugar to prevent glass foaming, 2) multiple feed nozzles, 3) acidification by nitric acid to a melter feed pH of ~5, 4) frit as the GFC source, 5) bubbling, and 6) operation at an average glass temperature of 1200°C. Results from these multi-day tests are presented in Figure 4. In every case, production rates failed to achieve or exceed the equivalent baseline rate of 1500 kg/d. Production rates were only 50% to 80% of the baseline rate. With the exception of the higher-temperature operation (1200°C vs 1150°C), none of the variables had a marked affect on production rate.

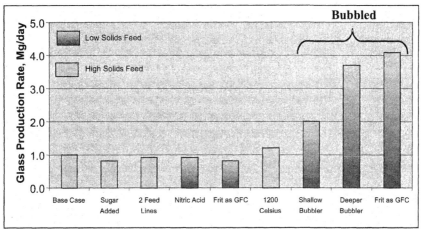

Figure 4. DM1200 Testing to Assess Effect of Testing Conditions on Production Rate (Results adjusted to WTP HLW Melter Equivalent Rates)

Three DM1200 tests were performed using bubblers. The more dilute AZ101 feed (approximately 0.4 kg oxide equivalent/L of melter feed) was used in the bubbled tests in order to obtain a conservative estimate based on a more realistic solids concentration from the Pretreatment facility of between 15 and 20 wt.% solids. Production rates were increased significantly by the use of bubblers. Two bubblers were placed in the DM1200 and tested at two depths, ~0.3 m and ~ 0.5 m below the nominal glass surface. These tests, using AZ101 and glass former chemicals, indicated an increase in production rate by a factor of 4.5 and 5. With the deeper bubbler position combined with using frit instead of glass former chemicals production rates increased a factor of five to a normalized WTP glass production rate of 4100 kg/d. Based on the results, the Project incorporated bubblers in the HLW melter design from the outset.

HLW Melter Glass Description: Four HLW tanks are being characterized and throughput rates demonstrated as part of the WTP scope of work. The tanks are designated AZ101, AZ102, AY102/C-106 and AY101/C-104. Four glasses were defined by VSL based on Project estimates in 2002 of the waste feed compositions after pretreatment, contract requirements, VSL glass formulation methodology and confirmatory laboratory testing. The four glasses are presented in Table II. The first three glasses are quite similar after GFCs are blended with the waste feed. The forth glass, AY-101/C-104 actually would contain just over 4% ThO_2 and almost 2.5% UO_2 had radioactive compounds been permitted. Possible surrogates, such as cerium, neodymium, and hafnium, were tested. However, heat-treatment tests showed significantly greater crystallinity in the three variations tested. Therefore, a simple renormalization of the waste composition, without uranium and thorium, was determined to be the most practical approach. Key properties of the glass are presented in Table III.

Table II. Glass Compositions Developed for Four HLW Feeds

Constituent	AZ-101	AZ-102	AY-102/C-106	AY-101/C-104
Ag_2O	—	—	—	—
Al_2O_3	5.19%	5.58%	5.29%	3.58%
As_2O_3	—	—	0.19%	—
B_2O_3	11.90%	12.51%	9.39%	10.81%
BaO	0.02%	—	—	—
CaO	0.28%	0.23%	0.30%	0.48%
CdO	0.06%	0.11%	—	—
Cl	—	—	0.11%	—
Cr_2O_3	—	—	0.08%	0.06%
Cs_2O	0.00%	0.05%	0.05%	0.05%
CuO	0.03%	—	0.04%	0.03%
F	0.04%	—	—	0.12%
Fe_2O_3	12.21%	12.51%	12.58%	9.54%
I	0.10%	0.10%	0.10%	0.10%
K_2O	0.03%	0.03%	—	—
La_2O_3	0.41%	0.37%	0.24%	0.15%
Li_2O	3.52%	3.26%	3.01%	3.31%
MgO	0.11%	0.07%	1.17%	—
MnO	0.17%	0.36%	4.00%	1.52%
Na_2O	11.65%	12.02%	11.83%	11.49%
Nd_2O_3	0.31%	0.16%	0.15%	0.11%
NiO	0.61%	0.45%	0.17%	0.47%
P_2O_5	—	0.03%	0.09%	0.04%
PbO	0.03%	0.07%	0.14%	0.12%
Sb_2O_3	—	—	0.25%	—
SeO_2	—	—	0.37%	—
SiO_2	47.40%	48.26%	47.04%	46.39%
SO_3	0.07%	0.04%	—	—
SrO	0.03%	—	0.92%	—
TiO_2	—	—	0.14%	0.02%
ZnO	2.01%	2.01%	2.07%	2.16%
ZrO_2	3.80%	1.77%	0.26%	9.43%
TOTAL	100%	100%	100%	100%

Table III. HLW Glass Properties

Tank I.D.:	AZ-101	AZ-102	C-106/AY-102	C-104/AY-101
Glass I.D.:	HLW98-77	HLW98-80	HLW98-86	HLW98-96
Waste Loading	25 wt.% oxides	24 wt.% oxides	25 wt.% oxides	35 wt.% oxides
Viscosity @ 1 150°C	5.0 N·s/m^2	5.1 N·s/m^2	4.4 N·s/m^2	5.7 N·s/m^2
Electrical Conductivity @ 1 150°C	0.36 S/cm	0.36 S/cm	0.36 S/cm	0.28 S/cm
Crystallinity @ 950°C	<0.5 volume %	<0.1 volume %	Zero volume %	<0.2 volume %

HLW Throughput Testing With Bubblers: In 2002 ORP directed the Project to incorporate the necessary design and process modifications to achieve a baseplate capacity of 6000 kg/d of HLW glass product. This change required two HLW melters, each having an instantaneous throughput capability of 3000 kg/d. Based on results described above, the use of five bubblers per melter were considered sufficient to achieve the new baseline requirement. Pilot testing in 2002 and early 2003 was performed to determine the bubbler air requirements necessary across the range of expected feed concentrations. Beginning with the AZ101 composition, 9-day tests were performed to determine the effect of feed concentration and bubbler air rate on throughput. The results of the first three tests are presented in Figure 5. Three feed concentrations were tested based on an expected range from pretreatment. The three feeds were based on undissolved solids (UDS) concentrations of 10, 15 and 20 weight percent prior to glass former addition. Feed property data are provided in Table IV.

As is evident from Figure 5, there is a marked effect on throughput due to the water content in the melter feed. Two bubblers were employed in the DM1200 to approximate the number of bubblers in the plant melter. The bubblers were placed 0.5 m below the glass surface. The maximum air rate of 65 slpm was determined based on visual observation of the bubbles at the glass surface. Above 65 slpm, the bubbling became less effective. It was concluded that at higher rates an air channel was forming from the bubbler to the glass surface. This would eliminate the mixing effectiveness of the bubblers and interactions with the cold cap at the glass surface. The data show that the required rate can only be met at high solids concentrations and high bubbler air rate. Interpolation of the data indicate that the required rate of 3000 kg/d would likely not be met below a HLW feed concentration of 17 wt% undissolved solids.

Testing was subsequently conducted for the remaining three HLW compositions at the higher concentration. The results of these tests are plotted in Figure 6 along with the AZ101 test data previously discussed. The data indicate that the four HLW feed compositions have similar processing rates as a function of bubbler air rate when feed concentrations are similar. Maximum rates were measured to be 900 to 950 kg/m^2/d.

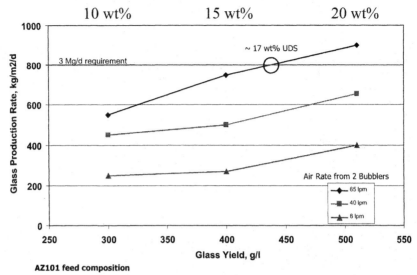

Figure 5. DM1200 Throughput Rate Results at Three Concentrations

Table IV. HLW Feed Properties

	AZ101			AZ102	AY102/C106	AY101/C104
Wt.% UDS:	10	15	20	20	20	20
Wt.% Water	72	64	55	54	54	56
Specific Gravity	1.24	1.33	1.39	1.42	1.42	1.40
Glass Yield, g-oxide/L	290	405	534	546	553	528
pH	10.2	10.5	10.6	10.4	10.2	10.9

These data were further reviewed against the design bases assumptions for scaling pilot data to the WTP melter. Two key assumptions were subsequently revised. Firstly, estimates of dilution water from transfer line flushes, feed line flushes, vessel vent demister flushes, etc. were now available to factor in to HLW feed concentration estimates. The steady-state flowsheet model estimated the solids concentration delivered to HLW was expected to vary between 14 and 17 weight percent UDS. Based on this fact alone, the testing already described indicated that

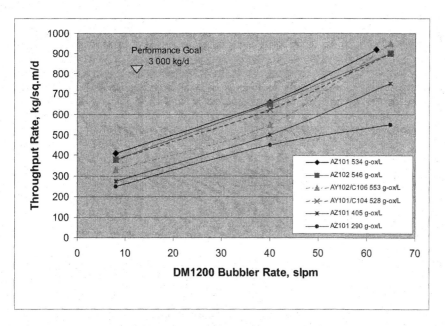

Figure 6. DM1200 Throughput Rate Results for Each HLW Tank Composition

the HLW melter design with bubblers could very well fall short of meeting its design requirements. Secondly, the scale-up requirements were revised to account for the number of bubblers used in the pilot melter compared to the plant melter. The DM1200 glass surface area is 32% of the size of the WTP HLW plant melters. The ratio of melter glass pool surface areas has historically been used as a melter scale-up factor. However, this simple approach should only be considered applicable when the number of bubblers per unit area is also equivalent. This is not the case. The DM1200 has two bubblers compared to five in the WTP melter. On a unit area bases this equates to 1.7 bubblers/m^2 in the DM1200 and 1.3 bubblers/m^2 in the WTP melter. This 30% increase in bubbler "density" needs to be accounted for in any scale up projections. Therefore, the DM1200 pilot target goal was revised to require, at a minimum, throughput rates 30% greater than 800 kg/m^2/d in order to provide confidence the WTP melter will attain 3000 kg/d (800 kg/m^2/d). This equates to ~1050 kg/m^2/d with feeds having concentrations based on 15 wt.% UDS prior to glass former addition.

Testing in 2003 was performed to determine the number of bubblers and their position required to achieve a production rate of 1050 kg/m^2/d or greater. To account for scale-up uncertainties and differences between simulated feeds and actual feeds; a goal of 1300 kg/m^2/d, or 125% of the target was established. Options for increasing the number of bubblers in the melter include providing more than one bubbler port within the existing bubbler design or adding bubblers using one or more of the spare nozzles on the melter. To prevent significant design impacts there was an incentive to not require additional bubbler ports or to require additional air supplies and controls to the existing five bubblers. Therefore, a multi-port bubbler design, which could operate with a single air supply, was emphasized. The final variable was to extend the bubbler to the floor of the melter.

Figure 7 presents the additional test data combined with the data shown previously in Figure 6. The per-square-meter DM1200 production rate data has been multipled by the WTP melter glass surface area of 3.72 m^2 to represent expected WTP melter throughput rates. Unless otherwise noted in the figure legend, the feed concentrations represented HLW feed at 15 wt.% UDS prior to glass former addition. It was also estimated that the HLW melter lid nozzle design could accommodate a bubbler design having an elongated foot with bubbler ports positioned 0.2 and 0.36 m apart. Therefore, testing was performed with bubbler assemblies used to represent bubbler ports separated by 0.2 and 0.36 m. Both multiple bubbler assembly options were successful in achieving rates above the adjusted required minimum of 3900 kg/day, (1 050 kg/m^2/d · 3.72 m^2). Simply placing two single port bubblers to the floor of the DM1200 resulted in a throughput rate of approximately 4100 kg/day. However, the margin between this rate and the minimum target is considered too small to account for testing variability, uncertainties in scaling and possible differences between actual wastes and test surrogates. Therefore a double port bubbler has been recommended for use.

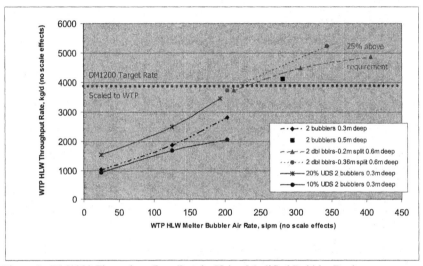

Figure 7. DM1200 Throughput Rate Results Using Modified Bubbler Designs

Overall, DM1200 testing evaluated bubbler depth at three depths, 0.3, 0.5 and 0.6 m (11.5, 19 and 25 in.). Each test attempted to achieve maximum production rates. In all cases, the bubbling rate reached a maximum above which no further increase in production rate could be achieved. These limited bubbling rates were 50, 68 and 90 liters per minute for depths of 0.3, 0.5 and 0.6 m, respectively. Depth alone appears to increase production rate at the same bubbling rate; while also allowing higher bubbling rates. This combined effect results in achieving even higher production rates. For example, at the same bubbling rate, the production rate is estimated to increase by 33% as the bubbler was moved to the deepest position. At the deeper position the bubbling rate was able to be increased by 38%. The combined increase in production rate 46%. These data are based on very limited results but are encouraging in predicting even higher throughput rates in the WTP HLW plant where the bubbler depth should be on the order of 1.1 m (42 in.).

Testing, physical modeling and engineering activities are currently underway to provide additional data and confidence that five bubblers with multiple ports will meet the WTP throughput requirements. These activities will be completed in 2004.

LAW MELTER THROUGHPUT

Melter Description

The LAW melter, shown conceptually in Figure 8, is designed to have a five-year operating life and baseplate production capability of 15 000 kg/d. The glass tank dimensions are 4.93 m long (on the electrode wall) by 2.03 m wide by 0.76 m deep. This results in a glass surface of 10 m^2 and a glass inventory of approximately 20 000 kg. The glass contact refractory is very similar to the HLW melter described earlier. The lid is constructed using courses of various alumina-based refractories. Nickel-based alloy electrodes are placed along the entire length and 2/3rd the depth of the melter providing up to 1500 kW of power at an average bulk glass temperature of 1150°C. Eighteen bubbler assemblies will be extended through the roof of the melter to within a few inches of the tank floor to provide agitation. Melter feed slurry will be brought into the melter through three feed nozzles at a nominal rate of 800 to 900 L·h^{-1}. Glass will be discharged from the tank through two discharges. A bubbler placed in the riser will be used to lift glass into the pour trough during the periodic glass discharges. Silicon carbide U-bar heaters will provide supplemental heat in each discharge.

Figure 8. LAW Melter Schematic

Low-Active Waste Melter Testing

The LAW melter technology has benefited from several years of progressive design evolutions by Duratek. Based initially on the U.S. DOE HLW melter designs the design has evolved to account for the much large size, throughput requirements, and Duratek innovations. A key objective of LAW melter research and testing was to demonstrate production rates and glass formulations that maximized waste loading while prohibiting the formation and accumulation of molten salts. To achieve these objectives, the project conducted nearly continuous operation of a pilot LAW melter at Duratek's Columbia, Maryland facility from May 2001 through May 2003. Since startup of the facility in 1999, the facility produced over 3 million kilograms of glass product and processed over 3.5 million liters of feed slurry. Glass formulation development and small-melter testing support were performed by the Catholic University of America's VSL under contract to Duratek, Inc. Compared to the plant melter, the pilot plant melter has 33% of the glass surface area and 110% of the depth. The pilot melter glass tank is essentially a 1/3rd segment of the plant melter. Therefore, unlike the DM1200 for HLW testing, DM3300 test results are assumed to extrapolate linearly to the WTP LAW facility melters. Aside from prototypic feed pumps that were used for the last 6 months of operation, no other pilot facility slurry handling or off-gas equipment was prototypic.

LAW Melter Glass Description: Eleven LAW tank compositions are being characterized and throughput rates demonstrated as part of the WTP scope of work. The eleven tanks were grouped into seven sub-envelopes based on common composition characteristics. The resulting sub-envelopes are described in Table V. The compositions are principally distinguished by their concentrations of sulfate relative to sodium and potassium. Glass solubility constraints for sodium and sulfur control the waste loadings of these glasses. The compositions and properties of the test glasses developed for melter testing in 2003 are presented in Tables VI and VII, respectively. Glass former chemicals and sugar are blended with the LAW solution to achieve the melter feed. The sugar serves to partially reduce the nitrates and nitrites in the waste resulting in a well-behaved cold cap melting process and reduced NO$_x$ emissions to the off-gas system.

Feed Variation Testing Results: Two distinct testing phases were conducted during the LAW pilot melter operational period. Testing in the first 15 months (2001 - 2002) was performed to determine the production rate of each sub-envelope composition. The ratio of waste to glass formers was also varied by $\pm 15\%$ to assess the sensitivity of production rate and the potential for sulfate accumulation to occur from feed batching uncertainties. The glass compositions used for these tests were quite similar to the compositions in Table VI with some minor adjustments made for updated tank characterization and plant recycle stream data. Test results are presented in Figure 9. Each test segment lasted 9 to 12 days; during which time 50 000 kg to 80 000 kg of glass product was produced. With the exception of the A2 nominal feed segment, the minimum rate requirements of 1500 kg/m^2/d were met. Subsequently to this test, increases in bubbler air rate and sugar addition were evaluated and determined to result in production rates above the minimum requirement. Dip sampling through several access nozzles was typically performed every three to four days to monitor for molten salt accumulations. None of the tests resulted in salt accumulations that could be characterized as significant. Many times small accumulations would be detected in the corners of the tank but these did not persist. Based on these results it was concluded that salt accumulations over time were not likely.

Table V. LAW Waste Characteristics Grouped into Seven Sub-envelopes

Envelope	A			B		C	
Sub Envelope	A1	A2	A3	B1	B2	C1	C2
Tanks applicable to Sub Envelope	AN-105 SY-101 AN-103	AP-101 AW-101	AN-104 possibly AP-108	AZ-101	AZ-102	AN-107	AN-102 possibly SY-102
Na_2O wt%		20%	14.8%	6.5%	~3%	~14%	11.2%
K_2O wt%	0.3 to 0.7 %	~2%	~0.3%	~0.2%	~0.2%	~0.3%	~0.2%
SO_3 wt%	0.1 to 0.2 %	0.1 to 0.2%	~0.35%	0.75%	~1%	~0.35%	~0.45%

Table VI. LAW Glass Properties

Sub-envelope:	A1	A2	A3	B1	B2	C1	C2
Waste Loading	25 wt.% oxides	24 wt.% oxides	19 wt.% oxides	6.4 wt.% oxides	3.7 wt.% oxides	14 wt.% oxides	12 wt.% oxides
Constituent	Glass Designation I.D.						
	LAWA44	LAWA126	LAWA137	LAWB83	LAWB96	LAWC22	LAWC35
Al_2O_3	6.13%	5.62%	6.06%	6.18%	6.17%	6.05%	6.07%
B_2O_3	8.89%	9.84%	9.93%	10.01%	10.02%	10.01%	9.43%
CaO	1.98%	1.99%	5.04%	6.77%	6.77%	5.09%	7.36%
Cr_2O_3	0.02%	0.02%	0.02%	0.03%	0.03%	0.01%	0.01%
Fe_2O_3	6.91%	5.54%	5.37%	5.28%	5.29%	5.41%	3.60%
K_2O	0.44%	3.81%	0.33%	0.18%	0.12%	0.07%	0.09%
Li_2O	-	-	2.48%	4.30%	4.30%	2.50%	3.26%
MgO	1.98%	1.48%	1.48%	2.98%	2.97%	1.51%	1.49%
MnO_2	-	-	-	-	-	0.03%	-
Na_2O	20.33%	18.46%	14.66%	5.53%	5.48%	14.57%	11.99%
NiO	-	-	-	-	-	0.03%	0.02%
SiO_2	44.09%	44.08%	46.16%	48.55%	48.74%	46.54%	47.30%
TiO_2	1.98%	1.99%	1.14%	1.40%	1.39%	1.15%	1.08%
ZnO	2.94%	2.94%	3.05%	4.84%	4.85%	3.06%	3.99%
ZrO_2	2.96%	2.97%	3.01%	3.16%	3.17%	3.01%	3.00%
Cl	1.18%	0.42%	0.79%	0.02%	0.01%	0.14%	0.39%
F	0.004%	0.35%	0.01%	0.08%	0.02%	0.34%	0.11%
P_2O_5		0.08%	0.11%	0.04%	0.01%	0.13%	0.16%
SO_3	0.19%	0.40%	0.37%	0.65%	0.65%	0.38%	0.63%
Sum	100%	100%	100%	100%	100%	100%	100%

Table VII. LAW Glass Properties

Sub-envelope:	A1	A2	A3	B1	B2	C1	C2
	Glass Designation I.D.						
	LAWA44	LAWA126	LAWA137	LAWB83	LAWB96	LAWC22	LAWC35
Viscosity @ 1150°C	6.9 N·s/m²	6.0 N·s/m²	3.5 N·s/m²	5.3 N·s/m²	5.3 N·s/m²	4.6 N·s/m²	3.5 N·s/m²
Electrical Conductivity @ 1150°C	0.52 S/cm	0.35 S/cm	0.29 S/cm	0.20 S/cm	0.20 S/cm	0.32 S/cm	0.27 S/cm

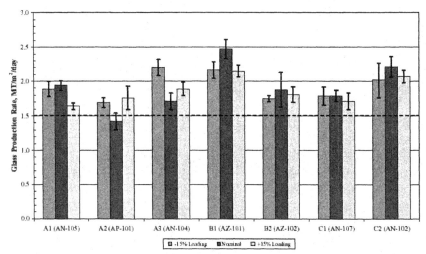

Figure 9. DM3300 LAW Melter Feed Variation Testing Results

Feed Changeover Testing Results: The last nine months of DM3300 testing assessed the effect of transitioning from one LAW feed composition to the next feed composition. Key variables were again glass production rate and sulfate accumulation as the glass composition in the tank converted between glass compositions. The production rate results of these tests are presented in Figure 10. The results are compared to the nominal feed variation testing results to allow the effect of changes in feed and glass compositions to be estimated. As stated earlier, the glass compositions were adjusted to reflect more complete data on waste and recycle stream compositions. Of significance were increased quantities of chloride, fluoride and sulfate which are believed to enhance cold cap melt rate.

Within the measurement tolerances of the data it is possible to say that statistically there is no difference in throughput rate for four out of six of the sub-envelopes. Only the A2 and A3 compositions showed marked improvement in production rate. Each of the sub-envelopes successfully exceeded the minimum production rate requirement of 1500 kg/m²/d. Consistent with the earlier test results, molten salt accumulations were observed in only trace quantities and were not increasing over time.

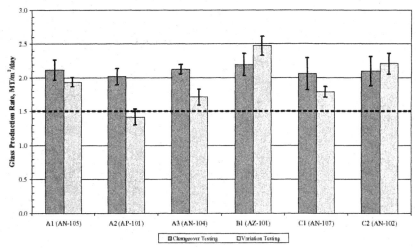

Figure 10. DM3300 LAW Melter Feed Changeover Testing Results

CONCLUSIONS

Testing of HLW and LAW pilot vitrification melter systems has been underway since 2001. During this period melter feed and glass compositions have been developed to provide representative surrogates for testing. HLW melter testing determined that bubblers will be required to meet the facility requirement of 6000 kg/d of glass product per melter. Testing further defined minimum bubbler design and operating requirements. LAW melter testing successfully demonstrated each of seven LAW glass and feed compositions during long-term campaigns. Final production rates exceeded facility requirements of 15 000 kg/d of glass product per melter. The absence of anything but trace molten salt accumulations also validated the glass formulation strategies that balance maximizing waste loading while avoiding salt accumulations.

ACKNOWLEDGEMENTS

The results presented here are based on work funded by the U. S. Department of Energy under contract DE-AC27-01RV14136. Testing was performed by Duratek and their subcontractor Catholic University of America. The Duratek Pilot Melter testing is led by Mr. Glenn Diener. The Catholic University of America testing is led by Drs. Ian Pegg, Isabella Muller and Keith Matlack.

BUBBLING AS A MEANS TO ENHANCE JOULE HEATED CERAMIC MELTER PRODUCTION RATES FOR VITRIFYING RADIOACTIVE WASTES

Bradley W. Bowan, II, Richard Meigs and Eric C. Smith
Duratek, Inc
10100 Old Columbia Road
Columbia, Maryland 21046
bbowan@duratekinc.com, esmith@duratekinc.com, rmeigs@duratekinc.com

ABSTRACT

Three vitrification programs that incorporated bubbled agitation systems with joule heated ceramic melters have demonstrated that this technology can achieve high glass production rates. The programs include one commercial process that remediated mixed waste at the Savannah River Site's M-Area, and two pilot-scale processes that are part of an integrated development and test program in support of the Department of Energy's River Protection Project Waste Treatment Plant (RPP-WTP). These programs have shown that glass production rates between 1 and 2 MT/m^2/day are achievable through melt pool agitation, nearly a 2 to 4 fold increase over the historical rates of joule heated ceramic melters (JHCM) without bubbling.

The technology was developed by Duratek and its partners in the early 1990s, and patented under the name DuraMelter. In 1997, the DuraMelter™ was successfully deployed at the M-Area, where 2,270,000 liters of mixed waste were converted into about 1,000,000 kg of durable glass. The process was centered on a 5 m^2 surface area JHCM, and achieved production rates of about 1 metric ton (MT) of glass/m^2 (surface area)/day. The technology has since been adopted for use at what will become the world's largest vitrification plant, the RPP-WTP, which will be used for the vitrification of low activity waste (LAW). In support of this pioneering facility, a pilot LAW vitrification facility (3.3 m^2 surface area JHCM) was built and operated from 1999 until 2003, producing nearly 3,300,000 kg of glass. This facility demonstrated that sustained production rates of 2 MT/m^2 /day are achievable with this technology. More recently, bubbling agitation has been demonstrated at a high-level waste (HLW) pilot test facility for the RPP-WTP. In that test, agitation was applied to a 1.2 m^2 surface area JHCM, yielding production rates in excess of the plant design requirement of 0.8 MT/m^2/day of glass.

INTRODUCTION

Since 1991, Duratek, Inc., and its long-term research partner, the Vitreous State Laboratory of The Catholic University of America, have worked to continuously improve joule heated ceramic melter vitrification technology in support of waste stabilization and disposition in the United States. From 1993 to 1998, under contract to the U.S. Department of Energy (DOE), the team designed, built, and operated a joule-heated melter (the DuraMelter™) to process liquid mixed (hazardous/low activity) waste material at the M-Area, Savannah River Site (SRS), South Carolina. This melter produced 1,000,000 kilograms of vitrified waste, achieving a volume reduction of approximately 70 percent and ultimately producing a waste form that the U.S. Environmental Protection Agency (EPA) delisted from the hazardous waste inventory.

The team built upon its M-Area experience by developing state-of-the-art melter technology that will be used at the DOE's Hanford site in Richland, Washington. Since 1998, the DuraMelter™ has been the reference vitrification technology for processing both the high level waste (HLW) and low activity waste (LAW) fractions of liquid HLW waste from the U.S. DOE's Hanford site (River Protection Project Waste Treatment Plant RPP-WTP). Process innovations have doubled the throughput and enhanced the ability to handle problem constituents in LAW. For HLW vitrification at the RPP-WTP, the technology has been adapted to accelerate HLW processing. This paper reviews these projects, and the throughput enhancement of JHCM technology achieved through bubbled agitation.

The M-Area Sludge Stabilization Project at the U.S. Department of Energy Savannah River Site resulted in the design, manufacture, and operation of the largest joule-heated melter for waste vitrification in the United States. During the one year of operation of the facility, the melter converted nearly 2,500,000 liters of mixed waste into about 1,000,000 kilograms of durable glass. The rate, at which the waste at M-Area was processed, exceeded the average state of the art by about a factor of two. This rate was achieved through Duratek's patented bubbled agitation system (Refs. 1-3). The waste was ultimately delisted from the site's hazardous waste inventory by the EPA.

The M-Area project became the first commercial project in the United States to stabilize a large volume of mixed wastes through vitrification. In the process, the M-Area project provided relevant, operational experience that underpinned the further development of vitrification technology for future use at other DOE facilities, particularly the River Protection Project Waste Treatment Plant (RPP-WTP) at the Hanford site in Washington State. Importantly, the M-Area project illustrated the significant contribution of the bubbled agitation system with respect to JHCM production rates, giving a starting point for further enhancement for the Hanford RPP-WTP program. The Hanford site currently stores about 210,000,000 liters of HLW tank wastes in underground storage tanks. The DOE seeks to vitrify a large portion of both the HLW and LAW as part of the RPP-WTP. This plant, using DuraMelter™ technology, will be the world's largest radioactive waste vitrification plant when commissioning begins in 2009. RPP melters are designed to produce waste glass three to five times faster than technologies developed in the early 1980s technologies by the DOE for the West Valley Demonstration Project in New York State and the Defense Waste Processing Facility at the Savannah River Site.

Since completion of the M-Area project, additional research and development of the DuraMelter vitrification technology has been underway at Duratek's LAW Pilot Plant in Columbia, Maryland, and its sister HLW Pilot Plant at the Vitreous State Laboratory of The Catholic University of America (CUA-VSL). The LAW Pilot Plant produced over 3,300,000 kilograms of glass during four years of testing — more than any JHCM waste glass melter of its type has ever generated. Meanwhile, the HLW Pilot Plant has produced over 230,000 kilograms of glass since testing started in 2001. These plants have provided a variety of data that will be used to support the successful startup and operation of production facilities at

RPP-WTP, inclusive of demonstrating that melter design target production rates for RPP-WTP will be achievable.

THE SRS M-AREA SLUDGE STABILIZATION PROJECT
The Project

DOE awarded a contract to stabilize 2,500,000 liters of mixed (radioactive and hazardous) waste stored on site at the Savannah River Site's M-Area in November 1993. These wastes had been generated from an electroplating process that was one step in the preparation of materials as part of the Cold War effort in the 1980s. The waste sludges were listed under RCRA as F-006 (electroplating wastes) and contained a total of about 14 curies of four uranium isotopes (^{233}U, ^{235}U, ^{236}U, and ^{238}U). The majority of the material occupied six 130,000-liter tanks and three 1,900,000-liter tanks. In addition, about one hundred twenty-five 200-liter drums of waste material (similar composition to the tank wastes) were treated. The waste sludges were predominantly spent filter aid (pearlite and diatomaceous earth) generated from treating electroplating wastewaters.

The Plan

Duratek and the CUA-VSL developed a plan to consolidate and blend the waste sludges to the extent existing storage capacity would permit, and to mix these wastes sequentially in batches with glass-forming chemicals in preparation for vitrification. The plan included pumping batches of blended waste continuously and feeding them through a stabilization process that used a single joule-heated DuraMelter™ with a 5-m^2 melt pool surface area. In addition to providing the feed preparation and delivery system, the plan also included attendant subsystems that would be coupled to the DuraMelter™: a multistage off-gas treatment train and a molten glass handling system that would convert vitrified waste into flattened marbles, called "DuraGems." The M-Area process would become the first commercial-scale deployment of vitrification technology to stabilize mixed waste streams within the DOE complex, and it would represent an approximate doubling of the world's glass making capacity for vitrifying radioactive wastes.

Waste glass produced by the M-Area vitrification facility was required to meet the Toxicity Characteristic Leaching Procedure-Universal Treatment Standards (TCLP-UTS) with high statistical confidence. The major TCLP metal present in the M-Area waste stream was Ni, which was typically 0.5 wt. percent in the glass produced by vitrification.

At project completion, about 2,860 drums containing 270 liters each of DuraGems were produced from the original M-Area waste inventory, reducing the overall waste volume by nearly 70 percent. The SRS was ultimately successful in delisting vitrified M-Area glass from the hazardous waste inventory through an agreement with the U.S. EPA, thereby allowing on-site disposal as low-level waste.

More than 1,000 metric tons of glass were produced over the life of the project (Fig. 1), and plant throughput reached the level of 4.5 metric tons of glass a day (design capacity). Sustained production periods were achieved, where nearly 5 metric tons of glass per day were produced. Additionally, weekly plant availability was sustained at about 80 percent from February 1998 until the project was completed. Overall plant availability was between

75 and 80 percent (the cumulative average of weekly availability) from the middle of April, 1998, until the end of testing.

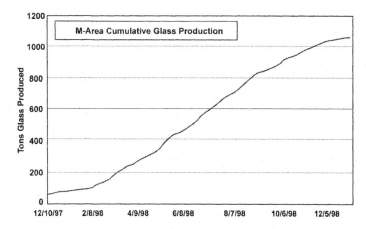

Fig. 1. M-Area plant cumulative glass production versus calendar time

The largest impact change during the restart phase was modifying the NOx control strategy. This was done by incorporating sugar (sucrose) into the feed and allowing its oxidation to destroy NOx (generated from the thermal decomposition of nitrate salts contained in the waste) within the melter plenum to relatively high efficiency. This had a high impact on reducing process control upsets due to a previously used chemical treatment and scrubbing technique. Additional success in plant availability was attributed to an improved particulate removal system installed in the off-gas system quencher. All plant airborne emissions were less than permitted release limits.

At the completion of waste processing, 2,860 drums (square cross section, 270-liter capacity) of acceptable glass product had been produced, satisfying the desired production performance design. In addition, the glass product was routinely monitored for the presence of melter corrosion products but no abnormal levels were observed.

M-Area waste treatment activities were completed early in the first quarter of 1999, followed by deactivation of the facility and closure of the initial waste storage tanks. This project was the first and largest application of joule-heated vitrification technology for commercial stabilization of mixed wastes, as well as one of the first "privatized" style waste treatment procurements for the U.S. Department of Energy. Plant performance of this first-of-a-kind program exceeded overall plant availability at design production levels. Examination of the DM 5000A following production indicated little wear on the internal melter components, and suggested a much longer life expectancy of the melter should operations have continued beyond the 15-month period. The M-Area melter achieved long term daily throughputs slightly less than 1 MT glass/m^2/day.

RPP-LAW PILOT PLANT
LAW Pilot Melter - Background
Near the end of the M-Area project in 1998, Duratek was commissioned to design, build, and operate a pilot melter facility to underpin and optimize key design assumptions for LAW melters for Hanford's RPP-WTP. The RPP is responsible for the treatment of radioactive waste material contained in 177 underground tanks at the Hanford Site in Richland, Washington.

The LAW Pilot Melter (DuraMelter™ 3300, or DM 3300) was designed to represent a one-third model of the full-scale design for the RPP LAW Melter. It was designed with a melt pool surface area of 3.3 m^2 compared to the 10 m^2 for the full-scale melter. (Note: Vertical melters are sized based on available melt pool surface area, through which heat is transported to a reacting feed pile.) The pilot melter design simulates the LAW melter key dimensions between opposing electrodes and maintains LAW melter design concepts and process parameters. The DM 3300 also used the same refractory design and cooling system as the M-Area DuraMelter™ 5000A, but altered the arrangement of the Inconel electrodes (all electrodes positioned against the walls, as opposed to having a center electrode).

A secondary objective of the LAW Pilot Melter was to select the optimum design details associated with the glass discharge channels. For this reason, three separate glass discharge systems were built into the design of the LAW Pilot Melter. The off-gas train and feed systems for the LAW pilot plant were custom designed and built to accommodate the support requirements of the DuraMelter™ 3300. One notable departure from the overall M-Area system design was that the LAW Pilot Melter incorporated a selective catalytic reduction (SCR) unit for the treatment of NOx emissions. Another significant departure was that gem-forming machines were not incorporated into the RPP flow sheet, and hence were omitted from the design of the LAW Pilot Melter.

Pilot Melter Testing Program
The LAW Pilot Melter test program evolved into three distinct phases. The initial phase was to validate key processing assumptions of the privatization contract. First, would the melter produce glass at the rate demonstrated at M-Area, i.e., 1 MT/m^2/day? Secondly, would the melter exceed a minimum life expectancy of one year? Upon completion of the first year of testing, the answers to both of these business and technology questions were a resounding "yes".

The program then entered its second phase of testing, which was to demonstrate that higher levels of sulfate could be incorporated into the LAW glass composition without adverse consequences to process rates or melter integrity. This second mission resulted from the project's need to find an alternative means of handling high levels of sulfate in the LAW feed, which were originally intended to be removed by a pretreatment step that proved ineffective during testing.

The third phase of the program resulted from contractual changes on the project, and the need to underpin new, higher production capacity (the required production capacity for the plant was increased), that would require demonstration during plant commissioning following construction.

In addition, the third phase of the testing program involved enhancing the life expectancy of the melter's bubblers. This would enhance circulation in the melt pool and increase the transport of heat to the surface of the pool thereby increasing the melter's production rate. In the earlier stages of testing, the melter's bubblers demonstrated a mean life expectancy of about 8 weeks. The project team set a target of 17 weeks for the life of the bubbler system on the LAW Pilot Melter, consistent with the design expectations for the full-scale plant.

Pilot Melter Production Capacity

Initially, the DM 3300 set out to demonstrate the ability to replicate the production rates of the M-Area's DM 5000A melter: ~ 1 MT/m^2/day. The concern here was that the Hanford LAW feeds were in the same chemical compositional ranges as the M-Area feeds, but not identical. Therefore, one might expect the Hanford LAW feeds to process at different rates than M-Area feeds, potentially jeopardizing the plant's ability to meet production targets (and the contractor's ability to recover its investment under the privatization contracting approach).

After an operational optimization test period using simulated Hanford LAW feeds, the DM 3300 proved that the system could meet the production goal of 1 MT glass/m^2/day. Testing continued to demonstrate that across three major LAW composition ranges, glass production rates in excess of 1 MT/m^2/day were achievable. The three chemical composition ranges were referred to as waste envelopes A, B, and C.

In early 2000, the project team decided to direct sulfate-rich material in the waste tanks directly into the vitrification process without pretreatment. The high sulfate feeds also exceeded the production benchmark and, in fact, they proved that the system could consistently exceed a production rate of 1.5 MT/m^2/day. Throughout the next several years of testing the system, all major waste compositional types, as well as variations of each (simulations of plant makeup errors), and the transition from one compositional target to others, demonstrated that production rates in excess of 1.5 MT/m^2/day were achievable and, in most cases, 2 MT/m^2 could be efficiently processed in a day (Fig. 2). The redesigned and enhanced system achieved a nearly two-fold increase in the production rate over that of the system used at the M-Area, which was itself a two-fold increase over previous state-of-the-art systems developed for the West Valley Demonstration Project (WVDP), and at the Savannah River Site's Defense Waste Processing Facility (DWPF) (Fig. 3). (Figures 2 and 3 are shown on the next page.)

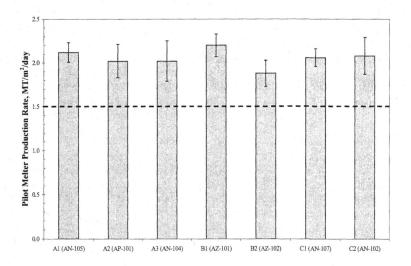

Fig. 2. LAW pilot melter production rates for the major compositional types. •MT/m^2/day= Metric tons glass produced/m^2 melt pool surface area (melter size)/day

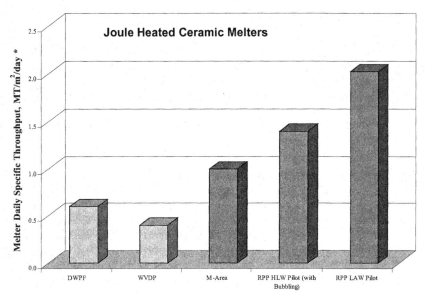

Fig. 3. HLW and LAW pilot melter glass production rates compared to M-Area, DWPF, and West Valley. MT/m^2/day= Metric tons glass produced/m^2 melt pool surface area (melter size)/day

LAW Pilot Melter Life Expectancy

At the time the LAW Pilot Melter program was conceived, there were no U.S.-based demonstrations of joule-heated ceramic melters for waste vitrification that had surpassed an operational period of much more than a year, and none had processed glass at the capacities necessary for the RPP-WTP to meet its mission on schedule. Therefore, it was a chief objective of the DM 3300 to exceed the one-year benchmark with waste materials having Hanford-relevant compositions. To satisfy this objective, the DM 3300 needed to operate continuously and around the clock for a significant enough duration to prove an extended operational life. The original business case for the project was that the full-scale LAW melter have an operational life of not less than one year. After one year of operating the pilot melter, and an examination of the interior of M-Area melter, the life of the system was estimated to be at least 3 years. At 4.25 years of continued operational testing (following the initial 2-3 month startup phase), more than 3,300,000 kilograms of glass have been produced (more than any joule-heated ceramic melter of its type), the DM 3300 has shown that a life expectancy approaching 5 years can be achieved with this technology. Not only did the DM 3300 show that it can continue to produce glass longer than the business case model and at the required rate of production, it also did so with higher levels of sulfate than the design specifications required.

LAW Pilot Melter Bubbler Life Optimization

The final development objective of the LAW Pilot Melter was to increase the life expectancy of the consumable bubbler devices. The bubblers, a patented enhancement to the joule-heated melter technology (see three referenced patents [1,2,3]), had an operationally demonstrated life expectancy of between 8 and 9 weeks, based upon the M-Area and early LAW Pilot Melter experience. Under the expected operational protocols of the RPP-WTP, each day of operational downtime will be a very costly event. Accordingly, the challenge was to increase the operational life of the bubblers to greater than 17 weeks. Over a nearly three-year period, Duratek engineers evaluated a menu of alloys, claddings, and design alterations to improve the life expectancy of these critical components. Ultimately, a solution yielded a 25.5-week life expectancy. For the proposed operating life of the LAW melter at the RPP-WTP, this technological advance alone will improve the operational availability of the system by about 5 percent.

HLW PILOT MELTER – DURAMELTER 1200

In January 2001, the sister HLW Pilot Plant (using the DuraMelter 1200 technology) was commissioned at the Vitreous State Laboratory of The Catholic University of America. This facility, also integral to the overall RPP-WTP research and testing program, also focused on improving the processing rates. In addition to being a one-third scale RPP-WTP HLW Melter (the DuraMelter 1200 – 1.2 m^2 melt pool surface area), this facility also incorporated a prototypic off-gas treatment system with a submerged bed scrubber (SBS) and a wet electrostatic precipitator (WESP).

Over the course of the next two and a half years, the HLW Pilot system produced data that support mass balance and production calculations for the system.

In the early stages of testing, it became apparent that simulated RPP HLW feeds were processed slower than other HLW feed types (DWPF and WVDP simulants). Accordingly, the testing focused on efforts to increase the processing rates [4]. A variety of options were reviewed, including altering feed solids content, adding frit instead of glass forming chemicals to the feed, and higher sugar concentrations (used as a feed reductant). With these, and other alternatives failing to provide sustainable production rates of greater than 0.8 MT glass/m^2 /day, a modified version of the LAW bubbling system was tested for efficacy with RPP HLW feed simulants.

Fig. 4 shows the results of some of these tests, where only tests that incorporated the bubbling system yielded throughputs in excess of the 0.8 MT glass/m^2/day. Here again, the technological success of the M-Area melter was incorporated into the RPP-WTP HLW process, and thus far, shows the ability to meet heightened production requirements without compromising routine processing operations.

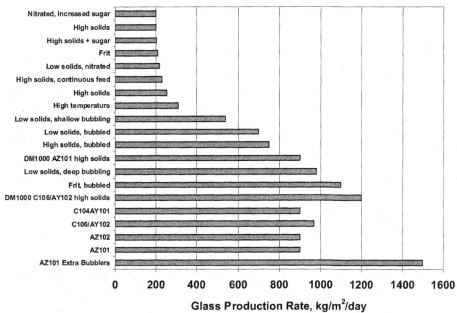

Fig. 4. HLW pilot melter production rate test results with and without bubbling [4]

SUMMARY

The operational throughput demonstrated during operation of the stand-alone waste vitrification facility at the Savannah River Site's M-Area was further enhanced and applied to the RPP-WTP Pilot Melter Test Program, collectively producing operational test data that will underpin the commissioning of the world's largest radioactive waste vitrification plant to be built at the Hanford Site (RPP-WTP). This underpinning includes the ability to meet 1) specific glass production rates for both LAW and HLW processing, 2) melter life expectancy, 3) high glass sulfate concentration, and 4) high process availability (through low rates of bubbler changeouts).

The M-Area project was the first commercial scale application of vitrification for mixed waste treatment, and the largest scale application of joule-heated ceramic melter technology in the United States. At almost 2-3 times the glass throughput of the nearest existing joule heated radioactive waste vitrification process, the M-Area project demonstrated a substantial increase in the state of the art and afforded an excellent platform for additional scaling input to the LAW and HLW melters for Hanford's RPP-WTP. The facility met all its design objectives, produced more than two thousand eight hundred sixty 270-liter drums of vitrified DuraGems (approximately 1,000,000 kg of glass), which satisfied established TCLP Universal Treatment Standards. The first of a kind DM-5000A Melter appeared relatively unchallenged by its 15-month operating campaign, suggesting that ample underutilized operating life expectancy remained.

The LAW Pilot Plant for the RPP-WTP located at Duratek's headquarters in Columbia, Maryland, and the HLW Pilot Plant at the Vitreous State Laboratory of The Catholic University of America built upon the M-Area experience to design, build and operate dedicated test facilities that support the ongoing mission of the RPP-WTP. The DM 3300, although somewhat smaller than the M-Area melter, produced more glass (3,300,000 kg), than any melter of its type and at higher sustained production rates (near 2 $MT/m^2/day$). The LAW Pilot Plant operated with few interruptions from January 1999 until April 2003. Since 2001, the DM 1200 has produced more than 180,000 kg of glass, which increased confidence that the design can be applied successfully to the RPP-WTP waste streams.

ACKNOWLEDGMENTS

The author gratefully acknowledges the efforts of the two hundred scientists, engineers, operators, and technicians of the Duratek, Inc. and the Vitreous State Laboratory of The Catholic University, whose diligent, conscientious, and tireless efforts over the past twelve years have resulted in tremendous advancements in joule-heated ceramic melter technology summarized in this manuscript.

REFERENCES

[1]Si Yuan Lin, U.S. Patent No. 5,868,814, "Apparatus for Recirculating Molten Glass."

[2]P. B. Macedo, R. K. Mohr, U.S. Patent No. 5,188,649, "Process for Vitrifying Asbestos Containing Waste, Infectious Waste, Toxic Materials and Radioactive Waste."

[3]P. B. Macedo, R. K. Mohr, U.S. Patent No. 5,340,372, "Process for Vitrifying Asbestos Containing Waste, Infectious Waste, Toxic Materials and Radioactive Waste."

[4]"Tests on the DuraMelter 1200 HLW Pilot Melter System Using AZ-101 HLW Simulants," K.S. Matlack, W.K. Kot, T.Bardakci, W. Gong, N.A. D'Angelo, T.R., Schatz, and I.L. Pegg, VSL-02R0100-2, Rev. 1, Vitreous State Laboratory, The Catholic University of America, Washington, DC, February 2003.

HIGH LEVEL WASTE PROCESSING EXPERIENCE WITH INCREASED WASTE LOADINGS

Carol M. Jantzen, Alex D. Cozzi, and Ned E. Bibler
Savannah River National Laboratory
Aiken, South Carolina 29808

ABSTRACT

The Defense Waste Processing Facility (DWPF) Engineering requested characterization of glass samples that were taken after the second melter (Melter #2) had been operational for ~5 months. After the new melter had been installed, the waste loading[†] had been increased to ~38 wt% after a new quasicrystalline liquidus model had been implemented. The DWPF had also switched from processing with refractory Frit 200 to a more fluid Frit 320. The samples were taken after DWPF observed very rapid buildup of deposits in the upper pour spout bore and on the pour spout insert while processing the high waste loading feedstock. These samples were evaluated using various analytical techniques to determine the cause of the crystallization. The pour stream sample was homogenous, amorphous, and representative of the feed batch from which it was derived. Chemical analysis of the pour stream sample indicated that a waste loading of 38.5 wt% had been achieved. The data analysis indicated that surface crystallization, induced by temperature and oxygen fugacity gradients in the pour spout, caused surface crystallization to occur in the spout and on the insert at the higher waste loadings even though there was no crystallization in the pour stream.

INTRODUCTION

Characterization of three glass samples that were taken from the first DWPF replacement melter (Melter #2) were performed. The samples were taken after the second melter had been in operation for ~5 months, after the waste loading had been increased to ~38 wt%, after the new quasichemical liquidus model [1] had been implemented, and after DWPF switched from processing with Frit 200 (high in SiO_2) to a more fluid Frit 320. The DWPF had observed a very rapid buildup of deposits in the upper spout bore and on the pour spout insert while processing the high waste loading feedstock. Rapid deposition in these locations had not occurred prior to processing the high waste loaded feeds and stopped after waste loading was decreased.

The sample analyses performed included chemical composition (including noble metals), crystal content, and REDuction/OXidation (REDOX) expressed as the $Fe^{+2}/\Sigma Fe$ ratio. The three glasses consisted of the following:

- a pour stream sample taken while filling canister S01859 during processing of Sludge Batch 2 (SB2),
- a sample that was scraped from the 2 inch upper pour spout bore (Figure 1) using a 1-7/8" diameter rotating drill bit while the melter was hot; the material had grown/accumulated at or just below the transition from the riser to the pour spout,
- a sample of glass adhering to a Melter #2 Type I insert (Figure 1) that had spalled from the interior of the insert after it had cooled.

[†] The waste loading (w) is defined thoughout this paper on a consistent basis, e.g. $w=100*[1-(g_{Li}/f_{Li})]$ where g_{Li} = grams of Li in 100 grams of glass and f_{Li} = grams of Li in 100 grams of frit

The samples were taken on August 28, 2003. The glass being processed at that time corresponded to DWPF batch 245.

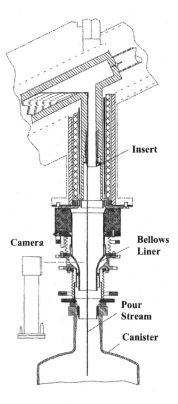

Figure 1. DWPF pour spout schematic showing location of the upper bore (section above the insert), the insert, the pour stream, and the canister.

BACKGROUND

The Savannah River Site has operated various pilot scale melters in support of the design of DWPF (Table I) since 1978. All have had prototypic risers and pour spouts. Two of the pilot scale melters experienced crystalline buildups in the pour spout risers and nozzles, and these case studies were examined for relevance to the current deposition problems.

The Project 1941 melter was ~½ the DWPF melt pool surface area and was initially dry fed a calcine oxide sludge and later slurry fed. The melter produced 74 tons of simulated waste glass. During the dry calcine feeding, a crystalline layer of ~9 inches thick formed at the bottom of the melter [2]. This material formed primarily during a 58 day idle at 1050°C. Upon probing the melter after the idle period, the deposits on the bottom of the melter floor were determined to be very dense ("hard") and comprised 4 inches of spinel deposits [2]. A less dense spinel layer comprised the remaining 5 inches of deposits. When the 1941 large scale melter was shut down, it was dismantled to evaluate its service life [3]. Additional crystalline deposits up to approximately 1 inch thick were found to have formed on the walls of the melter and in the riser and nozzle. The crystalline deposits found in the riser and nozzle of the 1941 melter were found to have almost completely plugged the riser [3]. When enough crystalline material had accumulated to fill the bottom of the 1941 melter, the deposits began to come out of the melter with the waste glass [3].

Table I. Pilot Scale Melters Operated In Support of DWPF Design.

Melter Designation	Melt Pool Surface Area (ft^2) *	Years of Operation	Pour Spout/Riser Pluggages	Crystal Buildup on Floor of Melter During Operation
Large Scale Project 1941 Melter	12	1978-1979	Yes	9"
Small Cylindrical Melter (SCM)	1.3	1979-1982	No	Several inches per campaign
Large Slurry Fed Melter (LSFM)	12	1982-1985	No	1/16-1/2"
Scale Glass Melter (SGM)	12.6	1986-1988	Yes	None
Integrated DWPF Melter System (IDMS)	3.14	1988-1994	No	None

* DWPF Design Basis of 228 lbs/hr or 8lbs/hr(ft)^2 times melt pool surface area (ft^2)

The large accumulations of three chemically distinct layers that eventually blocked the riser in the 1941 melter were attributed to the following:

- the bottom deposits were composed of Cr enriched spinels, an unidentified silicate, and a glass matrix enriched in alumina – this was attributed to corrosion of the K-3 refractory
- the second layer was composed of spinels, acmite, a glassy matrix, and entrapped waste glass of a normal (not alumina enriched) composition
- the top layer was similar to the second layer but did not contain entrapped waste glass.

Since the DWPF Melter #2 had not be idled at 1050°C nor had it been fed calcine feeds, this potential mechanism for pour spout crystallization was not considered relevant to the current accumulation of crystalline material.

The Scale Glass Melter (SGM) produced ~90 tons of glass in two years based on Frits 165, 168, and 200 and a reducing (formic acid) flowsheet. Crystalline deposits were not found in the SGM when the melter bottom was probed after the 5th melter Campaign (SGM-5) [4]. When the SGM was bottom drained after the 9th Campaign (SGM-9), no significant accumulations of deposits were observed [5]. However, in-situ formation of crystalline deposits did occur in the SGM melter pour spout and the deposition was similar to those experienced by DWPF Melter #2.

Deposition was experienced during the SGM-1 campaign due to a cool pour spout tip [6]. Temperatures were 980°C on the glass contact side near the glass disengagement point, 1040°C closer to the pour spout bore and 600°C on the opposite side of the pour spout bore during steady state pouring [7]. Crystallization of these deposits resulted in frequent channel pluggages and reduced production rate ~50%.

Pluggages formed in the SGM-1 pour spout discharge tube on 10 different occasions. The pluggages were composed of visibly crystallized glass. Batch pouring aggravated plugging; pluggages appeared to become worse as the length of time between pours increased. Characterization of the size, composition, and volume fraction of the crystallized (devitrified) phases in the SGM-1 pour spout was used to interpret the thermal history of the glass [8]. The identification of acmite crystals indicated that the affected sections of the pour spout were as cool as ~700°C while the thermocouples indicated that this region was hotter. The large size of the crystals indicated that the crystals had formed in the pour spout at these low temperatures over long time periods.

The SGM-1 pour spout pluggage and associated interpretation of the thermal history [8] led to a redesign of the pour spout, e.g. better insulation and relocation of the thermocouples to accurately profile pour spout temperatures. After this redesign, the SGM did not experience severe pour spout pluggages even when processing waste loadings up to 42 wt%.[‡] This suggested that the temperature profiles in the DWPF Melter #2 pour spout were relevant to the current accumulation of crystalline material.

PILOT SCALE MELTER WASTE LOADINGS AND LIQUIDUS TEMPERATURES

The glass composition data from the SCM-2, the LSFM, and the SGM was compiled into a database [9]. Compositional data for the LSFM and SCM-2 was compiled from analyzed feed and frit compositions. Compositional data for the SGM campaigns was from analyzed glasses taken from full sized canisters. Compositional data for the Project 1941 melter could not be found. The analyzed glass composition data from the IDMS melter campaigns was available but the glass analyses from these melter campaigns was suspect for liquidus temperate predictions due to Cr_2O_3 contamination from the grinders used [10]. Since Cr_2O_3 content has a large impact on the liquidus temperature calculated from the quasicrystalline model [1], this data was not included in comparison of pilot scale melters and DWPF operational history, e.g. predicted liquidus and waste loading.

The data in reference 9 was used to calculate the DWPF liquidus temperature from the DWPF historic liquidus model [11], the DWPF liquidus temperature from the newly implemented quasicrystalline model [1], and the waste loading achieved based on the Li_2O content of the frit.[‡] A comparison of the predicted liquidus temperature model calculated two different ways is shown in Figure 2 while the comparison of the quasicrystalline liquidus temperature to waste loading is shown in Figure 3. The ordinary least squares equation of best fit for the data shown in Figure 2 is

Quasicrystalline Liquidus (°C) = -590.476 + 1.5276 (Historic Liquidus,°C).   (1)

with an adjusted R^2 of 0.82 and a Root Mean Square Error of 34.58. Figure 2 shows the following:

- there is a linear correlation between the historic and quasicrystalline liquidus models
- there is approximately a 36°C offset in the historic liquidus (1050°C) and the corresponding quasicrystalline liquidus (1013°C) at the DWPF liquidus temperature limit of 1050°C

[‡] The waste loading (w) is defined thoughout this paper on a consistent basis, e.g. $w = 100*[1-(g_{Li}/f_{Li})]$ where g_{Li} = grams of Li in 100 grams of glass and f_{Li} = grams of Li in 100 grams of frit

- pilot scale melters such as the LSFM operated at lower liquidus values than the other pilot scale melters, e.g. SGM and SCM-2

A combination of the data displayed in Figure 2, the data given in reference 9, and the operating experiences summarized in Table I shows the following:

- the lower liquidus values experienced during the LSFM campaigns was associated with lower waste loadings in the range of 20-32 wt% when calculated from the frit Li_2O values
- the scale glass melter (SGM-6), which had a DWPF prototypic pour spout and ran reducing flow sheets, ran some very high waste loadings (up to 42 wt%) and no pour spout pluggages were experienced once the pour spout was insulated and the thermocouples relocated after the SGM-1 campaign.

EXPERIMENTAL

Elemental Analyses

The samples from Melter #2 were prepared for chemical analyses by pulverizing a portion of each in a Wig-L-Bug using agate balls and vial. The pulverized sample was sieved to <100-mesh (149 μm) and dissolved by two different dissolution methods to account for all the elements of interest.

A standard reference glass, Approved Reference Glass #1 (ARG-1), was analyzed at the same time as the unknowns. Each standard and each unknown was dissolved in quadruplicate and one replicate analysis of each sample was performed. The quadruplicate analyses were averaged to create the data in Table II. The dissolutions were analyzed by Inductively Coupled Plasma – Emission Spectroscopy (ICP-ES) and ICP Mass Spectroscopy (ICP-MS). The peroxide fusion analyses are reported in Table II preferentially because undissolved solids were found in some of the mixed acid digestions. Details are given elsewhere [9].

In order to provide a representation of the expected composition of the pour stream sample, the analysis of Slurry Mix Evaporator (SME) batch 245 and Melter Feed Tank (MFT) batch 245 were converted to oxides using the DWPF Product Composition Control System (PCCS). Table II gives the composition of the three DWPF samples as well as the composition of the Savannah River National Laboratory (SRNL) Tank 40 qualification glass sample and the measured composition of SME Batch 245 and MFT batch 245.

The composition of the pour stream sample resembles the SRNL Tank 40 qualification glass made with Frit 320 in the Shielded Cells Facility in Al_2O_3, MgO, B_2O_3, CaO, CuO, Li_2O, MgO, MnO, P_2O_5, and U_3O_8 (Table II) as it should, i.e., the Tank 40 sample was representative. The pour stream sample resembles most of the major components as determined by analysis of SME batch 245 (Table II). This is verified in Table III as the ratios of most components in SME batch 245 divided by the concentrations in the pour stream sample (PC0033) are close to 1.0 as they should be. The SME batch 245 analyses appear to be biased high or the pour stream samples biased low for Al_2O_3, B_2O_3, CaO, Fe_2O_3, NiO and U_3O_8 by 15-20%. The MFT batch 245 analyses divided by the concentrations in the pour stream sample (PC0033) are ~1.0 except for the Al_2O_3 and B_2O_3 analyses which are biased by ~20% (Table II).

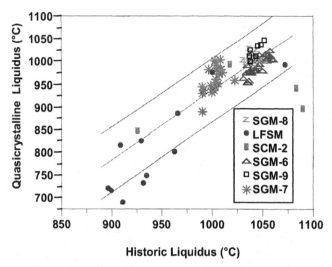

Figure 2. Comparison of the liquidus temperatures for various pilot scale melters calculated with the DWPF historic and quasichemical modes.

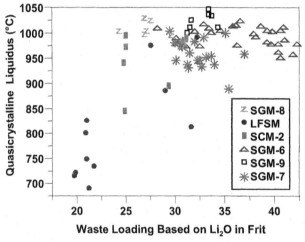

Figure 3. Comparison of the liquidus temperatures and waste loadings for various pilot scale melters calculated with the newly implemented DWPF quasicrystalline model.

The compositions of the analyses reported in Table II were used to calculate a predicted glass viscosity, liquidus temperature, and waste loading. The predicted viscosity based on the vitrified MFT product (43.48 poise) and the pour stream sample (46.05 poise) was in agreement to within 2.5 poise. The predicted liquidus based on the vitrified MFT 245 product (987°C) and the pour stream sample (997°C) was in agreement to within 10°C. The waste loadings predicted from the MFT (34.74 wt%) and the pour spout analyses (38.53 wt%) agree to within 3.79 wt%.

The composition of the upper pour spout bore sample (PC0006) was very different from the pour stream sample (PC0033) and very different from the MFT and SME analyses. The pour spout bore sample was deficient in Al_2O_3, B_2O_3, CaO, Li_2O, Na_2O, U_3O_8 and SiO_2 (Table III). Based on the relative Li_2O content of the pour stream sample (PC0033) to the bore sample (PC0006) it is estimated that the upper pour spout bore sample was ~62 wt% glass (Table IV). Based on the relative SiO_2 content of these samples it is estimated that the upper pour spout bore sample is ~52 wt% glass (Table IV). Compared to the pour stream, the upper pour spout bore sample was enriched in Cr_2O_3 over the pour stream sample by 35.9X, enriched in Fe_2O_3 by only 2.6X, and enriched in NiO by 15.4X. Based on the analyzed compositions given in Table II and the X-ray diffraction spectra, the number of moles of NiO, Cr_2O_3, and Fe_2O_3 over those reported in the pour stream were calculated. Based on these molar compositions it could be determined that 0.11 moles of $NiFe_2O_4$ and 0.034 moles of $NiCr_2O_4$ spinels comprised the remainder of the bore samples. When converted to weight percent, a mass balance indicated that the deposits were ~62 wt% glass, 25.78 wt% $NiFe_2O_4$ (trevorite) and 7.7 wt% $NiCr_2O_4$ (Table IV). The upper pour spout bore is also highly enriched in noble metal components; especially Rh_2O_3 and RuO_2 (Table V).

The insert sample (PC0031) was enriched the most in Cr_2O_3 over both the pour stream sample (PC0033) and upper pour spout bore sample (PC0006). The insert sample contains less Fe_2O_3 and NiO than the upper pour spout bore. This indicates that less $NiFe_2O_4$ spinels are accumulating in this area, e.g. the mass balance analyses indicate ~16.4 wt% $NiFe_2O_4$ (see Table IV). However, the high Cr_2O_3 content supported by the XRD identification of a Cr_2O_3 only phase indicates that there may be a reaction occurring with the hot glass and the Cr in the Inconel® 690 alloy insert. That is, an oxidized film of Cr_2O_3 may be forming to which some molten glass adheres. Since the sample received had "spalled off" the insert as it cooled, it is likely that the sample contained a good deal of the oxidized Cr_2O_3 film from the Inconel® 690. The mass balance indicates that ~21 wt% Cr_2O_3 comprises the analyzed deposits. Table III presents the ratios of the pour stream sample major components to the other compositions from Table II. The insert sample is also enriched in noble metals (Table V).

Noble Metal Analyses

The solutions that resulted from the peroxide fusion of the three samples were analyzed by ICP-MS for noble metals. Concentrations in weight percent along with the respective concentrations measured in the SRNL Tank 40 qualification sample are given in Table V. The data from the SB2 pour stream and insert are also given in Table V for comparison. The peroxide fusion (PF) was used to ensure that all the refractory spinels in the pour spout and insert samples were dissolved because the noble metal, Ru as RuO_2, is most often in the center of an insoluble spinel crystal where it has acted as a nucleating site. Based on the PF data in Table V, the noble metals, Ru and Rh were 56 to 81 times more concentrated in the upper pour spout bore than in the pour stream. The Ru and Rh were 10 and >17 times more concentrated in the insert glass.

Table II. Measured Compositions of the DWPF Melter #2 Samples Compared to SRNL Tank 40 Glass and DWPF SME Batch 245 and MFT Batch 245 (in Oxide Wt.%)

Oxide	Dissolution/ Analysis Methods	Pour Stream (PC0033)	Upper Pour Spout Bore (PC0006)	Insert (PC0031)	SRNL TK 40 Glass[12]	SME Batch 245	MFT Batch 245
Ag_2O	PF/ICPMS	0.0072	BDL	0.006	NM	NM	NM
Al_2O_3	PF/ICPES	4.33	2.40	2.54	4.93	5.59	5.04
B_2O_3	PF/ICPES	4.33	<2.93	<2.68	4.87	5.21	5.26
BaO	PF/ICPES	0.03	0.02	0.01	NM	NM	NM
CaO	PF/ICPES	1.48	0.95	0.87	1.53	1.69	1.26
CdO	PF/ICPES	0.05	0.05	0.05	NM	NM	NM
Cr_2O_3	PF/ICPES	0.15	5.38	21.22	0.31[a]	0.12	0.11
CuO	PF/ICPES	0.09	0.07	0.03	0.14	0.03	0.03
Fe_2O_3	PF/ICPES	13.64	36.03	25.19	16.16	15.69	14.07
K_2O	PF/ICPES	BDL	BDL	BDL	NM	0.1381	0.10
La_2O_3	PF/ICPES	0.04	0.04	0.05	NM	NM	NM
Li_2O	PF/ICPES	4.75	2.93	3.48	5.30	4.85	5.16
MgO	PF/ICPES	1.28	1.22	1.48	1.30	1.31	1.10
MnO	PF/ICPES	1.67	2.35	3.29	1.84	1.73	1.54
Na_2O	MA/ICPES	11.01	6.30	6.26	12.61	11.39	11.02
NiO	PF/ICPES	0.60	9.25	5.84	0.74	0.71	0.63
P_2O_5	PF/ICPES	0.60	<0.63	<0.63	0.74	<0.22	NM
SiO_2	PF/ICPES	46.87	24.26	24.55	51.2	45.67	48.11
SnO_2	PF/ICPES	0.21	0.20	0.19	NM	NM	NM
SrO	PF/ICPES	0.30	0.20	0.20	NM	NM	NM
Rh_2O_3	PF/ICPMS	0.0033	0.271	0.058	NM	NM	NM
RuO_2	PF/ICPMS	0.030	1.712	0.296	NM	NM	NM
TiO_2	PF/ICPES	0.04	0.03	0.11	NM	0.05	0.05
U_3O_8	PF/ICPES	3.45	2.63	2.45	4.06	4.22	3.80
ZnO	MA/ICPES	0.09	0.16	0.17	NM	NM	NM
ZrO_2	MA/ICPES	0.07	0.03	0.03	NM	BDL	0.09
SUM (w/o <)		95.12	96.48	98.37	105.73	98.4	97.37
Calculated Viscosity @1150°C (poise)		46.05	N/A	N/A	38.99	35.91	43.48
Calculated Liquidus (°C)		997	N/A	N/A	1056	1038	987
Calculated WL (Li_2O)		38.53[b]	N/A	N/A	36.10	39.32[b]	34.74[b]

(NM-Not Measured; BDL-Below Detection Limit; N/A-Not Applicable; MA-Mixed Acid dissolution; PF- Peroxide Fusion dissolution; WL-Waste Loading)

[a] Sample prepared in stainless steel grinder for Sludge Batch 2 (SB-2) qualification with Frit 320.

[b] Calculated using a value for Li_2O in the normalized frit of 8.13 wt% based on a weighted average of Frit 320 Lots 5, 8, and 13 in the ratio of 1:1:2.

Table III. Ratio of Major Components of the Pour Stream Sample to the Upper Pour Stream Bore Sample, to the Insert Sample, and to the SME and MFT 245 Analyses

Oxide	Insert /Pour Stream	Upper Pour Spout Bore /Pour Stream	SME batch 245/ Pour Stream	MFT batch 245/ Pour Stream
Al_2O_3	0.59	0.55	1.29	1.16
B_2O_3	N/A	N/A	1.20	1.21
CaO	0.59	0.64	1.14	0.85
Cr_2O_3	141.47	35.87	0.8	0.73
Fe_2O_3	1.85	2.64	1.15	1.03
Li_2O	0.73	0.62	1.02	1.09
MgO	1.16	0.95	1.02	0.86
MnO	1.97	1.41	1.04	0.92
Na_2O	0.56	0.57	1.03	1.01
NiO	9.73	15.42	1.18	1.05
SiO_2	0.52	0.52	0.97	1.03
U_3O_8	0.71	0.76	1.22	1.10

Table IV. Mass Balance for Samples Based on Data in Table II.

Calculated (Wt%)	Pour Stream (PC0033) Frit 320	Upper Pour Spout Bore (PC0006) Frit 320	Insert (PC0031) Frit 320
Glass (Based on Li_2O in Pour Stream	100	62	73
$NiFe_2O_4$ Spinel	0	25.78	16.4
$NiCr_2O_4$ Spinel	0	7.70	0
Cr_2O_3	0	0	21.07
$RuO_2 + Rh_2O_3$	0.03	1.98	0.36
SUM	100.03	97.46	110.83

Table V. Comparison of the Noble Metals (wt.%) of the SRNL Tank 40, Pour Stream, Upper Pour Spout Bore, and Insert Glasses.

Isotope	SB2 Pour Stream Glass (August 2002)	SB2 Insert Glass (August 2002)	SB2 Pour Stream Glass (PC0033)	SB2 Upper Pour Spout Bore (PC0006)	SB2 Insert (PC0031)	Ratio Insert/ Pour Stream[b]	Ratio Upper Spout Bore/ Pour Stream
Ru	NA	NA	0.023	1.3	0.23	10	56.52
Rh	NA	NA	0.0027	0.22	0.047	17.41	81.48

[b] This ratio should be ~0.47 because of dilution of the glass components by material from the insert (see text).

REDOX Analyses

Portions of the pour stream sample (PC0033) were pulverized using a non-metallic Wig-L-Bug in the SRNL Shielded Cells. The Environmental Assessment (EA) glass, a REDOX standard, was prepared by grinding it in a Tekmar grinder outside the SRNL Shielded Cells. The EA glass standard has a REDOX, expressed as ($Fe^{2+}/\Sigma Fe$), of ~ 0.18 [13]. Dissolutions of triplicate samples were performed using <100 mesh material prepared in the same manner as the material for compositional analysis. The dissolution was performed in the SRNL Shielded Cells. The spectrophotometric analysis were performed in a radiohood.

Table VI gives the results of the triplicate REDOX analyses of the pour stream glass and the EA glass standard. The measured REDOX of the EA glass was greater than expected. Microscopic investigation of the EA glass standard indicated that it was contaminated with metal filings from the Tekmar grinder. This did not affect the REDOX of the pour stream sample since it was ground with a non-metallic Wig-L-Bug.

A second pour stream sample and EA glass standard were both ground with the Wig-L-Bug to prevent metal contamination and the REDOX was re-measured in triplicate in the SRNL Shielded Cells (see Table VI). This set of analyses gave a blank corrected EA glass standard value of $Fe^{+2}/\Sigma Fe=0.24$ versus the reported standard value of 0.18 indicating that the problems with iron contamination had been avoided by using the Wig-L-Bug. The average remeasured values for the triplicate (PC0033) samples were 0.20 in agreement with the previous values determined in Table VI. The predicted REDOX of SME batch 245 was 0.17 based on the {[F]-[3N]} REDOX correlation [14] and 0.15 based on the new Electron Equivalents REDOX correlation [15]. This indicates that the REDOX model [1,15] predictions and batching in the DWPF SRAT and melter are working correctly and are on target.

Table VI. REDOX of Pour Stream Glass Prepared in the SRNL Shielded Cells.

Set	Sample	$Fe^{+2}/\Sigma Fe$ Standard Value	Target $Fe^{+2}/\Sigma Fe$ of Pour Stream	Blank Corrected Average ($Fe^{+2}/\Sigma Fe$)
1	EA Standard	0.18	N/A	0.41*
1	Pour Stream	N/A	0.15-0.17	0.20
2	EA Standard	0.18	N/A	0.24
2	Pour Stream	N/A	0.15-0.17	0.20

* iron or steel contamination observed

Contained Scanning Electron Microscopy (CSEM)

The <200 mesh crushed samples from the chemical analyses were used for Contained Scanning Electron Microscopy with Energy Dispersive Spectroscopy (CSEM/EDS). The CSEM analysis of the pour stream sample revealed uniformity across the entire sample and showed no crystallization. The EDS spectrums of various pour stream samples also indicated homogeneous DWPF-type glass.

The insert sample viewed at 500x appeared to have more surface texture than a typical glass sample. A single grain was examined that had a coating of a Fe and Cr rich material compared to the right hand side of the same grain that had an EDS spectra typical of glass. The coating appeared to be a portion of the Inconel® 690 to which the once molten glass had adhered.

The CSEM/EDS of the insert sample revealed copious amounts of spinels. Some of the spinels were more enriched in Cr than others. The non-crystalline portion of the sample gave a spectra typical of glass and demonstrated that the U component is in the glassy phase and does not participate in the crystallization. The EDS spectra did not indicate that the spinels were associated with RuO_2.

Contained X-ray Diffraction Analysis (CXRD)

Three different portions of the pour stream were analyzed by CXRD. The XRD pattern of the pour stream samples was typical of a borosilicate glass and free of any indicators of crystalline matter except for potential stainless steel (ss) contamination from grinding.

The CXRD analysis of the insert sample indicated the presence of glass, spinel, and chrome oxide (eskolaite). The spinel phase most likely resembles trevorite with chromium partially substituting for iron and iron partially substituting for nickel, based on the CSEM analyses.

The CXRD analysis of the upper pour spout bore sample indicated the presence of glass, spinel, and RuO_2. This is consistent with the chemical analyses of the upper pour spout bore sample being enriched in RuO_2 compared to the pour stream glass (See Table II and Table V). Likewise, there is an 81% enrichment of Rh_2O_3 (Table V). Therefore some mechanism is causing the RuO_2 and Rh_2O_3 to accumulate in the upper pour spout bore area over and above the amount of accumulation of the spinel forming components.

The spinel identified in the upper pour spout bore sample (PC0006) is a mixture of ~26 wt% trevorite ($NiFe_2O_4$) and ~8 wt% $NiCr_2O_4$ (see Table IV). The Cr_2O_3 is enriched in these deposits ~36X (Table III) relative to the amount in the pour stream sample (PC0033). Corrosion (oxidation) of the Inconel® 690 lining of the bore in this oxidizing environment on the pour spout is the most likely primary source of the Cr^{+3}. Therefore, the molten glass and the Inconel® 690 pour spout lining are chemically interacting in this hot oxidizing environment. That is to say that the oxygen fugacity in the upper pour spout bore is more reducing than that of the insert (air $\log f_{O_2} = -0.68$) but considerably more reducing than that of the melt pool (melt pool $\log f_{O_2} = -5.5$). Therefore, the region of the upper pour spout bore experiences large gradients in both temperature and oxygen fugacity which can induce spinel crystallization, e.g. the measured activation energy for spinel crystallization in DWPF type waste glass in an oxidizing atmosphere ($Fe^{+2}/\sum Fe\sim0$ at a $\log f_{O_2} = -0.68$) is 17.7 kcal/mole while the activation energy for spinel crystallization from a reducing glass ($Fe^{+2}/\sum Fe\sim0.5$ at a $\log f_{O_2} = -7$) is only 2.9 kcal/mole [16]. Therefore, crystallization of spinel is more rapid in the oxidizing atmosphere of the upper pour spout bore and insert than in the melt pool.

DISCUSSION

Volume Versus Surface Crystallization

Volume crystallization§ can involve rapid nucleation of the melt pool. Once formed, the type of $NiFe_2O_4$ spinel crystals that occur in DWPF waste glass melts are refractory (reported melt temperature of 1660±10°C [17]) and cannot be redissolved into the melt pool at the DWPF operating temperature of 1150°C. Therefore, the DWPF liquidus temperature model focused on preventing heterogeneous volume crystallization rather than preventing surface crystallization [18].

---

§   crystal growth begins from either homogeneous or heterogeneous nucleation sites with a melt [27]; volume crystallization of the spinel primary liquidus phase has been shown to be heterogeneous forming on melt insolubles in the waste such as RuO_2 [11,13].

Surface crystallization[**] has not been considered to be problematic in nuclear waste glass melters since spinel precursors ($NaFe_2O_4$ [19,20] and $LiFe_2O_4$ [21]), which can redissolve in the melt pool, have been found to form at the melt-atmosphere interface rather than insoluble $NiFe_2O_4$ spinels. Moreover, waste glass melts have been found to form a protective layer along the refractory walls which minimizes spinel formation in the melt pool from the refractory surfaces [22, 3], as long as the melt pool agitation or bubbling does not directly impinge on the melter walls. Surface crystallization can, however, be problematic where metallic materials of construction contact glass at temperatures lower than the liquidus temperature.

Pour Stream Sample: Volume Crystallization

Visual observation of the pour stream sample (PC0033) showed the sample to be typical of a DWPF-type glass (opaque and reflective). Compositional analysis by ICP-ES demonstrated a correlation between the pour stream sample and the MFT batch 245. The MFT analysis predicted a glass viscosity within 2.5 poise of the pour stream sample calculated viscosity. The MFT 245 analysis predicted a liquidus within 10°C of the pour stream calculated liquidus. The SME and pour stream calculations for waste loading agree to within 0.35 wt%, while the MFT and the pour stream calculations for waste loading agree to within 3.79 wt%. The measured REDOX was $Fe^{+2}/\Sigma Fe$ = 0.2 while the SME 245 target based on the {[F]-[3N]} correlation was 0.17 and the target based on the Electron Equivalents model was 0.15. Therefore, the viscosity [11], liquidus [1], and REDOX [14,15] models based on feed analyses appear to be adequately controlling the DWPF glass properties.

The pour stream sample, analyzed by CXRD in triplicate, contained no crystals and was totally amorphous. Therefore, there is no crystallization of spinel in the melt pool, which means that the liquidus model [1] is predicting and preventing volume crystallization in the melt pool as it was intended to do.

Pour Spout and Bore Samples: Surface Crystallization

Noble Metals: The pour spout insert sample was enriched in Ru 10X and Rh 17X over that present in the pour stream. The Ru as RuO_2 and the Rh as Rh_2O_3 could be acting as nuclei for the crystallization of the 16.4 wt% $NiFe_2O_4$ spinel, but during CSEM no RuO_2 or Rh_2O_3 were observed to be associated with or acting as nuclei for the crystallization of $NiFe_2O_4$ spinel. The role of the RuO_2 and Rh_2O_3 is indeterminate although an intermediate oxide compound $RhCrO_3$, is known to occur [23]. In addition, if RuO_2 is reduced to $Ru°$ locally in contact with the Inconel® 690 alloy, there are known solid solutions between $Ru°$ and $Ni°$ [24].

The upper pour spout bore contained 1.71 wt% RuO_2 and 0.27 wt% Rh_2O_3 in the deposits analyzed which is a 56X and 81X increase of these components in the pour spout bore samples compared to the pour stream. The CXRD analysis confirms the presence of an amorphous phase, a spinel phase, and RuO_2. In this case spinel and RuO_2 deposition could be synergestic.

Crystalline Deposits: Spinel and Cr_2O_3: The DWPF liquidus model was developed to prevent volume crystallization of the melt pool at the normal melter operational temperatures, e.g. between 1050-1150°C, at normal oxygen fugacities experienced in waste glass melters, e.g. between $\log f_{O_2}$ =-2 and $\log f_{O_2}$ =-9. Operation of the SGM melter, specifically SGM Campaign 6, at waste loadings in excess of 38 wt% (Figure 3), e.g. in the range in which DWPF

[**] crystal growth begins (i.e. nucleates) from the melt-atmosphere interface or the melt-container (melt-refractory) interface and grows perpendicular to the interface

experienced severe pour spout crystallization, is achievable if the pour spout is well insulated and kept hot.

As the glass flows up the riser, down the pour spout, and over the pour spout insert the following occurs:

- cooler temperatures are encountered, i.e., <1050°C which is below the liquidus of the glass being poured which enhances the kinetics of crystallization of spinel
- more oxidizing atmospheres (fugacities) are encountered, e.g. air $\log f_{O_2} = -0.68$ which enhances the kinetics of the crystallization of spinel [16] relative to the reducing atmosphere of the melt pool
- cooler Inconel® 690 surfaces are contacted that act as heat sinks inducing surface crystallization instead of bulk or volume crystallization
- cooler Inconel® 690 surfaces are contacted that are themselves being oxidized due to exposure to air and these surfaces release Cr_2O_3, which can further serve to nucleate spinels

Such crystallization in the melter riser, specifically in the tip of the pour spout channel had been observed during the first campaign of the DWPF pilot SGM. Crystallized deposits formed on ten separate occasions and were attributed to the fact that the tip of the pour spout channel lacked sufficient insulation which caused this region to be significantly cooler, ~800°C, than the thermocouples were indicating. Additional insulation and relocation of the thermocouples remediated the pluggage difficulties.

If the DWPF pour spout insert and upper pour spout bore are cooler than the liquidus temperature predicted by the DWPF Product Composition Control System (PCCS), e.g. a liquidus temperature of 987°C predicted for MFT 245 (Table II) and a liquidus of 997°C predicted from the pour stream analysis (Table II), then surface nucleation of crystals on these cooler surfaces is more likely to occur. This is because a higher waste loaded melt is closer to its crystallization temperature when it exits the melter than a lower waste loaded melt. Thus, unless a higher waste loaded melt is moved through the cooler region very rapidly, the glass crystallizes instead of "undercooling" to an amorphous state. In other words, the riser temperature profile is too steep (the bore is not hot enough) which allows spinels to form at higher waste loadings because the cooling rate is not fast enough in this region.

Heat Sink Induced Crystallization: In order to demonstrate the importance of cooling rate, dT/dt (where T is temperature in °C and t is time in seconds) calculations were performed assuming different temperatures for the upper pour spout bore ranging from 1100°C (the temperature of the LSFM bore), to 1040°C (the temperature of the SGM-1 bore), to 980°C (the temperature of the SGM-1 pour tip). The equation for the lowering of temperature with time for a finite body in contact with a heat sink (substrate) at a lower temperature may be written as follows [25]

$$\frac{T_t - T_s}{T_{t=0} - T_s} = \exp\left(t\left(\frac{K_m}{\rho \bullet C_p}\right)d^2\right)$$
(2)

T_t = temperature at time t (°C)
T_s = temperature of the substrate (°C)
K_m = thermal conductivity at the melting point (cal/cm sec °C)
C_p = specific heat (cal/g °C)
d = thickness (or radius if the cooling body is spherical) in cm
ρ = density in g/cm^3 at the melt temperature

Since DWPF glasses undergo Newtonian cooling, the term hd can be substituted for K_m where h is the heat transfer coefficient. Taking the logarithms of Equation 2 and differentiating with respect to t, the following expression for an instantaneous cooling rate (Q) can be calculated:

$$Q \equiv \frac{dT}{dt} = (T_t - T_s)\left(\frac{h}{\rho \bullet C_p}\right)d$$
(3)

where h = heat transfer coefficient (cal/cm^2 sec °C)
Note that the latent heat of fusion (ΔH_f) does not enter the calculation since crystallization only intervenes when the cooling rate is not great enough to prevent diffusional ordering from occurring [25]. The following parameters were used to approximate h, ρ, d, T_t, T_s, and C_p for DWPF type glass for a relative comparison of how substrate temperature can impact surface crystallization:

- radius of the DWPF pour stream ~0.25 cm
- $T_t = T_L$ (°C)
- T_s = varied from 900°C to 1150°C
- ρ = 2.47 g/m^3 for a high-Fe 131 glass [26,27] which is similar to the measured room temperature density in this report of 2.503 g/cm^3
- C_p = specific heat at T_s for a high-Fe 165 glass [28] which is similar to a high-Fe Frit 320 glass
- h = 1.7 cal/cm^{2}sec°C valid for SRL 131/Stage 1 waste glass above 900°C.

This allows Q ≡ dT/dt, a critical cooling rate to avoid crystallization to be estimated for DWPF glasses depending on the temperature of the Inconel® 690 substrate. This calculation is only approximate but serves to illustrate how much more rapidly the glass must be cooled as the Inconel® 690 substrate temperatures decrease, i.e., as the $\Delta T \equiv T_L - T_s$, the degree of undercooling increases. For example, at the liquidus temperature of 997°C for the DWPF pour spout sample (PC0003) analyzed in this report (see Table II) and a heat sink (pour spout) temperature of 980°C (the temperature of the pour spout tip near the disengagement point for SGM-1 from reference 7), a cooling rate of ~+9.7°C/sec is needed to prevent crystallization (see Table VII). At the pour spout bore temperature of SGM-1, e.g. 1040°C [7], a cooling rate of <1°C/sec is needed to prevent crystallization. If the pour spout were hotter, e.g. 1100°C, then a slower cooling rate is needed to prevent crystallization. This latter substrate temperature is consistent with the riser and pour spout temperatures reported during the 5th campaign of the LSFM melter, e.g. in the range of 1125°C ± 10°C and 1075°C ± 10°C, respectively, when no pour spout pluggages were observed [29].

For a melt with a liquidus of 1050°C, the cooling rates necessary to prevent crystallization at 980°C, 1040°C, and 1100°C become higher, e.g. 38°C/sec, +6°C/sec and <1°C/sec, respectively. So, more rapid cooling rates are necessary for the same substrate temperatures at higher waste loadings when the liquidus temperatures are higher. This can also be stated as more rapid cooling rates are necessary for larger undercoolings (ΔT), e.g. larger differences between T_L and T_s. It should also be noted that the larger the undercooling the more rapid the nucleation rate in glasses [30].

This suggests that if the DWPF pour spout bore or insert region is not sufficiently hot enough that the higher waste loaded glasses may be cooling off too slowly which allows surface nucleation of spinels on the inside of the upper bore. This, in conjunction with the more rapid nucleation of spinels in oxidizing environments and the availability of excess Cr_2O_3 from Inconel[®] 690 oxidation, has led to increased deposition in the spout and insert. This is consistent with the operating history of the LSFM which had a pour spout temperature of ~1075°C, poured lower waste loaded glasses, and did not have any pour spout pluggages. It is also consistent with the hotter spout designed for SGM after the 10 pour spout pluggages experienced when the pour spout tip was 980°C. Once the SGM was redesigned, it was able to pour glasses with calculated waste loadings up to ~42 wt% (see Figure 3). It should also be noted that the current DWPF melts have calculated viscosities in the range of the glasses melted during SGM campaign 6 during the pouring of canister six (SGM 6-6).

Table VII. Variation of Critical Cooling Rate with Heat Sink and Liquidus Temperatures.

Heat Sink Temperature (°C)	Melter Reference	Instantaneous Cooling Rate Needed for Glass with Liquidus of 997°C (°C/sec)	Instantaneous Cooling Rate Needed for Glass with Liquidus of 1050°C (°C/sec)
980	SGM-1 Tip of Pour Spout Before Insulation	9.7	38
1040	SGM-1 Bore Before Insulation	<1	6
~1100	LSFM Average in riser and bore	<<1	<1

Inconel[®] *690 Oxidation and Induced Crystallization:* Oxidation of Inconel[®] 690 to Cr_2O_3 rich oxide is evidenced by the mass balance of the pour spout insert samples analyzed in this report, e.g. the sample was ~73 wt% glass, 16.4 wt% $NiFe_2O_4$, and 21 wt% Cr_2O_3. At the temperature of the pour spout insert, and indeed anywhere between 800-1100°C, Inconel[®] 690 can rapidly oxidize to form a protective chrome oxide layer [31,32] even in the presence of Fe_2O_3 and FeO [33]. DWPF melts having an $Fe^{+2}/\Sigma Fe$ ratio=0.2 have a corresponding oxygen fugacity ($\log f_{O_2}$) of -5.5. At this oxygen fugacity and any oxygen fugacity more positive than

$\log f_{O_2} = -10$, Ni-rich alloys such as Inconel® 690 decompose to $NiCr_2O_4$ and NiO [32]. This "free" NiO further complexes with the Fe_2O_3 in DWPF glass forming $NiFe_2O_4$ which depletes the Inconel® 690 in NiO leaving an enrichment in Cr_2O_3 deposits. This is evidenced by the relative positions of the Inconel® 690 alloy composition (Figure 4 point A) to the insert deposit composition (Figure 4 point C) in mole percentage of Cr and Ni. Path AC in Figure 4 indicates that the insert deposits form by oxidation of Inconel® 690 and NiO depletion.

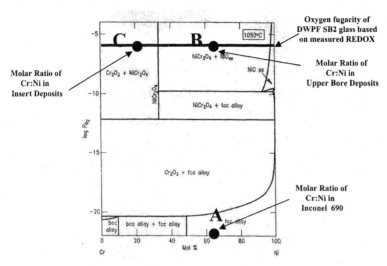

Figure 4. Binary phase diagram at 1050°C demonstrating the phases that are formed upon oxidation of a Ni-Cr alloy like Inconel® 690 [32].

The mass balance of the upper pour spout bore sample also indicated that the sample was Cr_2O_3 enriched, e.g. ~62 wt% glass, 25.8 wt% $NiFe_2O_4$, and 8 wt% $NiCr_2O_4$. Figure 4 readily shows that the Ni:Cr mole percentage of Inconel® 690 (point A) is the same as the ratio in the upper pour spout bore deposits (point B). The NiO released by Inconel® 690 oxidation further complexes with the Fe_2O_3 in DWPF glass forming $NiFe_2O_4$ as in the deposition of the insert deposits. These are the two main spinel components determined to be in the upper pour spout bore by mass balance (see Table III). This is evidenced by the relative positions of the Inconel® 690 alloy composition to the upper pour spout bore deposit composition to the insert deposit composition shown in Figure 4. These compositions represent the molar percentages of Cr and Ni in the deposits and in the alloy and indicate that the deposits form by oxidation of Inconel® 690 (path AB in Figure 4).

CONCLUSIONS
The Defense Waste Processing Facility (DWPF) Engineering requested characterization of three glass samples that were taken from Melter #2 after the waste loading had been increased and after rapid deposition had occurred in the DWPF pour spout region. The pour stream sample was determined to be homogenous, amorphous, and representative of feed tank chemistry from

which it was derived. This indicated that the DWPF viscosity, liquidus, and REDOX models are keeping the DWPF process in control.

The most likely mechanism by which the severe crystallization of the pour spout and insert occurred are the temperature and oxygen fugacity (oxidation) gradients in the DWPF pour spout in conjunction with the higher waste loadings. The DWPF liquidus model was developed to prevent volume crystallization of the melt pool at the normal melt pool temperatures, e.g. between 1050-1150°C, and at the normal oxygen fugacities experienced in waste glass melters, e.g. between $\log f_{O_2}$ = -2 ($Fe^{+2}/\Sigma Fe$=0.09) and $\log f_{O_2}$ = −9 ($Fe^{+2}/\Sigma Fe$=0.33). Operation of the SGM melter, specifically SGM Campaign 6, at waste loadings in excess of 38 wt% (Figure 3), i.e., in the range in which DWPF experienced severe pour spout crystallization, is achievable if the pour spout is well insulated and kept hot.

If the DWPF pour spout insert and upper pour spout bore are cooler than the liquidus temperature predicted by the DWPF PCCS, then surface nucleation of crystals on these cooler surfaces is more likely to occur. This is because a higher waste loaded melt is closer to its crystallization temperature when it exits the melter than a lower waste loaded melt. Thus, unless a higher waste loaded melt is moved through the cooler region very rapidly, the glass crystallizes instead of "undercooling" to an amorphous state. In other words, the riser temperature profile is too steep (the bore is not hot enough), which allows spinels to form at higher waste loadings because the cooling rate is not fast enough in this region.

Knowing that the DWPF pour spout bore and insert regions are more oxidizing than the melt pool and not sufficiently hot enough allows higher waste loaded glasses to cool too slowly. In other words, the degree of undercooling is too great, and the surface nucleation of spinels on the inside of the upper bore, spout, and insert can occur. The surface nucleation of spinels in the cooler more oxidizing regions of the pour spout is further enhanced in oxidizing environments because the activation energy of spinel nucleation is more rapid (17.7 kcal/mole) than in reducing environments (2.9 kcal/mole). In addition, the oxidative corrosion of Inconel® 690 provides excess Cr_2O_3 nuclei that can act as heterogeneous nuclei for spinel growth.

REFERENCES

[1] K.G. Brown, C.M. Jantzen and G. Ritzhaupt, **"Relating Liquidus Temperature to Composition for Defense Waste Processing Facility (DWPF) Process Control,"** WSRC-TR-2001-00520, Westinghouse Savannah River Co., Aiken, SC (October 2001).

[2] M.J. Plodinec, **"Long-Term Waste Management Progress Report Small-Scale Electric Meter, IV. Effects of Feed Mixing and Segregation on Glass Melting,"** U.S. DOE Report DPST-79-227, E.I. duPont deNemours & Co., Savannah River Lab., Aiken, SC (January 1979).

[3] W.N. Rankin, P.E. O'Rourke, P.D. Soper, M.B. Cosper, and B.C. Osgood, **"Evaluation of Corrosion and Deposition in the 1941 Melter,"** U.S. DOE Report DPST-82-231, E.I. duPont deNemours & Co., Savannah River Lab., Aiken, SC (March 1982).

[4] C.M. Jantzen, **"Lack of Slag Formation in the Scale Glass Melter,"** U.S. DOE Report DPST-87-373, E.I. duPont deNemours & Co., Savannah River Lab., Aiken, SC (April 1987).

[5] M.R. Baron and M.E. Smith, **"Summary of the Drain and Restart of the DWPF Scale Glass Melter,"** U.S. DOE Report DPST-88-481, E.I. duPont deNemours & Co., Savannah River Lab., Aiken, SC (May 1988).

[6] A.F. Weisman, **"Run Summaries from SGM-1, SGM-3, and SGM-4,"** U.S. DOE Report DPST-86-862, E.I. duPont deNemours & Co., Savannah River Lab., Aiken, SC (1986).

[7] A.F. Weisman, J.L. Mahoney, M. Rodman, K.R. Crow, D.M. Sabatino, and G.A. Griffin, **"Scale Melter Startup Review,"** U.S. DOE Report DPST-86-361, E.I. duPont deNemours & Co., Savannah River Laboratory, Aiken, SC (April 1986).

[8] C.M. Jantzen, **"Devitrification of Scale Melter Glass in Riser Heater,"** U.S. DOE Report DPST-86-461, E.I. duPont deNemours & Co., Savannah River Lab., Aiken, SC (May 1986).

[9] C.M. Jantzen, A.D. Cozzi, and N.E. Bibler, **"Characterization of Defense Waste Processing Facility (DWPF) Glass and Deposit Samples from Melter #2,"** U.S. DOE Report WSRC-TR-2003-00504, Rev. 0, Westinghouse Savannah River Co., Aiken, SC (April 2004).

[10]M.K. Andrews and J.R. Harbour, **"Chromium Levels in Feed and Glass for DWPF Startup Melter Campaigns,"** U.S. DOE Report WSRC-TR-95-0368, Westinghouse Savannah River Company, Aiken, SC (September 1995).

[11]C.M. Jantzen, **"Relationship of Glass Composition to Glass Viscosity, Resistivity, Liquidus Temperature, and Durability: First Principles Process-Product Models for Vitrification of Nuclear Waste,"** Proceed. 5th Intl. Symp. Ceram. in Nucl. Waste Mgt., G.G. Wicks, D.F. Bickford, and R. Bunnell (Eds.), Am. Ceram. Soc., Westerville, OH, 37-51 (1991).

[12]A.D. Cozzi, N.E. Bibler, T.L. Fellinger, J.M. Pareizs, and K.G. Brown, **"Vitrification of the DWPF SRAT Cycle of the Sludge-Only Flowsheet with Tank 40 Radioactive Sludge Using Frit 320 in the Shielded Cells Facility,"** U.S. DOE Report WSRC-RP-2002-00022, Rev. A, Westinghouse Savannah River Co., Aiken, SC (January 2002).

[13]C.M. Jantzen, N.E. Bibler, D.C. Beam, C.L. Crawford, and M.A. Pickett, **"Characterization of the Defense Waste Processing Facility (DWPF) Environmental Assessment (EA) Glass Standard Reference Material"**, WSRC-TR-92-346, Rev. 1, Westinghouse Savannah River Co., Aiken, SC (June 1994).

[14]K.G. Brown, C.M. Jantzen, and J.B. Pickett, **"The Effects of Formate and Nitrate on Reduction/Oxidation (Redox) Process Control for the Defense Waste Processing Facility (DWPF),"** WSRC-RP-97-34, Westinghouse Savannah River Co., Aiken, SC (February 1997).

[15]C.M. Jantzen, J.R. Zamecnik, D.C. Koopman, C.C. Herman, and J.B. Pickett, **"Electron Equivalents Model for Controlling REDuction/OXidation (REDOX) Equilibrium During High Level Waste (HLW) Vitrification,"** U.S. DOE Report WSRC-TR-2003-00126, Westinghouse Savannah River Co., Aiken, SC (May 9, 2003).

[16] C.M. Jantzen, D.F. Bickford, and D.G. Karraker, **"Time-Temperature-Transformation Kinetics in SRL Waste Glass,"** Adv. in Ceramics, 8, Am. Ceram. Soc., Westerville, OH, 30-38 (1984).

[17]A.E. VanArkel, E.J.W. Verwey, and M.G. VanBruggen, **"Ferrites I,"** Rec. Trav. Chim. 55, 331-339 (1936).

[18]C.M. Jantzen and K.G. Brown, **"Quasicrystalline Approach to Liquidus Temperature Prediction in Nuclear Waste Glasses,"** in preparation for J. Non-Crystalline Solids.

[19]M.J. Plodinec, **"Long-Term Waste Management Progress Report Small–Scale Electric Melter: II. Slag Formation,"** U.S. DOE Report DPST-78-453, E.I. duPont deNemours & Co., Savannah River Laboratory, Aiken, SC (August 1978).

[20]I.E. Grey and C. Li, **"New Silica-Containing Ferrite Phases in the System $NaFeO_2$-SiO_2,"** J. Solid State Chemistry, 69 [1], 116-125 (1987).

[21]J.D. Vienna, personal communication, Pacific Northwest National Laboratory (2002).

[22]C.M. Jantzen, K.G. Brown, K.J. Imrich, and J.B. Pickett, **"High Cr_2O_3 Refractory Corrosion in Oxidizing Melter Feeds: Relevance to Nuclear and Hazardous Waste Vitrification,"** Env. Issues and Waste Management Technologies in the Ceramic and Nuclear

Industries, Vol. IV Ceram. Trans., V. 93, J.C. Marra and G.T. Chandler (Eds.), Am. Ceram. Soc., Westerville, OH, 203-212 (1999).

[23]I.S. Shaplygin, I.I. Prosychev, and V.B. Lazarev, Zh. Neorg. Khim 26 [11] 3081-3083 (1981).

[24]H. Baker (Ed.), et. al, **"ASM Handbook, V. 3 Alloy Phase Diagrams,"** ASM Intl., 1992).

[25]P.T. Sarjeant and R. Roy, **"A New Approach to the Prediction of Glass Formation,"** Mat. Res. Bull., 3, 265-280 (1968).

[26] Technical Data Summary for the Defense Waste Processing Facility Sludge Plant, U.S. DOE Report DPSTD 80-38-2, E.I. duPont deNemours &Co., Aiken, SC (September 1982).

[27]J.P. Mosley, **"Calculated Physical Properties of SRP Waste Glasses Using Frit 131,"** U.S. DOE Report DPST-80-724, E.I. duPont deNemours &Co., Aiken, SC (December 1980).

[28]P.D. Soper and D.F. Bickford, **"Physical Properties of Frit 165/Waste Glasses,"** U.S. DOE Report DPST-82-899, E.I. duPont deNemours &Co., Aiken, SC (October 1982).

[29]W.P. Colven, D.M. Sabatino, J.L. Kessler, H.C. Wolf, **"Summary of the Fifth Run of the Large Slurry-Fed Melter,"** U.S. DOE Report DPST-82-890, E.I. duPont deNemours & Co., Savannah River Lab., Aiken, SC (September, 1984).

[30]D.R. Uhlmann and H. Yinnon, **"The Formation of Glasses"** in Glass Science and Technology, V.1, 1-47 (1983).

[31] K.L. Luthra, **"A Comparison of the Mechanism of Oxidation of Ti- and Ni-base Alloys,"** Environ. Eff. Adv. Mater., (R.H. Jones and R.E. Ricker, Minerals, Metals and Materials Society, Warrendale, PA, 123-131 (1991).

[32]A.D. Pelton, H. Schmalzried, and J. Sticher, J. Phys. Chem. Solids, 40 [12], 1103-1122 (1979) see also Phase Diagrams for Ceram. Figure 6270, Vol VI, Am. Ceram. Soc., Westerville, OH (1987).

[33]A. Muan and E.F. Osborn, **"Phase Equilibria Among Oxides in Steelmaking,"** Addison-Wesley Publ. Co., Reading MA (1965).

DWPF GLASS AIR-LIFT PUMP LIFE CYCLE TESTING AND PLANT IMPLEMENTATION

*Michael E. Smith, Allan B. Barnes, Dennis F. Bickford, Kenneth J. Imrich, Daniel C. Iverson, and Hector N. Guerrero
Westinghouse Savannah River Company
Savannah River Technology Center
*Building 773-42A
Aiken, SC 29808

ABSTRACT

Due to the accelerated cleanup at the Savannah River Site (SRS), efforts are underway to increase the glass melt rate and hence the high level waste processing throughput at the SRS Defense Waste Processing Plant Facility (DWPF). One of the proposed process/equipment improvements is a glass air-lift pump. The use of a glass air-lift pump to increase melt rate in the DWPF Melter has been investigated via several techniques including lab scale testing on various melters. The final test before implementation in DWPF was a long-term life cycle test (several months in duration) on a full size pump. The air-lift pump was successfully tested and no major problems were found. Based on this test a unit was designed and fabricated for DWPF and was installed in the DWPF Melter in February 2004.

INTRODUCTION

A DOE Tanks Focus Area program to access possible means of increasing the Defense Waste Processing Facility (DWPF) Melter melt rate was initiated in 2001. A lumped parameter comparison of DWPF data with earlier pilot plant scale data indicated that melt capacity for a given feed was limited by overheating of the glass immediately under the reacting feed (cold cap). Pumps were considered as a means of increasing glass circulation and opening a vent hole in the cold cap to allow increased electrode power, and thus increased melter total power. Limited locations for a pump in the DWPF melter top head, and glass pumping limitations of traditional pumps lead to the development of a system utilizing air-lift pumping.

The air-lift concept was tested with glycerin and an Inconel proof-of-principle air-lift was tested in molten glass and found to be an effective pump. In addition, small scale air-lift pumps were tested in the Slurry Fed Melt Rate Furnace (SMRF) to evaluate the overall behavior of the cold cap with an air-lift pump. Details of these tests and others are given elsewhere[1,2]. Prior experience with traditional pumps in glass, and evaluation of this performance with the lumped parameter heat transfer model has indicated that significant melt rate increases were possible from a single air-lift pump unit[1].

Due to the success of these tests and modeling work, a full-scale Inconel 690 unit was fabricated and installed in a glass hold tank at the Clemson Environmental Technologies Laboratory (CETL) facility in 2002. The main purpose of this test was to evaluate the expected unit life. To the extent possible, the test was performed to also provide information on air-lift design details and foam collapse rate. Design details of glass discharge height and nozzle design were also tested, as they require full scale testing in molten glass. This pump was designed to operate with a gas flow rate up to 850 standard liters/hour. In addition, it was used to provide initial indication of the interactions with the cold cap during a one day slurry feeding test[2].

Problems with the initial glass hold tank (glass foam buildup due to small glass hold tank diameter and heater failures) required a new test stand. Testing was completed in August 2003. A total of 72 days of operation (non-slurry feeding except for one day) were completed before the pump was removed. Most of the test time (48 days) was in the new test stand.

Anticipated benefits to DWPF of the air-lift pump were:
- Enhanced melt rate from direct action of increased overall glass circulation rates improving transfer of electrode power to the bottom of the cold cap. This may be the result of increased overall glass velocity or improved venting of cold cap gases trapped under the cold cap. This is the mode of melt rate improvement of traditional pumps operated with modest gas flow rates. However in the present case, the efficiency of the pumping action is improved, so that lower gas flows are required, and the gas is not forced to accumulate under the cold cap. (Gas bubbles generated by the melting process still have to vent from under of the cold cap.)

- Additional increases to melt rates from enhanced power available to the cold cap and slurry indirectly by heat transfer from the pumped glass to the melter plenum. It increases total power available to the melter by allowing additional electrode power to be applied without overheating of the glass under the cold cap.

- More uniform glass pool temperatures, making it easier to stay within temperature operating limits at the top and bottom of the glass pool.

- For Sludge Batch 2 feed processed until early in 2004, evaluation of the heat transfer across the surface of the glass in DWPF suggested that a barrier layer may form. The pumping action of the hot glass from an air-lift may push aside or raise the temperature of viscous layers or un-dissolved material floating on top of the glass pool, causing them to dissolve or dissipate.

This report discusses the full scale air-lift pump tests at the CETL. This includes the non-destructive and destructive evaluations performed on the pump after the test. It also discusses the implementation of the air-lift pump in the DWPF Melter.

LIFE CYCLE TEST DESCRIPTION
A full-scale Inconel 690 air-lift pump was fabricated at SRS for the performance testing in the Pour Spout Test Stand (PSTS). The pump has an outer diameter of 8.9 cm and an inner diameter of 6.4 cm (see Figure 1). The pump has two nozzles near the bottom that can either be run singularly or together. A thermocouple was installed in the pump to monitor glass temperature as well. The initial plan was to place it in the PSTS with simulated DWPF glass and then run it (without slurry feeding) for several months. Air flow of 566 standard liters/hour was targeted for one nozzle at the beginning of the test. The PSTS reservoir has an inner diameter of about 20.3 cm.

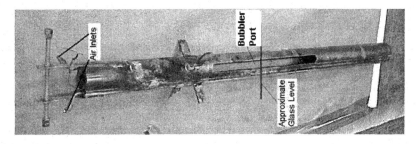

Figure 1. Air-lift pump

The pump was installed in the PSTS in September 2002. The glass temperature was maintained above 1100°C as indicated by the pump thermocouple. Due to excessive glass foaming over the top of the PSTS reservoir that was threatening to damage heaters, the pump was removed and the upper heated zone of the PSTS was extended to prevent this glass from foaming out. However, several other heater failures occurred during subsequent testing. Therefore, a new, more robust test stand was fabricated and used for the remainder of the test. It had 38.1 cm inner diameter Inconel 690 pot (127 cm tall) with more durable heaters.

The new test stand was charged with DWPF black startup frit (a frit designed for startup of the joule heated waste glass melters) and heated to an operational temperature of about 1120 °C. The pump was installed the next day. The glass level was 76 cm and the pump was submerged in the glass 61 cm. Unlike the previous test, both nozzles were operated at 283 standard liters/hour. These flow rates remained fairly constant for the remainder of the test. Glass temperature, as indicated by the pump thermocouple, was maintained at about 1120 °C for the duration of the test. No foaming buildup was observed and therefore the theory that the previous foamy buildup problems with the PSTS were due to the small PSTS inner diameter which allowed the foam to "climb up" the PSTS reservoir wall was confirmed.

Throughout the duration of the test the number of bubbles was counted for one minute each day. This number was in the range of 100 - 120 per minute. In addition, glass samples were taken weekly to determine if the glass composition was changing (alkalis being volatized). Subsequent analyses of these glass samples showed that the alkalis were not volatizing. Being a startup glass, the glass did not have corrosive components such as chlorides.

At the end of the test SRS personnel performed a short term slurry feeding test that was videotaped (see next section). The air-lift pump was removed the next day after a total of 72 days of operation time (48 days in new test stand). One of the air lines was disconnected prior to removing the pump from the molten glass. The second air line, which supplied air to the other nozzle, remained attached and functional until the pump was lifted out from the molten glass.

SHORT TERM SLURRY FEEDING EVALUATION
Air flow to the air-lift pump was maintained at 566 standard liters/hour. Simulated DWPF Macrobatch 2 melter feed was fed into the melter using a peristaltic pump. The feed rate was measured at 0.133 liters/minute and feeding was maintained for ~90 minutes. After feeding, the

pump continued to operate until the cold cap was completely melted. The two diametrically opposed exhaust ports on the pump will be referred to as ports A and B.

During the first 45 minutes of feeding, the feed tube was positioned so that the slurry flowed onto the glass exiting exhaust port B of the air-lift pump. Since the direction of the glass was from the center of the melt pool toward the wall of the Inconel pot, the slurry also flowed away from the pump tube and began forming a cold cap along the Inconel wall. Glass flow from the two exhaust ports of the pump remained uniform during the initial forming of the cold cap. As the cold cap grew, the glass flow from the pump exhaust port adjacent the slurry feed began to decrease, and glass flow from the pump became preferential to exhaust port A, 180° opposite the slurry feed, until exhaust port B became closed 25 minutes into the feeding cycle (see Figure 2).

The feed tube was then positioned so that the slurry was introduced 90° to the exhaust ports on the air-lift pump. Over the next 45 minutes of feeding, the cold cap continued to grow to a thickness of ~2 inches. The vent in the cold cap maintained by the glass flow from pump exhaust port A slowly closed and glass flow from the pump became preferential to exhaust port B, causing a vent to be re-formed in the cold cap at this location. The opening and closing of vents in the cold cap continued throughout the remainder of the feeding cycle requiring ~15 to 20 minutes for the glass flow from a given pump exhaust port to be blocked and re-opened. Once feeding stopped, the cold cap began to melt and the glass flow from the two air-lift pump exhaust ports became more uniform (see Figure 3). Throughout feeding, the pump air was readily vented to atmosphere and no adverse effects on cold cap stability or increased foaming were observed.

Figure 2. Air-lift pump exhaust port B (visible port on left) shown closed after 25 minutes of melter feeding at 0.133 liters/minute.

Figure 3. Glass flow out of air-lift pump exhaust port B (visible port on left) reopening cold cap vent (slurry feed had been stopped for ~2 minutes)

POST TEST NONDESTRUCTIVE AND DESTRUCTIVE EXAMINATIONS

Overall the air-lift pump was in excellent condition after 72 days of operation. Visual and metallurgical evaluation did not reveal any significant degradation, material loss or internal attack of the Inconel 690 portion of the pump. Even regions in high flow areas around nozzles or at the glass discharge point at the air/ glass interface did not reveal any observable degradation. Minimal internal attack, internal void formation and depletion of chromium in the near surface regions due to oxidation, were observed in the air passage and glass contact regions.

Degradation rate based on internal attack would predict a rate of approximately 1.27 centimeters per year, but movement of this interface would require material loss to occur. Material loss could not be estimated because significant local variations in wall thickness (from fabrication techniques employed) made comparison of ultrasonic thickness readings before and after the testing impossible. Welds were examined and did not show any significant molten glass attack. Minimal degradation of the Inconel 690 may have resulted because the glass chemistry (DWPF start-up frit) was not very corrosive. Because the data obtained is not fully representative of what would be expected in DWPF, degradation of the production pump should be carefully monitored during operation and a thorough metallurgical evaluation should be performed after it is removed from service.

Some glass has seeped into the thread gaps between the nozzles and the Inconel housing. Increasing the area of the sealing surface on the nozzles should help minimize glass intrusion into the thread gaps. There was also some oxidation of the Inconel 690 threads most likely resulting from air in leakage around the threads.

Spinel formation was noted at the inlet of each nozzle. The deposits were large enough to completely cover the air passage at the inlet of the nozzle. Although the deposits were extensive, they were porous and did not impede the air flow for the duration of the test.

Even minimal corrosion of the Inconel 690 in this system may have increased the concentration of transition metals in the glass because of the large surface (Inconel 690) to volume (glass) ratio. However, to minimize spinel formation during the operation of the DWPF production pump, argon should be used rather than air to minimize oxidation of the Inconel 690 air passage. Disruptions in gas flow should also be avoided in order to minimize glass intrusion into the air passage. These operational controls should minimize the formation of spinel deposit and extend the life of the pump.

IMPLEMENTATION OF AIR-LIFT PUMP IN DWPF

With the findings of the life cycle tests, an air-lift pump was designed and fabricated for use in the DWPF Melter. The center melt pool thermowell nozzle located in the middle of the top of the melter was chosen as the nozzle in which the pump was to be installed. The melt pool thermowell was removed and the new pump was put in place on February 10, 2004.

Initial observations indicate that the pump is helping to increase glass circulation. These observations include an increase in the lower melt pool temperature readings and the increase in total available electrode power. Due to unrelated off-gas instability problems, the overall impact on melt rate has not yet been determined. Process parameters are currently being systematically varied to determine the optimum operating conditions to maximize waste throughput in the DWPF Melter.

CONCLUSIONS

Based on the results of the life cycle air-lift pump tests, the one day slurry feeding test, and the subsequent metallurgical evaluation of pump, the following conclusions can be made:

1) The air-lift pump was successfully operated for a total of 72 days and provided sufficient flow to pump glass from the bottom of the melt pool without excessive foaming.
2) No problems were observed with the cold cap during the one day air-lift pump slurry feeding test.
3) No significant degradation of the Inconel 690 was observed even in the high flow regions or the air/melt interface.
4) Argon gas should be used with the Inconel 690 pump to minimize oxidation.
5) Significant spinel formation at the nozzle inlets was observed but because it was porous it did not disrupt the flow of air through the nozzles.
6) Operation of the pump in the DWPF Melter has increased glass circulation and increased melter bottom glass pool temperatures as expected and therefore should improve melt rate. The actual impact on DWPF melt rate has not yet been quantified.

REFERENCES

1. H. N. Guerrero and D. F. Bickford, "DWPF Melter Air-Lift Bubbler: Development and Testing for Increasing Glass Melt Rates and Waste Dissolution (U)", USDOE Report WSRC-TR-2002-00196, WSRC.
2. D. C. Witt, T. M. Jones, D. F. Bickford, "Airlift Mini-Bubbler Testing in the Slurry Fed Melt Rate Furnace (U)", USDOE Report WSRC-TR-2002-00494, WSRC.

CORROSION RESISTANCE OF METAL ELECTRODES IN AN IRON PHOSPHATE MELT

C.W. Kim, D. Zhu, and D.E. Day
Department of Ceramic Engineering and the Graduate Center for Materials Research, University of Missouri-Rolla, Rolla, MO 65409-1170

ABSTRACT

The corrosion resistance of Inconel 690 and 693 (both nickel-based alloy) has been investigated using samples submerged in an iron phosphate melt that contained 30 wt% of the simulated Hanford low activity nuclear waste (22.6% Na_2O, 20.0% Fe_2O_3, 52.2% P_2O_5, 2.9% SO_3, and 2.3% others, wt%) at 1050°C for 155 days. The weight loss for the submerged Inconel 690 and 693 samples was 14 and 8%, respectively. The overall corrosion rate, calculated from the initial and final dimensions of each sample, was 1.3 and 0.7 μm/day for the Inconel 690 and 693, respectively. The external surface of the corroded Inconel samples was depleted in nickel and the only corrosion product found by SEM-EDS and XRD on the external surface was (Fe, $Cr)_2O_3$. This layer appears to act as a chemically protective layer between the metal and iron phosphate melt. These preliminary results suggest that Inconel 690 and 693 have a good corrosion resistance in iron phosphate melts, with Inconel 693 having the better corrosion resistance.

INTRODUCTION

The U.S. Department of Energy (DOE) Hanford Site in Washington State has more than 55 million gallons of radioactive waste stored in 177 underground storage tanks [1]. It is planned to retrieve this waste from the tanks and separate it into a Low Activity Waste (LAW) and a High Level Waste (HLW) which will be separately vitrified in the Hanford Tank Waste Treatment Plant. Noteworthy compositional features of the Hanford LAW is that it contains significant amounts of sodium, sulfate, and phosphate [2-4], all of which could limit the maximum waste loading in a borosilicate glass, the only glass currently approved by DOE, to below 10 wt% [5]. The maximum LAW loading in a borosilicate glass is largely determined by the amount of sulfate in the waste [2,6-8]. If the LAW waste loading was not limited by the amount of sulfate, then the waste loading could be increased. If an iron phosphate glass was used, the amount of glass produced at Hanford could be reduced by as much as 43% [3].

Previous work [3,4,9] has suggested several advantages for vitrifying the Hanford LAW in an iron phosphate glass. The most important is the higher waste loading (up to 35 wt%) of the simulated LAW with no indication of sulfate salt segregation, as seen in most borosilicate glasses. It is unlikely that the LAW waste loading in iron phosphate glasses will be limited by the SO_3 content of the LAW. Another advantage is that iron phosphate glasses containing LAW simulant can be melted as low as 1000-1050°C. Iron phosphate melts also have a high fluidity (200 to 900 centipoise) at their melting temperature so they rapidly become chemically homogeneous, which reduces their melting times to only a few hours (< 4 h) compared to the > 48 h residence time for the borosilicate melts in the Joule Heated Melter (JHM) at the Defense Waste Processing Facility (DWPF), Westinghouse Savannah River Co.

Experience in melting iron phosphate glasses in the U.S. is limited. Several hundred iron phosphate compositions have been melted without problems in ordinary refractories, but the largest melts were less than 100 lbs. No attempts have been made to melt iron phosphate glasses on the scale expected for waste vitrification (e.g., 3 to 15 MT/day), primarily because these glasses are relatively new. However, phosphate glasses have been successfully melted in commercial size furnaces in the U.S. for several decades in the optical glass industry. Of more relevance to nuclear waste is that large quantities (2.2 million kg [10]) of a more chemically corrosive sodium-alumino phosphate glass has been successfully melted in Russia for up to six years in a JHM lined with commercial AZS refractories as part of their waste vitrification effort.

The present work was motivated by the potentially large savings in time and money that might be possible if the current JHM's, designed for melting borosilicate glass, could be used with minor modifications to vitrify the Hanford LAW in an iron phosphate glass. Consequently, the corrosion resistance of Inconel 690 and Inconel 693 in an iron phosphate melt containing 30 wt% of a simulated Hanford LAW has been studied to determine the suitability of these metals for electrodes in a JHM melting iron phosphate glass.

Inconel 690 was chosen since it is the current electrode material [11] being used in the JHM at DWPF to melt borosilicate glass. Inconel 693 was chosen since it is a modified version of Inconel 690, see Table I, that might offer a higher corrosion resistance and longer service life.

Table I. Composition and selected physical properties of Inconel 690 and 693 [12]

		Inconel 690	Inconel 693
Composition (wt%)	Nickel	58.0 min.	53.0 min.
	Chromium	27-31	27.0-31.0
	Iron	7-11	2.5-6.0
	Aluminum		2.5-4.0
	Niobium		0.5-2.5
	Manganese	0.50 max.	1.0 max.
	Titanium		1.0 max.
	Copper	0.50 max.	0.5 max.
	Silicon	0.50 max.	0.5 max.
	Carbon	0.05 max.	0.15 max.
	Sulfur	0.015 max.	0.01 max.
Physical property	Density (g/cm^3)	8.19	7.77
	Melting range ($^\circ$C)	1343-1377	1317-1367
	Electrical resistivity ($\mu\Omega\cdot$m)	1.148	1.168

EXPERIMENT
Preparation of Glass

An iron phosphate glass was prepared which contained 30 wt% of the high sulfate Hanford LAW simulant and 70 wt% glass forming additives Fe_2O_3 and P_2O_5. The overall LAW simulant and batch compositions are given in Table II. The appropriate amounts of the raw materials were mixed in a sealed plastic container for 30 minutes to produce a homogeneous mixture. This mixture was put in an alumina crucible and melted at 1050°C for 2 hours in an electric furnace in air. The melt was stirred 3 to 4 times with a fused silica rod over a period of 30 minutes to insure chemical homogeneity, and then poured onto the surface of a clean steel plate. The quenched glass was used for the corrosion tests.

Table II. Nominal composition (wt%) of Hanford LAW
simulant and batch containing 30 wt% LAW stimulant

	LAW simulant	Batch
Al_2O_3	4.4	1.3
Cl	0.6	0.2
Cr_2O_3	0.4	0.1
F	1.6	0.5
Fe_2O_3	0.0	20.0
Na_2O	75.3	22.6
P_2O_5	7.7	52.2
SiO_2	0.5	0.2
SO_3	9.5	2.9
Total	100.0	100.0

Corrosion Tests

Specimens (~14 mm × ~9 mm × ~7 mm) of Inconel 690 and 693 were cut from larger pieces of each metal using a diamond saw and then polished to 600 grit SiC paper. The dimensions and weight of each specimen were measured and recorded.

Fifty grams of the as-made iron phosphate glass was re-melted in an alumina crucible at 1050°C for 30 minutes and then one Inconel specimen, which had been preheated to 1050°C, was submerged in the melt. After a prescribed time, the crucible was removed from the furnace and the Inconel sample was removed from the melt and immediately quenched in water. After cleaning and drying the Inconel sample, its weight was measured and the sample again submerged in the melt in the crucible and the corrosion test continued. The iron phosphate melt was replenished every 7 days with new as-made glass to minimize any compositional changes over the entire period of the test.

The weight of each specimen (initial weight 7.07 and 6.29 g for Inconel 690 and 693, respectively) was measured every 7 days, from which the weight loss of the sample was calculated as a function of time. The ratio of the weight difference (initial weight minus weight of specimen after a certain time interval) divided by the initial weight was defined as a percent weight loss at a given testing time. The corrosion tests were terminated after 155 days, whereupon, the final dimensions of each specimen were measured.

RESULTS AND DISCUSSION

Weight Loss

The total weight loss after being submerged in the iron phosphate melt for 155 days was 14 and 8% for the Inconel 690 and 693 sample, respectively (Figure 1). The calculated weight loss rate for the two samples as a function of time is shown in Figure 2. The weight loss and its rate for Inconel 693 were about one half of that for Inconel 690, indicating that Inconel 693 was more chemically resistant to the melt.

The rate of weight loss for Inconel 690 and 693 was higher at the early stage of the corrosion test, and then gradually decreased with time. At the conclusion of the 155-day test, the weight loss rate was roughly two thirds and one third that of the initial rate for Inconel 690 and 693, respectively. This behavior is consistent with a chemically protective layer forming on the external surface of the metal during the test.

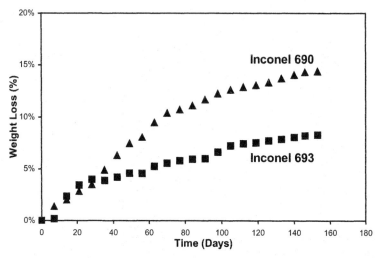

Figure 1. Weight loss (%) of the submerged Inconel 690 and 693 samples as a function of time in an iron phosphate melt containing 30 wt% Hanford LAW at 1050°C. Estimated error is represented by size of data points.

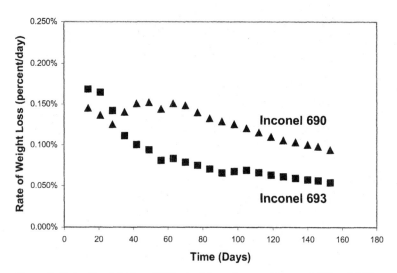

Figure 2. Rate of weight loss (%/day) of the submerged Inconel 690 and 693 samples as a function of time in the iron phosphate melt containing 30 wt% Hanford LAW at 1050°C. Estimated error is represented by size of data points.

The corrosion rate of Inconel 690 and 693 was also determined by measuring the initial and final dimensions of the samples after the 155-day test. The overall corrosion rate (calculated by assuming a constant rate from time = 0) was 1.3 and 0.7 μm/day for the submerged Inconel 690 and 693 sample, respectively. These values are significantly smaller than the 6.5 μm/day rate reported [11] for the Inconel 690 electrodes in a JHM being used to melt borosilicate glass at 1150°C in the DWPF at Savannah River. The typical appearance of the cross section of the submerged Inconel 693 sample after 155 days is shown in Figure 3.

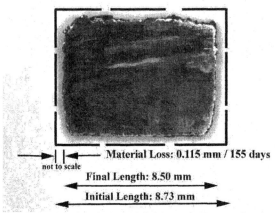

Figure 3. Cross sectional view of the Inconel 693 sample after the corrosion test at 1050°C for 155 days (0.115 mm/155 days = 0.7 μm/day). Note box around sample not to scale.

Corrosion Products

Only one alteration phase or corrosion product was found at the external surface of the Inconel 690 sample submerged for 155 days in the iron phosphate melt containing 30 wt% Hanford LAW simulant. SEM micrographs and EDS spectra for this Fe-rich solid solution of (Fe, Cr)$_2$O$_3$ phase are shown in Figure 4. Similarly, there was one main alteration phase at the surface of the Inconel 693 sample, but it was composed of crystals having two different morphologies (Figure 5). There were smaller grain-like crystals such as those in Figure 5(b) (Cr-rich solid solution of (Fe, Cr)$_2$O$_3$) and larger platy crystals, Figure 5(c), (Fe-rich solid solution of (Fe, Cr)$_2$O$_3$). The presence of only the Fe-rich phase in the Inconel 690 sample may be due to its higher iron content (7-11 wt%) compared to the lower iron content (2.5-6.0 wt%) of Inconel 693, see Table 1. This suggests that iron corrodes more easily than chromium and that substituting other components such as Al, Nb, and Ti for Fe in Inconel 693 (Table 1) enhances its corrosion resistance.

(a)

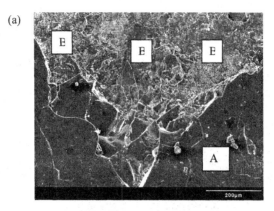

(b)

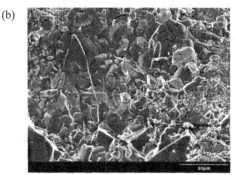

(bb)

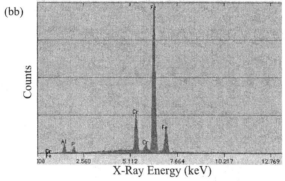

X-Ray Energy (keV)

Figure 4. (a) SEM micrograph of the external surface of the Inconel 690 sample after the 155-day corrosion test. A: residual glass attached to the surface; B: Fe-rich solid solution (Fe, Cr)$_2$O$_3$ alteration phase. (b) Close-up view of the Fe-rich (Fe, Cr)$_2$O$_3$ phase in area B. (bb) EDS spectra for the Fe-rich (Fe, Cr)$_2$O$_3$ phase.

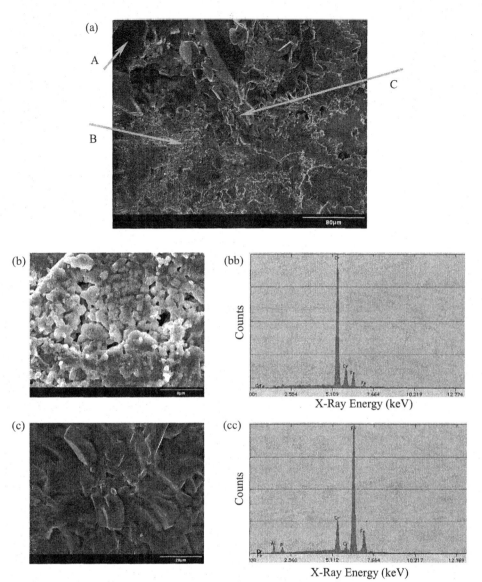

Figure 5. (a) SEM micrograph of the surface of the Inconel 693 sample after the corrosion test. A: residual glass attached to the surface; B: Cr-rich solid solution (Fe, Cr)$_2$O$_3$ alteration phase; C: Fe-rich solid solution (Fe, Cr)$_2$O$_3$ alteration phase. (b) Close-up view of the Cr-rich (Fe, Cr)$_2$O$_3$ phase in area B. (bb) EDS spectra for the Cr-rich (Fe, Cr)$_2$O$_3$ phase. (c) Close-up view of the Fe-rich (Fe, Cr)$_2$O$_3$ phase in area C. (cc) EDS spectra for the Fe-rich (Fe, Cr)$_2$O$_3$ phase.

The XRD spectra of Inconel 690 and 693 before and after the corrosion test are shown in Figure 6. The XRD Peaks for the Inconel 690 and 693 samples before corrosion were consistent with FCC alloys as expected. Diffraction peaks from the corroded samples of Inconel 690 and 693 corresponded to those for $(Fe, Cr)_2O_3$ (JCPDS-ICDD card # 02-1357). It should be noted that the diffraction peaks corresponding to the fresh Inconel alloys were also detected on the external surface of the corroded Inconel 693 sample (Figure 6(b)-II), while these peaks were absent on the corroded Inconel 690 surface. The thinner alteration layer (~20 μm) on the external surface of Inconel 693, compared to the ~40 μm thick layer on the Inconel 690 sample, indicates that the Inconel 693 was corroded less than 690, which is consistent with the presence of the diffraction peaks corresponding to the fresh Inconel alloy.

(a)

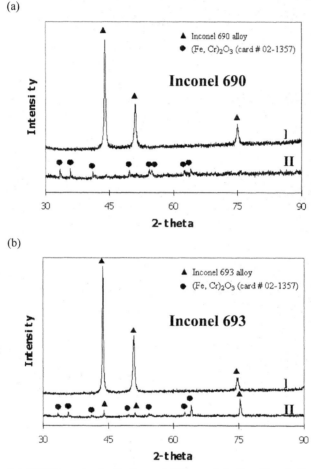

(b)

Figure 6. (a) XRD spectra for the Inconel 690 sample before (I) and after (II) the corrosion test. (b) XRD spectra for the Inconel 693 sample before (I) and after (II) the corrosion test.

No Ni-containing phase, such as $NiCr_2O_4$ (a Ni/Cr containing spinel) or $NiFe_2O_4$ (a Ni/Fe containing spinel), was detected on the surface of the corroded Inconel 690 and 693 samples, even though Ni was the main component (> 53 wt%) in each metal. However, Ni was detected by X-ray fluorescence in the glasses after corrosion tests. These results suggest that all the Ni lost from the Inconel samples go into the glass melt rather than forming an alteration phase on the metals.

CONCLUSIONS

The overall corrosion rate of Inconel 690 and 693 submerged for 155 days in an iron phosphate melt containing 30 wt% Hanford LAW stimulant at 1050°C was small, 1.3 and 0.7 μm/day for Inconel 690 and 693, respectively. The corrosion resistance of Inconel 693 was about twice that of Inconel 690 in the iron phosphate melt.

The surface of both Inconels at the completion of the corrosion test was depleted in Ni and composed of a layer of $(Fe, Cr)_2O_3$, which appeared to act as a chemically protective layer on the metals.

These preliminary results are encouraging since the iron phosphate melt did not corrode the Inconel 690 to any greater extent than what has been reported for Inconel 690 electrodes in the borosilicate melt in the JHM at DWPF. Furthermore, Inconel 693 may be an even better candidate for use in iron phosphate melts since its corrosion rate was only about one half that of Inconel 690.

ACKNOWLEDGEMENTS

This work was supported by the Environmental Management Science Program of the U.S. Department of Energy under contract FG07-96ER45618.

REFERENCES

[1] U.S. DOE, "Summary Data on the Radioactive Waste, Spent Nuclear Fuel, and Contaminated Media Managed by the U.S. Department of Energy", U.S. Department of Energy, Washington D.C. (2001).

[2] J.D. Vienna, W.C. Buchmiller, J.V. Crum, D.D. Graham, D.-S. Kim, B.D. MacIsaac, M.J. Schweiger, D.K. Peeler, T.B. Edwards, I.A. Reamer, and R.J. Workman, "Glass Formulation Development for INEEL Sodium-Bearing Waste", Pacific Northwest National Laboratory Report, PNNL-14050 (2002).

[3] D.S. Kim, W.C. Buchmiller, M.J. Schweiger, J.D. Vienna, D.E. Day, C.W. Kim, D. Zhu, T.E. Day, T. Neidt, D.K. Peeler, T.B. Edwards, I.A. Reamer, and R.J. Workman, "Iron Phosphate Glass as an Alternative Waste-Form for Hanford LAW", Pacific Northwest National Laboratory Report, PNNL-14251 (2003).

[4] C.W. Kim and D.E. Day, "Immobilization of Hanford LAW in Iron Phosphate Glasses", *Journal Non-Crystalline Solids,* **331** 20-31 (2003).

[5] Personal Communication, J.D. Vienna, Pacific Northwest National Laboratory (2004).

[6] H.D. Schreiber, C.W. Schreiber, E.D. Sisk, and S.J. Kozak, "Sulfur Systematics in Model Glass Compositions from West Valley", *Ceramic Transactions* **45** 349-358 (1994).

[7] G.K. Sullivan, M.H. Langowski, and P. Hrma, "Sulfate Segregation in Vitrification of Simulated Hanford Nuclear Waste", *Ceramic Transactions* **61** 187-193 (1995).

[8] H. Li, M.H. Langowski, and P.R. Hrma, "Segregation of Sulfate and Phosphate in the Vitrification of High-Level Wastes", *Ceramic Transactions* **61** 195-202 (1995).

[9] C.W. Kim, D. Zhu, D.E. Day, D.S. Kim, J.D. Vienna, D.K. Peeler, T.E. Day, and T. Neidt, "Iron Phosphate Glass for Immobilization of Hanford LAW", *Ceramic Transactions* (in press).

[10] JCCEM Workshop, "US/Russian Experiences on Solidification Technologies-Record of Meeting", Augusta, Georgia, September 4-5 (1997).

[11] Personal Communication, D.F. Bickford, Westinghouse Savannah River Company (2003).

[12] www.specialmetals.com.

FLUIDIZED BED STEAM REFORMING (FBSR) OF ORGANIC AND NITRATE CONTAINING SALT SUPERNATE

Carol M. Jantzen
Savannah River National Laboratory
Aiken, South Carolina 29808

ABSTRACT

A salt supernate waste (Tank 48H) generated at the Savannah River Site (SRS) during demonstration of In Tank Precipitation (ITP) process for Cs removal contains nitrates, nitrites, and sodium tetraphenyl borate (NaTPB). This slurry must be pre-processed in order to reduce the impacts of the nitrate and organic species on subsequent vitrification in the Defense Waste Processing Facility (DWPF). Fluidized Bed Steam Reforming (FBSR) is a candidate technology for destroying the nitrates, nitrites, and organics (NaTPB) prior to melting. Bench scale tests were designed and conducted at the Savannah River National Laboratory (SRNL) to demonstrate that bench scale testing can adequately reproduce the CO/CO_2 and H_2/H_2O fugacities representative of the FBSR process and form the appropriate product phases. Carbonate and silicate product phases that were compatible with DWPF vitrification were achieved in the bench scale testing and test parameters were optimized for a pilot scale FBSR demonstration.

INTRODUCTION

An In Tank Precipitation (ITP) technology was developed at the SRS to remove Cs^{137} from high level waste (HLW) supernates. During the ITP process monosodium titanate (MST) and sodium tetraphenylborate (NaTPB) were added to the salt supernate to adsorb Sr^{90}/Pu^{238} and precipitate Cs^{137} as CsTPB, respectively. This process was demonstrated at the SRS in 1983. The demonstration produced 53,000 gallons of 2.5 wt% Cs rich precipitate containing TPB, which was later washed and diluted to 250,000 gallons. This material is currently stored in SRS Tank 48H. The washed precipitate was to ultimately be disposed in borosilicate glass in the Defense Waste Processing Facility (DWPF).

Due to safety concerns, the ITP process was abandoned in 1998, and new technologies are being researched for Cs^{137} removal. In order to make space in the SRS Tank farm, the Tank 48H waste must be removed. Therefore, the Tank 48H waste must be processed to reduce or eliminate levels of nitrates, nitrites, and NaTPB in order to reduce impacts of these species before vitrification at the DWPF. Fluidized Bed Steam Reforming (FBSR) is being considered as a candidate technology for destroying the nitrates and the NaTPB prior to melting.

The FBSR technology is capable of destroying the organic sodium, potassium, cesium tetraphenyl borate phases, e.g. NaTPB, KTPB, and CsTPB, at moderate temperature and converting the organic phases to $(Na,K,Cs)_2O$ or $(Na,K,Cs)_2CO_3$ or $(Na,K,Cs)SiO_3$, CO_2 gas, and H_2O in the form of steam [1,2]. The high nitrate and nitrite content of the Tank 48H slurry will be converted to N_2 thereby minimizing NO_x emissions during processing. Any organics are oxidized to CO_2 instead of CO during processing. The FBSR can be electrically heated (pilot scale units) or operated in an auto-thermal mode, whereby the energy needs are supplied by the incoming superheated steam and by the oxidation of organics from the waste and carbon reductants. For production scale units, auto-thermal steam reforming is the preferred mode of operation. Since there is no open flame as in incineration and since the product emissions are CO_2 and N_2 instead of CO and NO_x, the FBSR process is Clean Air Act and Maximum

Achievable Concentration Technology (MACT) compliant. The solid oxide or mineral phases produced, e.g. $(Na,K,Cs)_2O$, $(Na,K,Cs)_2CO_3$, or $(Na,K,Cs)SiO_3$, are considered compatible with subsequent processing to borosilicate glass in the Defense Waste Processing Facility (DWPF) because their melting temperatures are $<1050°C$ and this should not create any melt rate related impacts.

The objectives of the current study were to demonstrate the following with a Tank 48H simulant:

- destruction of TPB with the FBSR process operating between 650-725°C
- destruction of nitrate at >99% with addition of sugar as a reductant
- destruction of anitfoam with the FBSR process operating between 650-725°C
- formation of Na_2CO_3 FBSR product to be compatible with mixing the FBSR product into a DWPF feed tank for subsequent vitrification
- formation of a Na_2SiO_3 or Na_4SiO_4 FBSR product to be compatible with mixing the FBSR product into a DWPF feed tank or as an addition to the Slurry Mix Evaporator (SME) in place of a portion of the frit
- assessment of the melting temperature of the Na_2CO_3 and Na_2SiO_3 FBSR products to evaluate impacts (if any) on melt rate
- optimization of the amount of reductant to ensure that excess reductant was not contained in the FBSR product that would alter the carefully controlled REDuction/OXidation (REDOX) equilibrium in the DWPF melter [3]
- optimization of test parameters for subsequent pilot scale testing of Tank 48H simulant at Idaho National Engineering and Environmental Laboratory (INEEL)
- demonstration that bench scale studies can duplicate the complex reactions in the FBSR process

BACKGROUND

Studsvik built and tested a commercial Low-Level Radioactive Waste (LLRW) FBSR Processing Facility in Erwin, TN, in 1999 [4]. In January 2000, commercial operation commenced [1]. The Studsvik Processing Facility (SPF) has the capability to process a wide variety of solid and liquid LLRW streams including: ion exchange resins, charcoal, graphite, sludge, oils, solvents, and cleaning solutions with contact radiation levels of up to 100R/hr. The licensed and heavily shielded SPF can receive and process liquid and solid LLRWs with high water and/or organic content.

The Erwin facility employs the THermal Organic Reduction (THOR[sm]) process, developed by Studsvik, which utilizes pyrolysis[*]/steam reforming technology. THOR[sm] reliably and safely processes a wide variety of LLRWs in the unique, moderate temperature (~700°C), dual-stage, pyrolysis/reforming, fluidized bed treatment system. The reforming process has demonstrated effectiveness in volatilizing/combusting organics and separating sulfur and halogens from inorganic waste materials. Of special relevance is the capability of the THOR[sm] technology to convert nitrates to nitrogen and sodium salts to sodium compounds that are suitable for direct disposal and/or subsequent vitrification.

In February 2002, THOR[sm] demonstrated the capability of producing sodium aluminosilicate waste forms for Hanford's sodium-bearing low activity waste (LAW) [2]. Other demonstrations performed by Hazen showed that LAW waste could be transformed into Na_2CO_3, $NaAlO_2$, or

---

[*] Pyrolysis chemically decomposes organic materials by heat in the absence of oxygen.

Na$_2$SiO$_3$ feed material for the LAW Hanford melter. Addition of no solid co-reactant yields a sodium carbonate product. Sodium combines with carbon dioxide in the reformer gases to provide a sodium carbonate product. Addition of a Al(OH)$_3$ co-reactant will provide an NaAlO$_2$ product, addition of SiO$_2$ will provide an Na$_2$SiO$_3$ product. Addition of kaolin clay will provide an NaAlSiO$_4$ product. The latter has been shown to perform well as a final waste form [2,5].

In November 2002, THOR^{sm} was contracted to demonstrate the FBSR technology to produce a carbonate waste solid for Idaho National Engineering and Environmental Laboratory's acidic and radioactive Sodium-Bearing Waste (SBW) [6]. This demonstration successfully converted the SBW to a Na$_2$CO$_3$ product that met the Waste Isolation Pilot Plant (WIPP) Waste Acceptance Criteria (WAC) for transuranic (TRU) waste. During the demonstration, data were collected to determine the nature and characteristics of the product, the operability of the technology, the composition of the off-gases, and the fate of key radionuclides (cesium and technetium) and volatile mercury compounds. The product contained a significant fraction of elemental carbon residues. Mercury was quantitatively stripped from the product but cesium, rhenium (Tc surrogate), and the heavy metals were retained. Nitrates were not detected in the product and NO$_x$ destruction exceeded 98% [6]. The steam reformer off-gas was monitored and it was determined that no O$_2$ was present. The off-gas was mostly (76%) H$_2$O (wet, N$_2$-free basis). CO levels averaged 1.3%, while the measured CH$_4$ levels averaged 0.1%.

In the THOR^{sm} FBSR process, a granular/particle bed material is fluidized with low pressure superheated steam. The lower zone of the fluid bed is operated in strongly reducing conditions to facilitate reduction of nitrates and nitrites to nitrogen gas. The upper zone of the fluid bed is operated under oxidizing condition by injection of oxygen. The oxidizing zone converts residual carbon reductants and organics into carbon dioxide and water vapor. The fluidized bed material can include ceramic media and/or reformed product granules. The incoming waste feed coats the fluidized particles and is instantly dried. The large active surface of dried nitrates readily reacts with the hot carbon reductant particles, carbon monoxide and hydrogen gases, and the reduced metal and metal oxides particles in the fluidized bed. Hydrogen and CO are formed when the steam oxidizes any solid carbon material (Equation 1) or any intermediate hydrocarbons (Equation 2). The CO in the gaseous mixture resulting from Equation 1 and Equation 2 can be made to react with more water vapor in the water gas shift reaction (WGSR) as shown in Equation 3. These reactions are important sources of heat that facilitate the dehydratrion, denitration, and reaction of the waste plus additive mixtures.

$$C(s) + H_2O(g) \rightarrow CO(g) + H_2(g) \tag{1}$$

$$CH_4(g) + H_2O(g) \rightarrow CO(g) + 3H_2(g) \tag{2}$$

$$H_2O(g) + CO(g) \rightarrow H_2(g) + CO_2(g) \tag{3}$$

Free hydrogen promotes radical generation and chain reaction propagation in the reformer bed. A significant benefit of the FBSR process is that it produces zero-liquid releases. All water is released as water vapor.

EXPERIMENTAL

A simulant of the Tank 48H solution was prepared according to Table I. This slurry has approximately 13.7 wt% solids. Antifoam (IIT Corp. B52) was added at 100 ppm antifoam per wt% solids [7]. Five wt% Fe$_2$O$_3$ was added as Fe(NO$_3$)$_3$•9H$_2$O to provide an indicator of the

REDuction/OXidation (REDOX) equilibrium that the sample experienced in sealed crucibles inside the oven. Having ~5 wt% Fe_2O_3 present enabled the $Fe^{+2}/\Sigma Fe$ ratio of the solid product sample to be measured from which the oxygen fugacity, $\log f_{O_2}$, $\log p_{H_2O}/p_{H_2}$, and $\log p_{CO_2}/p_{CO}$ of the reactions inside the sealed crucibles could be determined. All samples were analyzed for $Fe^{+2}/\Sigma Fe$ analysis by the Baumann method [8].

Table I. Tank 48H Simulant Recipe

Species	M/L
NaTPB	0.0728
NaOH	1.8425
NaNO$_2$	0.4709
NaNO$_3$	0.2753
Na$_2$CO$_3$	0.1295
NaAlO$_2$	0.1118
Na$_2$SO$_4$	0.0071
Na$_3$PO$_4$	0.0077
NaCl	0.0088
NaF	0.0059
KNO$_3$	0.0779

The T48H simulants were batched into stainless steel beakers. The slurry was carbonated with dry ice to convert the NaOH to Na_2CO_3 until a pH of ~9.5 was reached. This "acidification" from pH 13.3 to 9.5 also minimized foaming of the slurry. This ensured that once the carbonated material was put into a sealed crucible that a CO_2 atmosphere would be maintained. This served to duplicate the control of the atmosphere in the FBSR with CO_2 gas.

The reductant of choice was sucrose. A test matrix (Table II) was developed that varied three different levels of reductant based on the following stoichiometric equations:

$$C_{12}H_{22}O_{11} + 9.6NaNO_3 \rightarrow 7.2CO_2 + 11H_2O + 4.8Na_2CO_3 + 4.8N_2 \qquad (4)$$

$$C_{12}H_{22}O_{11} + 16NaNO_2 \rightarrow 4CO_2 + 11H_2O + 8Na_2CO_3 + 8N_2 \qquad (5)$$

Where the stoichiometric ratio of [C]:[N] for nitrate species is 12/9.6=1.25 (Equation 4) and 12/16=0.75 for nitrite species (Equation 5).

Three different levels of sucrose (none, ½X stoichiometric, and 1X stoichiometric) and three different reaction times (1/2 hour, 3 hours, and 48 hours) were tested. High purity (99.999%) Al_2O_3 crucibles were used to simulate Al_2O_3 bed material and to determine if the FBSR product was adhering to the simulated bed media. Temperatures of 650°C and 725°C were tested to see which levels of reductant optimized the WGSR (Equation 3) at which temperatures.

The known melt temperature of alumina containing FBSR products are >1280°C [9,10]. Hence, the lower melting Na_2CO_3 and Na_2SiO_3 FBSR products were targeted for study. Since the feed was already carbonated, nothing needed to be added to the samples to optimize the Na_2CO_3 product. Precipitated silica was added to the tests where the desired final FBSR product was Na_2SiO_3 or Na_4SiO_4.

The carbonated slurries were dried to peanut butter consistency to ensure that some H_2O remained in the sample to create steam for the WGSR. Alumina crucibles were sealed with nepheline ($NaAlSiO_4$) gel that melts at a temperature lower than the test temperature. This prevents air inleakage during reaction but allows other gases to escape by slow diffusion through the gel. The sealed samples were placed in a calibrated furnace at the test temperature designated in the test matrix. This generated a combined atmosphere of steam, CO from decomposition of the sucrose and CO_2 thus duplicating the WGSR species (Equation 3). The furnace was purged with 99.99% Ar to ensure that no O_2 mixed with any H_2 or CO escaped through the crucible seal.

Samples were analyzed by X-ray diffraction (XRD) to determine if the desired FBSR product was achieved. Samples were measured by High Pressure Liquid Chromatography (HPLC) to determine if the TPB was adequately destroyed by the FBSR reactions. Analyses were also conducted to determine if any secondary TPB reaction products were present, e.g 3PB and 2PB. Total carbon (TC), Total Inorganic Carbon (TIC), and Total Organic Carbon (TOC) were also analyzed.

Samples were measured by Ion Chromatography (IC) for NO_2, NO_3, F, Cl and SO_4 to determine the fate of these anions and the percent nitrate destruction. For those samples that the desired FBSR product was a silicate, samples were dissolved using a $LiBO_2$ fusion and the solution analyzed by Inductively Coupled Plasma – Emission Spectroscopy (ICP-ES) for Na, K, and Si to determine if the correct ratios of silica additive had been achieved during experimentation. Differential Thermal Analysis (DTA) was performed on selected products to determine the melting temperature. Details of all the analyses performed are given elsewhere [11].

DISCUSSION
Baseline Testing

A sample (T48-0) was tested as a baseline. The T48-0 sample was carbonated, anitfoam and $Fe(NO_3)_3 \bullet 9H_2O$ were added, and the sample was dried at 60°C. This sample was analyzed for TPB, anions, TC, TIC, TOC and REDOX as a baseline case (see Table II). These analyses demonstrated that there was 95,100 ug/g of TPB (Table III) present in the samples after the carbonation and drying steps. The presence of the TPB was also confirmed by the measurement of TOC, which showed 19,500 ug/g of organic carbon. Either the TPB or the anitfoam (an organic) may have reacted with the $Fe(NO_3)_3 \bullet 9H_2O$ because an all Fe^{+3} dried solution should have had a REDOX measurement of ~0 and the measurement was 0.44. This indicated that a considerable amount of Fe^{+2} was present or that the organics interfered with the REDOX measurement. Anion analysis of the base case indicated <100 ug/g of NO_2 and 163,000 ug/g or 16.3 wt% of NO_3. This number was used with the measured NO_3 data [11] to calculate the NO_3 destruction values given in Table III.

Tetraphenylborate Destruction

Samples were tested at two different temperatures, 650°C and 725°C (Table II). Tetraphenylborate (TPB) was completely destroyed in all the samples tested, i.e., the TPB, 2PB and 3PB were all <5ug/g indicating that the thermal treatment destroyed all the TPB and its derivatives. This was confirmed by the TOC analyses for all the samples thermally treated when <100 ug/g of TOC was observed. This indicates that FBSR is a viable technology for destruction of the organics in Tank 48H.

Table II Simulated T48H Steam Reformer Optimization Matrix

Test ID	Temp (°C)	Time (Hours)	Sugar Stoichiometry	Addition	Phase(s) Desired
T48-0	25	0	0	0	Na_2CO_3
T48-1	725	½	0	0	Na_2CO_3
T48-2	725	3	0	0	Na_2CO_3
T48-2B	650	3	0	0	Na_2CO_3
T48-3	725	3	½	0	Na_2CO_3
T48-4	725	½	1	0	Na_2CO_3
T48-5	725	3	1	0	Na_2CO_3
T48-5B	650	3	1	0	Na_2CO_3
T48-6	725	½	0	SiO_2	Na_4SiO_4
T48-7	725	3	0	SiO_2	Na_4SiO_4
T48-7B	650	3	0	SiO_2	Na_4SiO_4
T48-8	725	3	½	SiO_2	Na_4SiO_4
T48-9	725	½	1	SiO_2	Na_4SiO_4
T48-10	725	3	1	SiO_2	Na_4SiO_4
T48-10B	650	3	1	SiO_2	Na_4SiO_4
T48-11	725	48	0	SiO_2	Na_4SiO_4
T48-12	725	48	1	SiO_2	Na_4SiO_4
T48-13	650	48	1	0	Na_2CO_3
T48-14	725	48	1	SiO_2	Na_4SiO_4 + faujesite
T48-15	725	48	1	SiO_2	Na_2SiO_3 + faujesite

Carbonate FBSR Products

For all of the FBSR samples in which the desired product was Na_2CO_3 (samples T48-1 through T48-5B and T48-13), analysis by XRD indicated that a mixture of $Na_2CO_3 \bullet H_2O$ and Na_2CO_3 was formed regardless of temperature and residence time in the furnace (Table II and Table III). However, for the T48-13 sample that was heated at 650°C for 48 hours, the XRD analysis indicated no minor constituents. This meant that the minor constituents that had appeared in the same sample reacted for only 3 hours were due to incomplete reaction. Test T48-5B or T48-13 appeared optimal for making the Na_2CO_3 FBSR product at 650°C with 1X stoichiometric sugar and 3-48 hour residence time, as no minor phases were identified as incomplete reactants. Only the two primary phases, Na_2CO_3 and $Na_2CO_3 \bullet H_2O$, were present in the T48-5B and T48-13 samples.

Silicate FBSR Products

In the current study Na_4SiO_4 was chosen as the FBSR phase of choice because it melts at ~1120°C and can only coexist with a liquid phase down to temperatures as low as 1040°C. This choice was made to limit any potential liquid phase in the steam reformer that might cause bed agglomeration. Na_2SiO_3 may be acceptable for use as feed in the DWPF as its melting temperature is ~1080°C but it can coexist with a Na_2O-SiO_2 liquid phase that melts as low as 825°C and may cause FBSR bed agglomeration. Sodium silicate (Na_2SiO_3) was the silicate FBSR product phase made by THOR[sm] in their pilot scale studies with Hanford's high Na^+ containing Low Activity Waste (LAW).

Table III Simulated T48H Steam Reformer Analytic Results

Test #	Major Phase Desired	Major Phases Identified by X-Ray Diffraction	Minor Phases Identified by X-Ray Diffraction	NaTPB, 3PB, 2PB (ug/g)	Percent NO_x Destroyed
T48-0	Na_2CO_3	$Na_3H(CO_3)_2(H_2O)_2$, $Na(NO_3)$, $NaNO_2$, $Na_2CO_3 \cdot H_2O$	Na_2SiO_3, $KAl(SO_4)_2(H_2O)_{12}$	95,100 <5,<5	0
T48-1	Na_2CO_3	$Na_2CO_3 \cdot H_2O$, Na_2CO_3	$Al(OH)_3$ (?), $Ca_8Al_2Fe_2O_{12}CO_3(OH)_2 \cdot 22H_2O$	<5,<5,<5	30.1
T48-2	Na_2CO_3	$Na_2CO_3 \cdot H_2O$, Na_2CO_3	$Al(OH)_3$ (?), $Ca_8Al_2Fe_2O_{12}CO_3(OH)_2 \cdot 22H_2O$	<5,<5,<5	4.3
T48-2B	Na_2CO_3	$Na_2CO_3 \cdot H_2O$, Na_2CO_3	$Al(OH)_3$ (?), $NaNO_3$, $Ca_8Al_2Fe_2O_{12}CO_3(OH)_2 \cdot 22H_2O$	<5,<5,<5	24.5
T48-3	Na_2CO_3	$Na_2CO_3 \cdot H_2O$, Na_2CO_3	$Ca_8Al_2Fe_2O_{12}CO_3(OH)_2 \cdot 22H_2O$	<5,<5,<5	99.5
T48-4	Na_2CO_3	$Na_2CO_3 \cdot H_2O$, Na_2CO_3	Ca_2SiO_4 (?), $Al(OH)_3$(?)	<5,<5,<5	98.1
T48-5	Na_2CO_3	$Na_2CO_3 \cdot H_2O$, Na_2CO_3	$Ca_8Al_2Fe_2O_{12}CO_3(OH)_2 \cdot 22H_2O$ $Al(OH)_3$(?)	<5,<5,<5	97.5
T48-5B	Na_2CO_3	$Na_2CO_3 \cdot H_2O$, Na_2CO_3	NONE	<5,<5,<5	99.1
T48-6	Na_4SiO_4	Na_2CO_3, $Na_2CO_3 \cdot H_2O$, $K_{48.2}Al_{48.2}Si_{143.8}O_{384} \cdot 243H_2O$	Na_2SiO_3	<5,<5,<5	35.0
T48-7	Na_4SiO_4	Na_2CO_3, Na_2SiO_3, $Na_{7.89}(AlSiO_4)_6(NO_3)_{1.92}$	$Na_2CO_3 \cdot H_2O$ NaTPB (?)	<5,<5,<5	63.3
T48-7B	Na_4SiO_4	Na_2CO_3, Na_2SiO_3, $Na_{7.89}(AlSiO_4)_6(NO_3)_{1.92}$	$Na_2CO_3 \cdot H_2O$, $K_{48.2}Al_{48.2}Si_{143.8}O_{384} \cdot 243H_2O$	<5,<5,<5	41.4
T48-8	Na_4SiO_4	Na_2CO_3, $Na_2CO_3 \cdot H_2O$, Na_2SiO_3,	$KAlSiO_4$, $K_{48.2}Al_{48.2}Si_{143.8}O_{384} \cdot 243H_2O$	<5,<5,<5	95.2
T48-9	Na_4SiO_4	Na_2CO_3, $Na_2CO_3 \cdot H_2O$, Na_2SiO_3, $K_{48.2}Al_{48.2}Si_{143.8}O_{384} \cdot 243H_2O$	$KAlSiO_4$	<5,<5,<5	92.8
T48-10	Na_4SiO_4	$Na_2CO_3 \cdot H_2O$, Na_2SiO_3, $K_{48.2}Al_{48.2}Si_{143.8}O_{384} \cdot 243H_2O$	NONE	<5,<5,<5	98.7
T48-10B	Na_4SiO_4	Na_2CO_3, $Na_2CO_3 \cdot H_2O$, Na_2SiO_3, $K_{48.2}Al_{48.2}Si_{143.8}O_{384} \cdot 243H_2O$	$KAlSiO_4$	<5,<5,<5	94.1
T48-11 (see 6)	Na_4SiO_4	Na_2SiO_3, $K_{48.2}Al_{48.2}Si_{143.8}O_{384} \cdot 243H_2O$	$Na_2CO_3 \cdot H_2O$	<5,<5,<5	99.8
T48-12 (see10)	Na_4SiO_4	$Na_2CO_3 \cdot H_2O$ Na_2SiO_3	NONE	<5,<5,<5	98.8
T48-13 (see 5B)	Na_2CO_3	$Na_2CO_3 \cdot H_2O$, Na_2CO_3	NONE	<5,<5,<5	99.0
T48-14	Na_4SiO_4 faujesite	$Na_2CO_3 \cdot H_2O$, Na_2SiO_3	$Na(NO_3)$	<5,<5,<5	98.5
T48-15	Na_2SiO_3 faujesite	$Na_2CO_3 \cdot H_2O$, $Na_2Si_2O_5$	Na_2SiO_3, $Na(NO_3)$ $K_{48.2}Al_{48.2}Si_{143.8}O_{384} \cdot 243H_2O$	<5,<5,<5	99.4

For all of the simulated FBSR samples in which the desired product was Na_4SiO_4 with a $Na_2O:SiO_2$ ratio of 2:1 (T48-6 through T48-12), a potassium aluminosilicate zeolite phase known as faujesite ($K_{48.2}Al_{48.2}Si_{143.8}O_{384} \cdot 243H_2O$) was identified by XRD and a sodium silicate of a 1:1 $Na_2O:SiO_2$ stoichiometry had formed (Table III). Excess SiO_2 does not appear on the XRD pattern since the precipitated SiO_2 that was added to the sample is amorphous and will not give an XRD pattern. It was apparent that the faujesite was consuming some of the SiO_2 that was meant to form the 2:1 sodium silicate phase.

Subsequent testing (T48-14) was designed to compensate for the silica being consumed by the faujesite. X-ray Diffraction analysis (Table III) again indicated that the major phases in the T48-14 sample after a 48 hour residence time were still the faujesite and the 1:1 $Na_2O:SiO_2$ phase. Analysis of all the silicate FBSR products was performed to determine if the correct ratios of $Na_2O:SiO_2$ had been added during experimentation. These analyses indicated that some of the $Na_2O:SiO_2$ ratios measured were biased low by ~20%. This may be because the precipitated silica contains absorbed water and the exact amount of the absorbed water had not been measured.

In summary, if a silicate FBSR phase was desired, a silica FBSR phase was the major phase formed. Although the exact $Na_2O:SiO_2$ ratio of the desired FBSR silicate phase was never achieved due to incomplete reaction and silica deficient starting mixtures, this would not hinder the usage of any sodium silicate FBSR material made from Tank 48H slurry in DWPF.

Nitrate and Sugar Destruction

In the sample test matrix (Table II), samples with the designation of B indicate comparison of tests at the two different reaction temperatures. These were designed into the test matrix to test the optimum NO_x destruction at the various temperatures, e.g. optimize the WGSR. Hence samples T48-2B, 5B, 7B and 10B were tested at 650°C, while samples T48-2, 5, 7, and 10 were tested at 725°C.

For two of the pairs of samples tested at the different temperatures, T48-2 and 2B and T48-5 and 5B, the desired FBSR product was Na_2CO_3. Samples T48-2 and T48-2B had no sugar and samples T48-5 and T48-5B had 1X stoichiometric sugar. Comparison of the XRD spectra of the two tests without sugar demonstrates that the FBSR products in absence of sugar includes un-decomposed $NaNO_3$ which indicates that nitrate destruction is incomplete when sugar is absent at 650°C and 725°C. This was confirmed by nitrate analyses [11]. For the sample pair T48-5 and T48-5B sucrose was present at 1X stoichiometry. These samples had 99.1% and 97.5% NO_3 destruction at the 650°C and 725°C temperatures respectively (Table III). This indicates that the WGSR may be better optimized at 650°C than at the 725°C. Likewise, the TOC analyses [11] indicated no residual TOC in the form of sucrose in the samples and the XRD spectra did not indicate any residual $NaNO_3$ (Table III).

For two of the pairs of samples tested at the different temperatures, T48-7 and 7B and T48-10 and 10B, the desired FBSR product was a sodium silicate. Sample T48-7B had no sugar and Sample T48-10B had 1X stoichiometric sugar. Comparison of the XRD spectra of the two tests without sugar did not show any un-decomposed $NaNO_3$ but the nitrate analyses [11] indicated that considerable NO_3 remained in the samples without sugar. The nitrate destruction percentages given in Table III for these samples indicate that nitrate was only partially destroyed at either temperature for samples T48-7 and T48-7B. For the sample pair T48-10 and T48-10B sucrose was present at 1X stoichiometry. These samples had 98.7% and 94.1% NO_3 destruction (Table III). This again indicates that the WGSR may be better optimized at 650°C than at the 725°C. Likewise, the TOC analyses [11] indicated no residual TOC in the form of sucrose in the samples and the XRD spectra did not indicate any residual $NaNO_3$ (Table III).

The small amount of TOC measured in all the samples [11] indicated that at ½ to 1X sugar stoichiometry that most of the sugar added is consumed during denitration and that the FBSR product should not be overly reducing and thus compatible with DWPF processing of the FBSR product.

Particle Agglomeration to Simulated Al_2O_3 Bed Material

No adherence of the silicate or carbonate phases onto the Al_2O_3 crucibles was noted in any of the tests. Therefore, if the FBSR bed media used is Al_2O_3 there should not be any particle agglomeration with the bed material regardless of whether the FBSR product is carbonate or silicate. This also indicates that the Na_2SiO_3 phase that was produced most often as an FBSR product in this study appears to be an acceptable FBSR product phase in that it did not preferentially melt and react with the Al_2O_3 crucible.

FBSR Product Melt Temperatures

FBSR product samples T48-5B (Na_2CO_3 made at 650°C), T48-10 (mixed Na_2CO_3 and Na_2SiO_3 and faujesite), and T48-11 (Na_2SiO_3 and faujesite) were measured by Differential Thermal Analysis (DTA) to determine their melting temperature. The melt temperatures were 980°C, 1022°C, and 1049°C, respectively. These melt temperatures are all compatible with melting of these phases in the DWPF.

REDOX Measurements and the Water Gas Shift Reaction (WGSR)

Because the FBSR product is a mixture of oxide species, the Electro-Motive Force (EMF) REDOX series developed for DWPF glasses [12] was used to calculate the log f_{O_2} from the measured REDOX of the FBSR product [11]. Published correlations [13] between $-\log f_{O_2}$, temperature, log p_{H_2O}/p_{H_2}, and log p_{CO_2}/p_{CO} allows the log p_{H_2O}/p_{H_2} and log p_{CO_2}/p_{CO} to be determined for the atmosphere achieved in the sealed crucibles. The log log p_{H_2O}/p_{H_2} and log p_{CO_2}/p_{CO} are the partial pressures of the two half reactions for the WGSR given in Equation 3. The average REDOX ratio for the FBSR samples tested at 725°C and 650°C show that log f_{O_2} values of -9.69 and -10.75, were achieved respectively. These negative log f_{O_2} values mean that no oxygen was present during the FBSR reactions. The log f_{O_2} can be converted to log p_{H_2O}/p_{H_2} and/or log p_{CO_2}/p_{CO}. The log p_{H_2O}/p_{H_2} in the FBSR crucibles were between +5 and +6. The log p_{CO_2}/p_{CO} in the FBSR crucible studies were in the range of +4.5 to +5. The positive values for log p_{H_2O}/p_{H_2} and log p_{CO_2}/p_{CO} indicate that the conditions of the WGSR were adequately simulated.

Volumes of FBSR Product for DWPF

The 250,000 gallons of T48H slurry should make ~29,470 gallons of solid FBSR Na_2SiO_3 solid product or ~26,246 gallons of solid FBSR Na_2CO_3 solid product for subsequent treatment in the DWPF. This calculation used a measured FBSR product density of 1.46 g/cc [6]. It is calculated that 25,500 gallons of FBSR product (almost all of T48) could be added to the next 500,000 gallon DWPF sludge batch (at 18 wt% solids) if ~7 wt% Na_2O from the FBSR product is substituted for 7 wt% Na_2O in a given DWPF frit as done previously [14].

CONCLUSIONS

The purposes of the current study, organic destruction and downstream processing of T48H waste slurry, were fulfilled as documented by the following:

- TPB was destroyed in all 19 samples tested with the simulated FBSR process at operational temperatures 650-725°C; 650°C seemed to optimize the NO_3 destruction
- >99% destruction of nitrate was achieved with addition of sugar as a reductant at 1X stoichometry and TOC analyses indicated that excess reductant was not present in the FBSR product which ensures that the REDuction/OXidation (REDOX) equilibrium of the DWPF melter would not be adversely impacted

- destruction of anitfoam was also achieved at operating temperatures between 650-725°C based on measured TOC
- for all tests in which Na_2CO_3 was the desired FBSR product phase, Na_2CO_3 was produced, which has been shown to be compatible with the DWPF melt process as it melted at 980°C as measured by DTA
- for all tests in which Na_4SiO_4 or Na_2SiO_3 was the desired FBSR product a mixture of sodium silicates
  - this was determined to be a problem with water absorption by the SiO_2 additives used and the consumption of SiO_2 by a potassium aluminate zeolite (faugesite) that formed
  - formation of a sodium silicate (mixed with Na_2CO_3 or alone) is compatible with mixing the FBSR product into a DWPF feed tank or as an addition to the Slurry Mix Evaporator (SME) in place of some of the DWPF glass forming frit because the mixtures melted at temperatures of 1022°C and 1049°C, respectively, as measured by DTA
- the recommended test parameters for pilot scale testing of Tank 48H simulant at Idaho National Engineering and Environmental Laboratory (INEEL) were given by samples T48-5B and T48-14
- the sealed crucible studies demonstrated that bench scale studies can duplicate the complex reactions, especially the Water Gas Shift Reactions, and the associated $\log p_{H_2O}/p_{H_2}$ and $\log p_{CO_2}/p_{CO}$ atmospheres in the FBSR process.

REFERENCES

[1]J.B. Mason, J. McKibben, J. Ryan, J. Schmoker, " **Steam Reforming Technology for Denitration and Immobilization of DOE Tank Wastes,"** Waste Management 03 Conference (February 2003).

[2]C.M. Jantzen, **"Engineering Study of the Hanford Low Activity Waste (LAW) Steam Reforming Process,"** U.S. DOE Report WSRC-TR-2002-00317, Westinghouse Savannah River Co., Aiken, SC (July 2002).

[3]C. M. Jantzen, J. R. Zamecnik, D.C. Koopman, C.C. Herman, and J. B. Pickett, **"Electron Equivalents Model for Controlling REDuction-OXidation (REDOX) Equilibrium During High Level Waste (HLW) Vitrification,"** U.S. DOE Report WSRC-TR-2003-00126 (May 2003).

[4]J.B. Mason, T.W. Oliver, M.P. Carson, and G.M. Hill, **"Studsvik Processing Facility Pyrolysis/Steam Reforming Technology for Volume and Weight Reduction and Stabilization of LLRW and Mixed Wastes,"** Waste Management 99 Conference, (February 1999).

[5]B.P. McGrail, H.T. Schaef, P.F. Martin, D.H. Bacon, E.A. Rodriguez, D.E. McCready, A.N. Primak, and R.D. Orr, **"Initial Evaluation of Steam-Reformed Low Activity Waste for Direct Land Disposal,"** U.S. DOE Report PNWD-3288, Battelle Pacific Northwest Division, Richland, WA (January 2003).

[6]D.W. Marshall, N.R. Soelberg, K.M. Shaber, **"THOR[sm] Bench-Scale Steam Reforming Demonstration,"** U.S. DOE Report INEEL/EXT.03-00437, Idaho National Engineering & Environmental Laboratory, Idaho Falls, ID (May 2003).

[7]M.A. Baich, D.P. Lambert, and P.R. Monson, **"Laboratory Scale Antifoam Studies for the STTPB Process,"** U.S. DOE Report WSRC-TR-2000-00261, Westinghouse Savannah River Co., Aiken, SC (October 2000).

[8]E.W. Baumann, **"Colorimetric Determination of Iron(II) and Iron(III) in Glass,"** Analyst, 117, 913-916 (1992).

[9]E.M. Levin, C.R. Robbins, and H.F. McMurdie, **"Phase Diagrams for Ceramists,"** Vol I, American Ceamic Society, Westerville, OH, Figure 506, 601pp. (1964).

[10]E.M. Levin, C.R. Robbins, and H.F. McMurdie, **"Phase Diagrams for Ceramists,"** Vol II, American Ceamic Society, Westerville, OH, Figure 2284, 623pp. (1969).

[11]C.M. Jantzen, **"Disposition of Tank 48H Organics by Fludidized Bed Steam Reforming (FBSR),"** U.S. DOE Report WSRC-TR-2003-00352, Rev.1, Westinghouse Savannah River Co., Aiken, SC (March 2004).

[12]H.D. Schreiber, and A.L. Hockman, **"Redox Chemistry in Candidate Glasses for Nuclear Waste Immobilization,"** Journal of the American Ceramic Society, Vol. 70, No. 8, pp. 591-594 (1987).

[13] A. Muan and E.F. Osborn, **"Phase Equilibria Among Oxides in Steelmaking,"** Addison-Wesley Publ. Co., Reading, MA, 236pp (1965).

[14]C.M. Jantzen, **"Glass Compositions and Frit Formulations Developed for DWPF,"** U.S. DOE Report DPST-88-952, E.I. duPont deNemours & Co., Savannah River Laboratory, Aiken, SC (November 1988).

STEAM REFORMATION OF SODIUM BEARING WASTE: PROS & CONS

Darryl D. Siemer
Idaho National Engineering & Environmental Laboratory
PO 1625, Idaho Falls, ID 83415-7111
siemdd@inel.gov

ABSTRACT

Liquid reprocessing waste which contains a good deal of sodium nitrate cannot be efficiently calcined in a fluidized bed reactor unless a chemical reducing agent such as sugar is added. If the goal is to produce calcine capable of passing radwaste leach tests (a finished waste form), it is necessary to also add a mineralizing agent (clay) and run the reactor at a considerably higher temperature than is typical of radwaste calciners (e.g., 725°C instead of 500°C) This paper describes the outcome of a 5-day attempt to make such a "direct disposal" calcine via "steam reforming" and presents the author's opinions about that scenario.

INTRODUCTION

Most of the liquid reprocessing waste produced by the AEC/ERDA/DOE Complex was converted to hydroxy sludge/salty liquid slurries via neutralization with excess NaOH and then stored in underground steel tanks. The Idaho National Engineering and Environmental Laboratory (INEEL) was an exception because most of its waste was converted to a mix of sand-like granules and dust (fines) with fluidized bed calciners[1]. The rationale included: 1) calcines generally occupy less space than do either raw liquid wastes or neutralized slurries made from them; 2) dry powders are safer to store for indefinite periods than liquids; and, 3) calcination generally facilitates the subsequent implementation of technologies capable of producing products better suited for permanent disposal in a geological repository[a]. By the time that INEEL's fuel reprocessing facility shut down for the final time (1992), about 90% of its waste had been calcined.

The "technical" reason why INEEL didn't complete calcination is that its "sodium bearing waste" (SBW) contains a higher proportion of alkali metals (mostly sodium) relative to its other ash-forming components than did its other wastes.[b] Alkali nitrate salts do not "calcine" in purely thermal calcination systems operated at temperatures (e.g., 500°C) low enough to minimize the loss of "semivolatiles" such as chlorine, cadmium, ^{99}Tc, ^{106}Ru, ^{137}Cs, etc. These salts instead melt to form a mastic which quickly agglomerates a fluidized bed-type calciner's particulate reaction medium. A chemical solution to the problem was discovered at Argonne National Laboratory over forty years ago and repeatedly demonstrated in INEEL's pilot plants[5-7]; i.e., if a stoichiometric amount of a water soluble carbohydrate (e.g., ~38 grams of sucrose/mole of nitrate) is dissolved in the SBW, calcination will produce a free-flowing powder that consists primarily of sodium carbonate. In addition to facilitating calcination, the addition of sugar also serves to suppress

---

[a] For example, pretreatment via calcination generally enhances the productivity of a glass melter[2]. Calcination is absolutely necessary if a liquid waste is to be converted to a hot-isostatially-pressed ceramic ("Synroc"). It also both improves the quality and reduces the volume of any "grout"[3] that might be produced from most types of radwaste.

[b] If alkalis collectively constitute less than ~20 mole percent of the ash-forming cations in a nitrate-based waste, bed agglomeration is generally not encountered. Consequently, an inefficient solution to this problem is dilution - add a relatively large amount of low-sodium ash formers (e.g., aluminum nitrate) to the SBW before calcining it.

---

[106]Ru volatilization and ameliorate the "opacity issues" generated by the fact that purely thermal calcination converts most of the nitrate to NO_x.

INEEL's ~1 million gallons of SBW remains in liquid form because fluidized bed sugar calcination (FBSC) was deemed too dangerous to implement with the existing calciner[c]. A more recent rejection of vitrification as well has served to kindle a great deal of interest in a "new" way to calcine radwaste, "fluidized bed steam reformation" (FBSR). The fundamental difference between FBSC and FBSR is that the latter is performed at a higher temperature and a large excess of the chemical reductant (sugar, etc.) is added to further reduce NO_x production/emission. The most attractive FBSR technology offered by today's vendors is THOR's "mineralization" process because its product has been characterized as more "durable" than radwaste-type glasses[8-10]. The rationale for this is that the addition of clay converts the alkalis to water-insoluble aluminosilicate minerals (primarily nepheline and various sodalite analogs).[d]

At the time that the first draft of this paper was written, INEEL personnel had performed six steam reforming demonstrations with a DOE-owned fluidized bed reactor located at the Science Applications International Corporation (SAIC), Science and Technology Applications Research (STAR) Center in Idaho Falls. While most of them were done in ways that produced calcines consisting primarily of sodium carbonate, the one which serves as the subject of this paper set out to produce a "mineralized" product similar to that made by Hazen Research, Inc. and subsequently characterized by two DOE laboratories[9-11]. This paper presents an overview of what happened during INEEL's test and offers the author's opinions about what it means. For a more detailed description of the test itself, see Soelberg et. al.[12]

EXPERIMENTAL

SAIC's fluidized bed test system can either be manually operated or run automatically with a process logic controller (PLC) system featuring multiple human-machine interface (HMI) stations. Continuous emissions monitoring systems (CEMS) measure and record the concentrations of several important gases at various points. The primary components of the reactor include the reformer vessel, product collection systems, a feed system, the off-gas control system, and the PLC. It covers a footprint of about 12 m by 12 m. Equipment and piping are fabricated from 300-series stainless steel except for the reformer vessel, which is made from Inconel 800H. The main features of the reformer vessel include the fluidized bed-containing section, a freeboard (particle disengaging) section, and a gas distributor through which the fluidizing gas was introduced. The bed and freeboard sections were externally heated with electrical resistance heaters. The fluidized bed section is 15 cm in diameter (the same size as Hazen's reactor) and 76 cm tall. The freeboard section is 30 cm in diameter and 1.5 m high. Numerous ports in the bed and freeboard sections enable access for liquid and solid input streams and instrumentation. Interchangeable distributor and bottom receiver designs plus numerous ports make the test platform versatile and relatively simple to reconfigure.

A mix of superheated steam (approximately 7 kg/hr) and elemental oxygen (approximately 1 kg/hr) served to fluidize the bed particles. The purpose of the oxygen was to provide additional

---

[c] This conclusion was based upon an assumption that the sugar would be batch-mixed with SBW in quantities large enough to feed the reactor for several days (e.g., 10,000 gallons). The safety issues generated by that assumption could have been addressed by simply adding the sugar (as syrup) with an in-line mixer.

[d] This idea is almost as venerable as sugar-calcination – DORR-OLIVER recommended adding clay to its fluidized bed salt waste incinerators to prevent bed agglomeration over thirty years ago[13].

heat[e] - without it, the reactor's external heaters and the superheated steam could not provide sufficient energy to maintain the desired bed temperature (~720°C) at more than a nominal (> 2 kg/hour) slurry feed rate. A pneumatic nebulizer utilizing ~3 kg/hour of N_2 sprayed the feed slurry into the reactor immediately above its distributor plate.

Off-gas exiting this reactor passed through a cyclone which removed the larger gas-entrained particulates. A reversible auger situated at the base of the cyclone either returned its catch to a sample collection point (occasionally) or recycled it back into the base of the reactor (most of the time). The off-gas subsequently passed through a vessel containing seven 6.3-cm diameter, 61-cm long, sintered-metal blow back filters having a nominal pore size of 2 μm. Filtered-out dust dropped into a nitrogen-purged product collection drum. The filtered off-gas was diluted with sufficient air to oxidize the combustible gases (hydrogen, carbon monoxide, ammonia, methane, etc.) produced by water-gas/reforming reactions and then run into a large gas-fired thermal oxidizer which was operated at ~1,000°C. The gas was then quenched with a water spray and then scrubbed with a venturi scrubber. The amount of quench water and scrub solution temperature were controlled to maintain a water-neutral system – one which requires no makeup. This simplifies both the operation of the scrubber and the evaluation of results. The scrubbed off gas was then reheated and passed through a 3-stage bed of sulfur impregnated activated carbon which absorbed the mercury.

Hands-on operation of the equipment was done by SAIC personnel. Both the original run plan and on-the fly decision-making were jointly produced by INEEL and visiting THOR personnel.

Inputs

The waste simulant (Table I) is a non-radioactive version of the SBW contained in INEEL tank WM-180 (which represents about 30% of INEEL's remaining liquid reprocessing waste). Like INEEL's real SBW, it is a low viscosity liquid containing relatively little undissolved solids. The actual feedstream consisted of a thin slurry of this simulant with solid mineralizing additives (various amounts of finely powdered quartz plus high-grade kaolin, nominally $Al_2O_3.2SiO_2.2H_2O$) which was in-line mixed with 55 wt % sucrose-water syrup just before it entered the reactor. Sufficient clay/quartz were added to convert the alkalis to a mineral assemblage comprised entirely of the alkali aluminosilicate minerals normally associated with the process (nepheline, nosean, sodalite, etc.). The amounts added per liter of simulant ranged from 44 g of quartz plus 257 g clay at the beginning of the test to 322 g of clay-only at the end. Quartz was eventually omitted because it seemed to promote excessive bed agglomeration. Other inputs included ~0.05 cm-diameter, sintered alumina "starting bed", charcoal granules, iron oxide (purportedly, a NO_x reduction catalyst), and, on one occasion, elemental iron powder. While the absolute amounts of sugar and charcoal were both varied over wide ranges, there was always more than enough of them together to provide a strongly reducing reactor atmosphere.

---

[e] The oxygen provided heat by burning some of the reducing agents (elemental carbon and sugar) added to "reform" the waste.

Table I: INEEL tank WM-180 SBW simulant

	moles/L	g/L		moles/L	g/L
CATIONS					
Acid (H+)	1.1	1.13	Magnesium	0.012	0.292
Aluminum	0.66/	17.9	Manganese	0.016	0.775
Boron	0.012	0.133	Mercury	0.0014	0.271
Calcium	0.047	1.89	Nickel	0.0013	0.086
Cesium	0.0032	0.431	Potassium	0.20	7.66
Chromium	0.0034	0.174	Rhenium	0.0011	0.20
Copper	0.0007	0.044	Sodium	2.1	47.4
Iron	0.022	1.21	Strontium	0.0020	0.176
Lead	0.0013	0.274	Zinc	0.0011	0.069
ANIONS					
Chloride	0.03	1.06	Phosphate	0.029	2.74
Fluoride	0.047	0.90	Sulfate	0.070	6.72
Nitrate	5.3	330	Water	40.6	731

One liter of this simulant would produce 128 g of oxide-basis "ash"; or, if "sugar-calcined", 157 g of carbonate-calcine; or, if calcined with 290 g of kaolin, 385 g of mineralized calcine.

"Conservative PCT"

For reasons to be discussed later, the author applies a modified version of the PCT to any multiphasic waste form material (anything other than glass): It involves grinding the sample just enough to pass through a 100-mesh (150 micron) screen; putting everything into a tared Teflon™ PCT vessel along with 10x as much deionized water; putting that vessel into a desiccator containing lime water (to absorb CO_2) situated in a 90°C convection oven; removing it after 7 days; cooling to room temperature, filtering, and immediate analyzing of the leachate. Since raw PCT data are usually normalized with respect to surface area, this version assumes that the sample consists of impervious spheres just large enough to pass through a 100-mesh screen. For materials possessing the "skeletal" (Archimedian) density of typical alkali aluminosilicate minerals (~2.5 g/cm^3) this corresponds to an area:mass ratio of 160 cm^2/g.

OBSERVATIONS

- Total in-reactor gas flowrate was about 0.4 gram-moles/s [f] (~two-thirds water vapor)
- Typical NO$_x$ feedrate (moles nitrate/moles total gas) ~ 20,000 ppmv wet-basis
- Nominal in-reactor gas velocity @720°C & 0.84 atm (base of the reactor), ~136 cm/s
- Mean in-bed gas residence time @720°C & 0.84 atm, ~ 0.55 s

- Grab samples of the bed product generally contained mineralized agglomerates[g] which grew larger and more numerous as the run continued. The rate of lump formation decreased when quartz powder was left out of the feed slurry.

---

[f] About 15 scfm in the usual US engineering units.

[g] The water solubility and carbonate concentration of these lumps (both low) were similar to those of the non-agglomerated bed material.

- On two occasions those agglomerates grew sufficiently large and numerous to defluidize the bed causing immediate process shut down.
- Off gas samples taken upstream of the off gas burner typically contained several thousand ppm (by volume) of ammonia plus several ppm HCN. The concentration of ammonia correlated directly with total reductant (Σ charcoal + sugar) feed rate.
- NO_x destruction by the reactor itself ranged from 60% to 95% with strong dependence upon total reductant feed rate.
- The off gas burner (thermal oxidizer) destroyed ~60% of the NO_x and ~100% of the NH_3 reaching it. Overall nitrate-to-N_2 conversion over the entire run averaged ~93%.
- The off gas burner diluted the off gas by a factor of ~6.
- ~100% of the mercury ended up on the first stage of the sulfur-impregnated charcoal bed
- ~0.0009% of the simulant's sodium ended up in the scrub water. This means that the blowback filters did an excellent job of particulate removal.
- Because the off gas burner quantitatively destroyed ammonia, mineral acids derived from the SBW's anions eventually reduced the pH of the scrub water to ~2. Without that burner, the pH of the scrub would have become buffered at ~8 by a high and steady-state concentration of ammonium bicarbonate.
- The system's cyclone solids recycle rate was generally about the same as the total ash content of the feed stream (typically 1-2 kg/hr).

RESULTS

Product Characterization

Except as noted, the following observations are based upon analyses of composited samples taken from the drums in which both of the product fractions had accumulated throughout the entire five-day test. Consequently, neither fraction represented an "optimum" product.

Sufficient feed to generate ~115 kg of product produced ~74 kg of fines, ~3.6 kg of cyclone dust samples, and ~37 kg of bed product[h]. The gross compositions (major components only – alumina, silica, and soda) of both product fractions were similar and consistent with the amounts of "ash formers" (SBW salts, clay, silica, and iron) fed to the reactor.

Regardless of the total reductant feed rate (Σ charcoal + sugar), grab samples of either product fraction never contained more than 0.1% residual nitrate (or nitrite) and generally none at all. Grab samples of both product fractions generally contained substantial amounts of elemental carbon (~0 to 50 wt% lampblack-like dust in fines – 15% in the composite, 0 to 10% 2-5 mm chunks in bed samples – 6.7 wt% in the composite). The amount of "product carbon" in the grab samples depended upon the total reductant feed rate.

Ten-minute exposure to 100x as much ~90°C water would typically solubilize ~20% of the sodium in fines samples and ~4% of that in bed samples. This suggests that FBSR did indeed succeed in "mineralizing" most of the salt. The primary solute in such water-leachates is sodium accompanied by an equivalent amount of (in decreasing order of concentration) aluminate, carbonate (with fines, not bed samples), phosphate, sulfate, chloride, silicate, and fluoride ions.

---

[h] This figure (37 kg) does not include the total of ~63 kg alumina "starting bed" charged to the reactor on three different occasions. At a bulk density of 0.7 g/cm^3, 37 kg corresponds to ~4 bed turn-overs.

The "tapped bulk densities"[i] of the composited fines and bed samples were ~ 0.35 g/cm³ and ~ 0.7 g/cm³ respectively. The mean particle diameter of the fines fraction appeared to be on the order of ~1-2 μm (via optical microscopy). The mass mean particle diameter (MMPD) of the bed material was about 0.6 mm (via screening).

Several "semivolatile" materials (rhenium, cesium, chloride, and sulfate) were present at significantly higher concentrations in the fines product fraction (Tables II and III). The likely cause is that the temperature of the blowback filters was sufficiently lower (200-300 C-degrees) than that of the "reformer" to permit the accumulation of fines which covered them to act as efficient condensers.

Table II: Distribution of anions in both product fractions (ion chromatographic finish)

Anion/expected Conc. (ppm)	ppm in BED		ppm in FINES		% in scrubwater
	Water leach	Via fusion	Water leach	via fusion	
Chloride/2700	590/550	600/680/690	2500/2900	2200/2390	16
Sulfate/17500	1200/1670	4400/4370	2600/2990	5900/5500	5

"Fusion" entailed fusing sample powder wetted with 25wt% NaOH soln. plus 30% H_2O_2 in a Pt crucible in a 450°C muffle, transferring everything to a centrifuge tube with water rinses, bubbling sufficient CO_2 through it to convert most of the hydroxide to carbonate; heating the slurry @ 90°C for ~10 min.; and then diluting to a convenient volume with pure water. The slashes separate results of replicate analyses.

Table III: Distribution of cesium and rhenium (^{99}Tc surrogate) in solid products (ICPMS finish)

Element/expected Conc. (ppm)	%in scrub water	Bed analyses (ppm)			Fines analyses (ppm)		
		Water leach	Acid leach	$LiBO_x$ fusion	Water Leach	Acid Leach	$LiBO_x$ fusion
Cs/1100 ppm	0.035	60	82	110	480	820	1100
Re/519 ppm	0.14	120	82	80	330	460/500	125

"acid leach" means heating the sample with aqua regia followed by dilution with water.

The BET surface area of the bed product was ~83 m²/g..

X ray diffraction spectrometry indicated that the bed product was predominantly crystalline and that nepheline was its primary component. The fines fraction was predominantly amorphous and featured weak quartz and carnegieite peaks superimposed on a broad "hump."

The sum of all sulfur-bearing species (sulfate, sulfite, and sulfide) in both solid product fractions plus the scrub liquor accounted for just under one-third of the sulfur fed to the reactor.

There was under 1 ppm ammonium or cyanide in either solid product fraction.

DISCUSSION

The off gas produced by a FBSR reactor differs from that of a FBSC reactor primarily in that it contains reduced-form nitrogen species (ammonia, HCN, etc.) and other combustible gases in

---

[i] The bulk density of a powder sample is not an absolute number because its value depends upon the degree of consolidation (settling) achieved before volume is noted. The figures in this paper were obtained by dumping powder into a tared 10 cc glass graduated cylinder, tapping it for about 30 seconds, and then measuring its mass and volume. This much tapping typically reduces the volume of a dumped fines sample by about 50%.

addition to the NO_x, CO_x, and water vapor. It is *not* intrinsically easier to render such off gas harmless than those produced by either a "normal" calciner or glass melter.

One of the reasons why this paper paints a relatively unfavorable picture of this process is that its product proved to be less attractive than advertised. A key factor in judging radioactive waste forms is "durability" and today's chief test of this characteristic, the PCT (ASTM C1285-02), is apt to generate grossly misleading (optimistic) results with non-glass materials. For example, a quick water leach of either of the products of INEEL's test removes most of the chloride[j]. Since ASTM C1285-02 leaves sample powder-washing to the discretion of the analyst, an "official" test performed for the purpose of determining the normalized leach rate of chloride (NL_{Cl}) *might* be done on material that no longer contains chloride – which, in turn, means that its results *might* be interpreted to mean that FBSR had succeeded in immobilizing the chlorine (sodalite?).

The formal PCT's size-discrimination creates additional bias because different sized fractions of non-vitrified materials often exhibit quite different intrinsic solubilities. In this instance, 100% of the fines product fraction was under the PCT's lower particle size cut-off (75 μm) which means that two-thirds of what the reactor produced (the poorer-performing fraction) *could* "legally" be excluded from consideration. Furthermore, the fact that the concentrations of cesium and rhenium (a good technetium surrogate) were five times greater in the fines than in the bed product fraction means that a PCT performed solely on the latter could not gauge the process' ability to immobilize key radioisotopes.

Additional confusion is generated by the practice of normalizing fraction-leached results to surface area. For example, seven-day exposure of a 50:50 mass-wise mix of INEEL's fines & bed products to 10x as much 90°C water (PCT conditions) solubilized 12.5% of the sodium. If this is normalized to a "conservative" surface area (see EXPERIMENTAL), the waste form's NL_{Na} works out to 7.8 g/m^2 – somewhat higher (worse) than that of today's HLW sodium-leach standard[k] (NL_{Na} of EA glass ~6.7 g/m^2). If a reasonable estimate of just the *external* surface area of the sample particles is used for this calculation, NL_{Na} drops to a much more comforting ~0.03 g/m^2. Their BET area[l] would generate an even more persuasive ~0.0015 g/m^2!

The other reason for this writer's opinion about this process is that no amount of "tweaking" by the attending fluidized bed experts could make it operate continuously for more than a day or so without some sort of agglomeration-related "system crash".

Roughly two-thirds of the SBW's sulfate "disappeared" because reactor conditions rendered sufficiently reducing with excess charcoal to attain the test's NO_x abatement goal (90%) served to convert most of it to a gaseous form (probably SO_2) which an acidic scrub solution could not capture. (The concentration of SO_2 in the off gas may have as high as ~30 ppm - not enough to exceed typical regulatory or risk assessment limits.) Since only ~one-half of the sulfur retained by the reactor's solid products (Table II) was immediately water soluble, a rather small fraction of the sulfur in the feed (½ of (1 minus ⅔) or one-sixth of the total) may have indeed been converted to nosean[10].

INEEL's test generated a total of ~270 liters of fines, granules, and chunks from ~300 liters of SBW – a volume reduction of about 10%. *If* 100% of the product had been bed material (no

---

[j] The same observation and line of reasoning applies to the cesium in the fines product fraction.

[k] This figure is 4x worse than the standard that DOE's proposed low level glasses must meet (NL_{Na} <2 g/m^2)

[l] The folks championing concrete waste forms often normalize PCT results to BET areas to make their product seem more attractive. This is silly from a technical point of view because such huge surface areas render the concept meaningless: for example, any material possessing a surface area:mass ratio greater than 0.15 m^2/gram (1/6.7) can totally dissolve and still pass today's standard for high-level glasses (NL_{Na} < 6.7 g/m^2)!

fines), net volume reduction would have been on the order of 50%. On the other hand, previous INEEL experience suggests that sugar-calcination (or sugar-reformation) of the same simulant without clay would have resulted in a volume reduction of about 80%.

The author considers "full mineralization"[m] FBSR to be a poor way to pretreat INEEL SBW for vitrification. The product contains too much aluminum to be efficiently converted to a borosilicate glass in a normal-temperature (~1150°C) melter and its high and variable elemental carbon content would render redox control difficult. He also considers it non-optimal as a pretreatment for grouting. Its products' low bulk density, low water solubility, and high surface area cause them to behave like silica fume in a cement-mixer; i.e., excessive water is required to produce grout sufficiently fluid to readily fill the canister and it also tends to promote flash set[15]. The fines fraction is especially troublesome.

With respect to some of the vendor's other suggested waste treatment scenarios[8], experience gained during both this and prior INEEL tests[16] suggests that bed agglomeration would render it impossible to produce a product consisting primarily of sodium silicate with a normally-configured fluidized bed reactor. On the other hand, that same experience also suggests that a calcine consisting primarily of sodium carbonate[17] or high-alumina sodium aluminate (both of which are water soluble and therefore poor waste forms) would be relatively simple to produce.

Bed agglomeration is the bête noir of fluidized-bed calcination and it is apt to be especially problematic with full-mineralization FBSR: 1) it is especially likely to happen because sodium tends to form relatively low-melting (sticky) eutectics with silicates; and, 2) such incidents would be especially difficult to recover from because a "rocked-up" bed could not be simply dissolved-out with dilute nitric acid. INEEL's test reactor had to be repeatedly disassembled and rodded-out - prohibitively expensive with a fully remoted system. If weight-wise waste loading, product volume, and product quality didn't matter, bed agglomeration could be easily avoided by running the reactor like a fluidized bed salt waste *incinerator*[13]; i.e., add excess clay and increase gas flow so that ~100% of the product elutriates out as fines. Unfortunately, all three of the aforementioned characteristics *are* important.

CONCLUSIONS/RECOMENDATIONS

While much of the difference in product "durability" noted between this and previous reports can be attributed to how that characteristic was measured (which version of the PCT was used), some is due to the fact that Hazen's test[8] purportedly produced only the better-performing bed product, no fines. Because INEEL's test did not do this, it didn't really prove anything – it simply "raised issues". Consequently, any further repeats of INEEL's mineralized FBSR test should be done in a manner that also produces a bed-only product.

Hazen accomplished this by running the off gas exiting its cyclone directly into a scrubber (no blow black filters) and continuously recycling the scrub-slurry back to the feed tank[n]. Perhaps a

---

[m] As opposed to "partial mineralization"; i.e., SBW sugar-calcination done with sufficient added clay to suppress carbonate retention but not enough to produce an insoluble calcine. This is how the author batch-calcines SBW for his "hydroceramic" grouting experiments[14]. I don't know if it could be accomplished with a fluidized bed reactor – it definitely could be done with a rotary kiln equipped with breaker-bar or chain-type "deagglomerators."

[n] INEEL did not choose to do this because its experience with similarly-configured fluidized bed calciners (no blow back filters) suggested that the degree of solids recycle anticipated for this particular process would "choke" the scrubber. Excessive solids blow-over was expected because micron-sized clay particles would constitute about two-thirds of the ash forming material going into the reactor. The fact that elutriated solids could not be readily solubilized in the scrub tank presaged plugging-related operation problems with the scrub recirculation system.

better way for INEEL to implement total solids recycle would be to periodically dump the filter fines catch into the feed tank[o].

If such a revised system could be made to operate reliably[p], the retest would produce a bed-only calcine and therefore results more consistent with those already published[9-10]. However, this *still* sidesteps the fact that it is unreasonable to base a judgment about the relative worth of a calcine-producing technology and vitrification on PCT results – even results obtained with a "conservative" version of that test. The reason for this is that glass is intrinsically monolithic and calcine is intrinsically dust-like. Since leaching occurs at the surfaces and real waste forms will not be deliberately powdered before burial, the surface area of the *intact* waste form should count. For example, a one million-gram glass monolith (smaller than most real or proposed US glass waste forms) possesses a geometric surface area of about 3 m^2. Its "real" surface area isn't much greater than that because glass is basically just a super-viscous liquid. Roughly one gram (the exact amount depends upon how area is measured) of INEEL's mineralized "bed product" possesses that much surface area.

The underlying justification for INEEL's efforts was to investigate another way of rendering its remaining liquid waste road-ready – an activity which has constituted much of its raison d'être for the past twelve years. If one chooses to piece together relevant portions of other recent DOE Complex reports[18], it quickly becomes apparent that a similar-sized (15 cm diameter), forced convection-type, glass melter close-coupled with a thin-film evaporator could probably have converted the same 300 liters of SBW into roughly one-sixth as much product (~45 liters of glass vs ~270 liters of calcine) in about one-half the time (~2 days). This plus the fact that the time allotted for "studying" INEEL's options has run out suggests that decision makers should consider reviving the "direct vitrification" option identified in its final environmental impact statement.[19] This writer believes that a properly designed vitrification system would: a) be less troublesome to operate than FBSR; b) generate a considerably lesser volume of equally-easy-to-purify off gas (because melters don't require fluidizing or "reforming" gases); and c) produce a waste form more acceptable to both citizen-stakeholders and independent technical reviewers[20].

LITERATURE CITATIONS

[1]B.R. Wheeler, J.A. Buckham, and J.A. McBride, "Comparison of Various Calcination Processes for Processing High-level Radioactive Waste", IDO-14622, Idaho National Engineering and Environmental Laboratory, Idaho Falls, Idaho, April 1964.

[2]VECTRA, "Fluid Bed Calciner Test Report-Final" VECTRA GSI Report # WHC-VIT-03, WHC-SD-WM-VI-031, Aug. 1995 (work subcontracted to PROCEDYNE).

[3]A.H. Zacher et al., "Denitration of High Nitrate Salt Wastes with Reductants", DE 00006226, PNNL-12144, Pacific Northwest National Laboratory, Richland, Washington, May 1996.

[4]J.W. Loeding, E.L. Carls, L.J. Anastasia, and A.A. Jonke, "The Fluid-bed Calcination of Radioactive Waste", ANL-6322, Argonne National Laboratory, Idaho Falls, Idaho, May 1961.

[5]J.C. Petrie, "Report on Run 8, Twelve-Inch Diameter Calciner", NRTS letter Petr-2-65A to E. J. Bailey, dated June 4, 1965.

---

[o] While filter-fines could also be dry-augered directly into the base of the reactor, it seems rather likely that they would be immediately blown out again.

[p] There is no compelling reason to believe that implementing 100% dust recycle would ameliorate the operational problems caused by *excessive* agglomeration.

[6]J.C. Petrie, "Report on Run 12, Twelve-Inch Diameter Calciner", NRTS letter Petr-13-65A to E. J. Bailey, dated December 30, 1965.

[7]J. Pao and J.A. McCray, "Interim Status Report for Sugar-Additive Laboratory-Scale Mockup Calcination Testing", letter JHP-05-96/JAM-07-96, September 11, 1996.

[8]J. Bradley Mason, J. McKibbin, K. Ryan, and D. Schmoker, "Steam Reforming Technology for Denitration and Immobilization of DOE Tank Wastes", *WM '03 Conference*, Feb 23-27, 2003, Tucson, AZ.

[9]C.M. Jantzen, "Characterization and Performance of Fluidized Bed Steam Reforming (FBSR) Product as a Final Waste Form", WSRC-MS-2003-00595, Westinghouse Savannah River Company, Aiken, South Carolina, 2003.

[10]C.M. Jantzen, "Engineering Study of the Hanford Low Activity Waste (LAW) Steam Reforming Process, " WSRC-TR-2002-00317, Westinghouse Savannah River Company, Aiken, South Carolina, July 12, 2002.

[11]B.P. McGrail, H.T. Schaef, P.F. Martin, D.H. Bacon, E.A. Rodriguez, D.E. McGrady, A.N. Primak, and R.D. Orr, "Initial Evaluation of Steam-Reformed Low Activity Waste for Direct Land Disposal", U. S. DOE Report PNWD-3288, Battelle Pacific Northwest Division, Richland, WA, January 2003.

[12]N.R. Soelberg, D.W. Marshall, S.O. Bates, and D.D. Taylor, "Phase-2 THOR Steam Reforming Tests for Sodium-Bearing Waste Treatment, INEEL/EXT-04-01493, Idaho National Engineering and Environmental Laboratory, Idaho Falls, Idaho, January 2004.

[13]C.J. Wall, J.T. Graves, and E.J. Roberts (DORR OLIVER), "How to Burn Salty Sludges," *Chemical Engineering*, pp. 77-82, April 1975.

[14]D.D. Siemer, J. Olanrewaju, B.E. Scheetz, and M.W. Grutzeck, "Development of Hydroceramic Waste Forms for *INEEL* Calcined Waste", *Ceramic Transactions*, **119**, 391-398, (2001).

[15]D.D. Siemer, M.W. Grutzeck, and B.E. Scheetz, "Cementitious Solidification of Steam Reformed DOE Salt Wastes", ECI Alternative Nuclear Waste Forms, Girdwood, AL January 18-23, 2004.

[16]N.R Soelberg, D.W. Marshall, S.O. Bates, and D.D. Siemer, "SRS Tank 48H Waste Steam Reforming Proof-of-Concept Test Results," INEEL/EXT-03-01118, Idaho National Engineering and Environmental Laboratory, Idaho Falls, Idaho, September 15, 2003.

[17]D.W. Marshall and N.R Soelberg, "TWR Bench Scale Steam Reforming Demonstration," INEEL/EXT-03-00436, July 2003.

[18]Vitrification rate estimate is based on "Stir Melter" tests performed at Clemson University (see SRS doc. WSRC-TR-9900305, Rev 1, page 24) and SBW evaporation tests performed by the LCI Corp. (see INEEL doc. ICP/EXT-04-00172).

[19]"Idaho High-Level Waste & Facilities Disposition", FINAL ENVIRONMENTAL IMPACT STATEMENT, DOE/EIS-0287, September 2002.

[20]J.A. Gentilucci, J.E. Miller, R.L. Treat, and W.W. Schultz, "Technical Review of the Applicability of the Studsvik, Inc. THOR Process to INEEL SBW", doc. TRA-0101, March 2001.

ANSTO'S WASTE FORM RESEARCH AND DEVELOPMENT CAPABILITIES

E. R. Vance, B. D. Begg, R. A. Day, S. Moricca, D. S. Perera, M. W. A. Stewart, M. L. Carter, P. J. McGlinn and K. L. Smith
Materials and Engineering Science
Australian Nuclear Science and Technology Organisation
Menai, NSW 2234, Australia

ABSTRACT

Waste form development at ANSTO for mainly high-level radioactive waste is directed towards practical applications. Some longstanding misconceptions about titanate ceramics are dealt with. We currently focus on waste forms for problematic niche wastes that do not readily lend themselves to direct vitrification. Titanate-bearing waste form products we have developed or are developing are aimed at immobilization of tank wastes and sludges, U-rich wastes from radioisotope production from reactor irradiation of UO_2 targets, INEEL calcines and Na-bearing liquid wastes, Al-rich wastes arising from reprocessing of Al-clad fuels, Mo-rich wastes arising from reprocessing of U-Mo fuels, partitioned Cs-rich wastes, and ^{99}Tc. Other waste forms include encapsulated zeolites or silica/alumina beads for immobilization of ^{129}I. Waste form production techniques cover hot isostatic and uniaxial pressing, sintering, and cold-crucible melting, and these are strongly integrated into waste form design. In addition, building on previous work on speciation and leach resistance of Cs in cementitious products, we are studying geopolymers. Recently we have embarked on studies of candidate inert matrix fuels for Pu burning. Although we have a strong emphasis on the properties of candidate wasteforms for actual wastes, we have a considerable program directed at basic understanding of the waste forms in regard to crystal chemistry, dissolution behaviour in aqueous media, radiation damage effects and processing techniques.

INTRODUCTION

In 1978, Ringwood (1) suggested assemblages of titanate minerals could be used to incorporate HLW, as titanate minerals are much more water-resistant than those used in supercalcine (2), the then principal alternative to borosilicate glass. In titanate assemblages waste ions are only dilutely incorporated into the phases, whereas in the supercalcines fission products and actinides were the basis of the phases. Synroc-C is one of the early titanate assemblages designed to incorporate HLW from fuel reprocessing. Subsequently it has become the archetype from which waste forms for various applications have been derived at ANSTO. Table I shows the phase constitution of synroc-C, containing 20 wt% HLW, and the radionuclides can be incorporated in the various phases. This material can be consolidated into a dense ceramic by hot uniaxial or isostatic pressing at ~ 1150°C (3).

Table I. Composition and mineralogy of synroc-C

Phase	wt%	Radionuclides in lattice
Hollandite, $Ba(Al,Ti)_2Ti_6O_{16}$	30	Cs, Rb
Zirconolite, $CaZrTi_2O_7$	30	RE, An*
Perovskite, $CaTiO_3$	20	Sr, RE, An
Ti oxides	15	
Alloy phases	5	Tc, Pd, Rh, Ru etc.

*RE, An = rare earths and actinides respectively

The grain size is • 1 μm, to optimise mechanical properties and prevent subsequent radiation-induced microcracking. The alloy phases derive from elements which form metals under the reducing conditions prevailing during hot-pressing. The leach rates at 90°C in water from synroc-C of the most soluble elements, alkalis and alkaline earths, are typically < 0.1 g/m^2/day for the first few days, and they decrease asymptotically to values of ~ 10^{-5} g/m^2/day after 2000 days. Leach rates of other elements are much lower. Leach rates of 10^{-5} g/m^2/day correspond to a corrosion rate of ~ 1 nm/day.

In the 1980s, the inactive Synroc production process was scaled-up via the Synroc Demonstration Plant to produce ~ 50 kg monoliths containing 20 wt% of simulated HLW, with properties as good as those of gram-sized laboratory samples. The ceramic could tolerate waste loadings up to 30% HLW (neglecting radiogenic heat effects), with no changes in the phases, just their abundances.

SYNROC DERIVATIVES

In the early 1990s, we recognised that Synroc derivatives could be applied to a variety of radioactive wastes. Effort was also put into the development of alternative consolidation techniques to uniaxial hot pressing, such as hot isostatic pressing, sintering, and melting to expand the application of ANSTO's waste forms. ANSTO has subsequently undertaken work both on contract and collaboratively, in addition to its own initiatives, on a variety of waste streams as described below.

We are using our 25 years of experience in waste form development and the choice of the processing method is a key determinant in the final choice of waste form. A key advantage of hot isostatic pressing is that the calcined waste + precursor is sealed in a metal can before pressing, so that the major part of the waste form consolidation process involving the highest temperature produces essentially zero offgas; and the entire process produces offgas only in the calcination stage where temperatures are much lower than those in the hot-pressing. Hot isostatic pressing in waste form production has been validated at Argonne-West.

Partitioned Groups of Elements from Advanced Reprocessing Cycles

The French and Japanese have independently explored the possibility of separating (or partitioning) reprocessing HLW into several groups- the actinides, rare earths/Zr, Pd-group metals, and the heat-producers (such as Cs/Sr)- and disposing of the different groups separately. Given the long half lives of actinides in general, waste forms for actinides must survive for periods of •10^4 years. For actinide-rich wastes, a waste form was produced that contained 80 wt% zirconolite and only 5-10% each of other titanate phases. For ^{137}Cs/^{90}Sr, the half-life is only ~30 years, so hollandite/perovskite-rich materials might only have to last a few hundred years, but be capable of coping with elevated temperatures early in their disposal history. However this simple picture is complicated by the long-lived ^{135}Cs isotope. Immobilization of ^{99}Tc/Cs mixtures has also been demonstrated (4)-see also below.

Surplus US Weapons Pu

Zirconolite-rich waste forms were also initially explored for immobilization of surplus US weapons Pu. A ~ 400 g sample containing 50 g of Pu was successfully formed by hot isostatic pressing (5). However it emerged that substantial amounts of U would accompany the Pu. Substantial quantities of neutron absorbers (Hf and Gd) were also needed to suppress the potential for criticality excursions, particularly in any leaching plumes in a repository. It was finally decided to use a waste form primarily composed of a pyrochlore-structured phase ((CaAn)$_2$Ti$_2$O$_7$), that can incorporate ~10 wt% of Hf and Gd oxides, as well as 10 and 20 wt% respectively of Pu and U oxides. Pyrochlore and zirconolite have very similar structures.

The ideal "baseline" ceramic for pure Pu is composed of 95% pyrochlore-structured titanate plus 5% of Hf-doped rutile (5). After evaluating a range of processing techniques, sintering was found to be the preferred processing method due to its simplicity and fact that the only radionuclides to be immobilized were actinides, which are relatively non-volatile at elevated temperatures. Inclusion of "impurity" waste elements (other than U) however was an issue because the immobilization route was only mandated for surplus Pu containing significant amounts of impurities. These produced minor phases in addition to the Hf-doped rutile, and at the expense of the major pyrochlore-structured phase. These minor phases, such as zirconolite and brannerite (UTi_2O_6) were also durable.

This waste form for immobilization of surplus Pu was selected by a competitive process over all other candidate wasteforms by the US government in 1997. ANSTO assisted in the development of the baseline ceramic and investigated the crystal-chemical, leaching, and radiation damage characteristics of the individual constituent phases (eg. brannerite). It was later suggested that replacement of Ti by Zr may prove useful to minimize radiation effects in these ceramics, but such material is less tolerant to impurities in the Pu and the sintering temperature is significantly elevated (6). Although the immobilization option has been suspended in the US, ANSTO's interest in immobilization of Pu-bearing waste remains. This was the first time that a crystalline waste form was approved by regulatory authorities.

Other Wastes

Glass-ceramics have been devised for Hanford tank wastes and sludges (7). These glass-ceramics have high waste loadings, in the 50-70 wt% range, and consist of synroc-type titanate phases very largely incorporating the fission products and actinides, in an aluminosilicate glass matrix. They can be made by melting techniques to achieve a high throughput. Hot isostatic pressing is also another possible production technique. High-uranium wastes from radioisotope production at nuclear reactors can be immobilized in several kinds of titanate ceramics. These have waste loadings of up to 44 wt% and have very good leaching behaviour (8), comparable with that of synroc-C. For Al-rich wastes arising from reprocessing of Al-clad fuels, synroc-D devised by Ringwood (1) should be useful. This waste form was first directed towards the Al- and Fe-rich waste streams at Savannah River, SC, USA but after borosilicate glass was chosen over synroc by the Peer Review Panel (9) in 1982, no further work on this wasteform was done until 2002. These studies, using modern characterisation techniques, notably electron microscopy, are confirming the essential correctness of Ringwood's phase assemblage and lattice locations of radionuclides for Al-rich waste. The Al is mainly incorporated in a spinel phase, with the fission products incorporated in zirconolite, perovskite and hollandite, to yield good aqueous durability and high waste loadings. Some glassy material can also be accommodated in some versions of synroc-D (10).

Glass-ceramics have also been developed for the INEEL calcines and these have waste loadings of ~ 50 wt% depending on the precise composition and whether the alumina- or zirconia-rich calcines are being considered (11). Glass-ceramics are also viable for Na-bearing liquid wastes (unpublished work). Mo-rich wastes would arise from reprocessing of U-Mo fuels and present studies at ANSTO are focussing on sintered ceramics containing powellite ($CaMoO_4$) and sodium zirconium phosphate (NZP) as key immobilization phases. Waste loadings in these materials are typically 50 wt%. Following detailed leach studies of Cs-bearing barium hollandite (12), we have investigated a wide range of barium hollandite-type solid solutions which can be sintered in air to incorporate Cs and which have excellent leaching behaviour. In further applications, we have been able to melt in air Cs-bearing hollandite-rich samples without any loss of aqueous durability. These contain other synroc phases in addition to the major hollandite. ^{99}Tc is a volatile element and because of its long lifetime, it features heavily in performance assessments at around the 10^5 yr mark. It can be readily incorporated in titanate ceramics as a metal alloy with stainless steel under reducing conditions (5). The feasibility of incorporating Tc as Tc^{4+} in a number of titanate phases is being assessed, using hot isostatic pressing or sintering in neutral atmospheres (13).

Cements and geopolymers which are formed by polymerisation of aluminosilicates dissolved in an alkaline aqueous solution have been widely advanced for immobilization of intermediate-level waste. However studies have only rarely been made (14) on the speciation of radionuclides in these materials and we are utilising solid state nuclear magnetic resonance, leaching tests, and scanning microscopy to gain speciation information on Cs and alkalis-particularly as to whether the alkalis inhabit the matrix or the pore water- as well as on aluminium and silicon ions. The incorporation of ^{129}I into alumina or silica beads is well known (see (15)), but we are working on encapsulation strategies using a variety of matrices and processing options that avoid I volatilisation.

Fission Product Disposition in Inert Matrix Fuels

We have recently embarked on studies of inert matrix fuels for burning surplus Pu in reactors. Here the emphasis is to reach an optimum fuel in which the fission products produced by irradiation can find their way into refractory, leach-resistant phases, while preserving the necessary high melting point, high thermal conductivity and resistance to swelling from irradiation damage at elevated temperatures as far as possible. To this end, we are investigating synroc-D like ceramics in which the spinel phase gives improved thermal conductivity and a slight melting point improvement over synroc-C (16).

UNDERPINNING RESEARCH AT ANSTO

ANSTO's waste form development program has an extensive range of capabilities for waste form characterization. Crystal chemistry studies are targeted towards waste form design via solid solution behaviour, and establishment of the valences of actinides and other ions. Experimental techniques include X-ray diffraction, electron microscopy (including electron energy loss spectroscopy), X-ray absorption, diffuse relectance and X-ray photoelectron spectroscopies, and solid state nuclear magnetic resonance. Aqueous dissolution is directed at dissolution rates of different actinide valences and the effect of redox conditions, as well as standard MCC, PCT and Soxhlet testing. Radiation damage work involves structural effects of heavy-ions, using fast electrons to obtain displacement energies of cations and oxygen ions, and alpha-decay effects via ^{238}Pu doping. Waste form production techniques cover hot isostatic and uniaxial pressing, sintering, and cold-crucible melting.

Solid-state chemistry studies of inactive and actinide-doped samples involves electron microscopy/microanalysis, diffraction, X-ray absorption and photoelectron spectroscopies, optical spectroscopy, electron spin resonance, electrical resistivity, nuclear magnetic resonance and positron annihilation lifetime studies. In solid-state chemistry studies the main theme is the valence and site occupation of fission product and actinide ions incorporated in the host lattice. The conclusion is that both the prevailing oxygen fugacity and the crystal chemical forces exerted by the host (including the presence of appropriate charge compensators) control the final valence. An ionic model appears to be generally applicable for the ions in the synroc phases, especially under oxidizing conditions. Solid solution limits of fission products, actinides and neutron absorbers in the various phases are also of interest. Positron annihilation lifetime spectroscopy has been successful in showing directly the presence of cation vacancies in some of the doped titanate phases. For instance, Ce-doped zirconolites were deduced by other techniques to contain cation vacancies when they were heated in oxidising, but not in reducing atmospheres (17). This was borne out in observing a 20% increase in annihilation lifetime when the material was heated in an oxidising rather than a reducing atmosphere. Similar increases have been observed in vacancy-bearing perovskites, doped with La acceptors (unpublished work).

In the first instance, aqueous durability studies measure the release into solution of various species. We are careful to distinguish between elements in solution, those in colloidal form, and those adhering to the sides of the leach container or precipitated on the sample surface. Such distinction is especially important for actinides. Thus we utilize surface alteration studies via X-ray photoelectron spectroscopy (XPS), electron microscopy, energy recoil analysis of aqueous components, secondary ion mass spectroscopy (SIMS), and alpha-recoil spectroscopy, in addition to ICP analysis of dissolved species. XPS has been used to study the decalcification of perovskite at depths of < 5 nm after leaching in deionised water at 90 and 150°C. It was also shown that the presence of Ca in the leach liquid inhibited the decalcification. A cross-sectional transmission electron micrograph on a synroc-C surface leached for 1 year at 150°C in a pH = 4 aqueous buffer solution highlighted preferential leaching of the perovskite grains, with partial in-filling by anatase (18). There was virtually no evidence of leaching of the other synroc phases. The leached depth of the perovskite was approximately 0.2 μm in this experiment. Complementary results were obtained by SIMS. Energy recoil analysis of a synroc surface leached for 30 days in D_2O at 190°C, showed that the average penetration of D (probably as OD-) was only ~ 15 nm (19). Alpha-recoil spectroscopy has indicated some actinide buildup on leached synroc surfaces.

Radiation damage studies deal with diffraction studies of metamict (rendered X-ray amorphous) synroc-analogue minerals, and electron, neutron, and heavy-ion irradiated synthetic samples, including the use of incorporated ^{244}Cm and ^{238}Pu alpha-emitters (half-lives of 18 and 87 yr. respectively). Much information on these questions has been available since the 1980s, when it was agreed that the only significant and permanent damage processes to solid wasteforms arise from alpha-decay events, with the main damage arising from the recoil atom, not the alpha-particle itself. Strachan et al.(20) have recently shown by taking strenuous efforts to eliminate radiolysis effects, that there is no leach rate enhancement when the pyrochlore-based waste form for surplus Pu immobilization is amorphized. The doses of heavy ions such as 1.5 MeV Kr ions (simulating alpha-recoil nuclei) required to amorphize the different actinide-bearing phases vary only by less than a factor of 10 at temperatures where self-annealing is small (21). Recently, we have investigated the displacement energies of anions in perovskite and zirconolite using cathodoluminescence after fast electron irradiation (22) and cations in perovskite by HARECXS analysis (23). These experimental studies are being supplemented by computational work to calculate individual defect energies.

CONCLUSIONS AND FINAL REMARKS

ANSTO waste form research and development has broadened considerably in recent years. In addition to our well-known titanate ceramics for Purex waste and surplus US Pu, we now have a suite of ceramics targeted towards high-U wastes from radioisotope production, Al-rich wastes arising from the reprocessing of Al-clad reactor fuels, ^{99}Tc, high-Mo wastes and partitioned Cs wastes. Glass-ceramics for the immobilisation of US waste calcines and defense wastes are also under study, together with encapsulation of ^{129}I and modes of radionuclide incorporation in cements and geopolymers. Work on fission product disposition in inert matrix fuels for Pu burning is continuing. We also have experience of a wide range of processing technologies and extensive characterization facilities. Our key selling point is that we have an experienced team that can maximize waste loading and aqueous durability to a degree consistent with performance standards for a given waste stream. By integrating the waste form and process design we can produce a range of waste forms via a single process platform.

REFERENCES
(1) A.E. Ringwood, S.E. Kesson, N.G. Ware, W. Hibberson and A. Major, "Geological Immobilisation of Nuclear Reactor Wastes", *Nature*, **278**, 219-23 (1979).

(2) G.J. McCarthy, "High-Level Waste Ceramics: Materials Considerations and Product Characterization", *Nuclear Technology*, **32**, 92 (1977).

(3) A.E. Ringwood, S.E. Kesson, K.D. Reeve, D.M. Levins and E.J. Ramm,"Synroc"; p. 233 in *Radioactive Waste Forms for the Future*. Edited by R. C. Ewing and W. Lutze. Elsevier, North-Holland 1988.

(4) K.P. Hart, E.R. Vance, R.A. Day, C.J. Ball, B.D. Begg, P.J. Angel and A. Jostsons, "Immobilization of Separated Tc and Cs/Sr in Synroc"; p. 221 in *Scientific Basis for Nuclear Waste Management XIX*. Edited by W. M. Murphy and D. A. Knecht. Materials Research Society, Pittsburgh, PA 1996.

(5) E.R. Vance, A. Jostsons, S. Moricca, M.W.A. Stewart, R.A. Day, B.D. Begg, M.J. Hambley, K.P. Hart and B. B. Ebbinghaus, "Synroc Derivatives for Excess Weapons Plutonium"; pp. 323-9 in *Ceramics Transactions (Environmental Issues and Waste Management Technologies IV), Volume 93*. Edited by J. C. Marra and G. T. Chandler. American Ceramic Society, Westerville 1999.

(6) M.W.A. Stewart, B.D. Begg, E.R. Vance, M. Colella, K. Finnie, K.P. Hart, H. Li, G.R. Lumpkin, K.L. Smith and W.J. Weber , "The Replacement of Titanium by Zirconium in Ceramics for Plutonium Immobilization"; p. 311 in *Scientific Basis for Nuclear Waste Management XXV*. Edited by B. P. McGrail and G. A. Cragnolino. Materials Research Society, Pittsburgh, PA 2002.

(7) E.R. Vance, R.A. Day, M.L. Carter and A. Jostsons, "A Melting Route to Synroc for Hanford HLW Immobilization"; pp. 289-96 in *Scientific Basis for Nuclear Waste Management XIX*. Edited by W. M. Murphy and D. A. Knecht. Materials Research Society, Pittsburgh, PA 1996.

(8) E.R. Vance, M.L. Carter, S. Moricca and T. Eddowes, "Titanate Ceramics for Immobilization of U-Rich wastes", this volume.

(9) L.L. Hench, D.E. Clark and J. Campbell, "High-Level Waste Immobilization Forms", *Nuclear and Chemical Waste Management*, **5**, 149 (1984).

(10) A.E. Ringwood, S.E. Kesson and N.G. Ware, "Immobilization of U.S. Defense Nuclear Wastes using the Synroc Process"; pp. 265-72 in *Scientific Basis for Nuclear Wasre Management, Volume 2*. Edited by C.J.M. Northrup. Plenum, New York 1980.

(11) R.A. Day, J. Ferenczy, E. Draberek, T. Advocat, C. Fillet, J. Lacombe, C. Ladirat, C. Veyer, R. Do Quang, and J. Thomasson, "Glass-Ceramics in a Cold-Crucible Melter : The Optimum Combination for Greater Waste Processing Efficiency", ibid.; WM '03(CD-ROM, sess 26/26-04), WM Symposia, Inc., Tucson, AZ, USA (2003).

(12) M.L. Carter, E.R. Vance, D.R.G. Mitchell, J. V. Hanna, Z. Zhang and E. Loi, "Fabrication, Characterisation, and Leach Testing of Hollandite $(Ba,Cs)(Al,Ti)_2Ti_6O_{16}$", *Journal of Materials Res*earch, **17**, 2578, (2002).

(13) M.W.A. Stewart, E.R. Vance and R.A. Day, "Titanate Wasteforms for Tc-99 Immobilization", WM'04 Conference, Tucson, AZ (2004).

(14) J.V. Hanna, L.P. Aldridge and E.R. Vance, "Cs Speciation in Cements"; pp. 89-96 in *Scientific Basis for Nuclear Waste Management XXIV*. Edited by K. P. Hart and G. R. Lumpkin. Materials Research Society, Warrendale, PA 2001.

(15) D.S.Perera, B.D.Begg, R.L.Trautman, D.J.Cassidy, E.R.Vance and R.A.Day, "Hot Isostatic Pressing Method for Immobilising Radioactive Iodine-129", International Hot Isostatic Pressing (HIP) Conference, Moscow, Russia, 20-22 May, 2002.

(16) M.W.A. Stewart, E.R. Vance, P.M. Cook and E.R. Maddrell, "Directly Disposable Inert Matrix Fuel", IMF9 Workshop on Inert Matrix Fuel, Kendal, UK, Sept 10-11, 2003.

(17) J.H. Hadley, F.H. Hsu, Yong Hu, E.R. Vance and B.D. Begg, "Observation of Vacancies by Positron Trapping in Ce-doped Zirconolites", *Journal of the American Ceramic Society*, **82**, 203-5 (1999).

(18) K.L. Smith, G.R. Lumpkin, M.G. Blackford, R.A. Day and K.P. Hart, "The Durability of Synroc", *Journal of Nuclear Materials*, **190**, 287-94 (1995).

(19) N. Dytlewski, E.R. Vance and B. D. Begg, "Energy Recoil Analysis of Deuterium Incorporated in Synroc by Reaction with D_2O at 120 and 190°C", *Journal of Nuclear Materials*, **231**, 257-59 (1996).

(20) D.M. Strachan, R.D. Scheele, A.E. Kozelisky and R.L. Sell, "Effects of Self Irradiation from [238]Pu on Candidate Ceramics for Plutonium Immobilization", Pacific Northwest National Laboratory Report, PNNL-14232 (2003).

(21) G.R. Lumpkin, K.L. Smith and M.G. Blackford, "Heavy Ion Irradiation Studies of Columbite, Brannerite, and Pyrochlore Structure Types", *Journal of Nuclear Materials*, **289**, 177-87 (2001).

(22) R. Cooper, K.L. Smith, M. Colella, E.R. Vance and M. Phillips, "Displacement Energies in Perovskite ($CaTiO_3$)", *Journal of Nuclear Materials*, **289**, 199-203 (2001).

(23) K.L. Smith, N.J. Zaluzec, R. Cooper, M. Colella and E.R.Vance, "Cation and Anion Displacement Energies in Oxides: Review of Recent Experiments and Results", 10th International Conference Ceramics Congress and 3rd Forum on New Materials (Florence, July 2002), pp 15-22.

STABILIZATION OF ARSENIC-BEARING IRON HYDROXIDE SOLID WASTES IN POLYMERIC MATRICES

F. Rengifo, B. Garbo, A. Quach, W.P. Ela and A.E. Sáez
Dept. of Chemical and Environmental Engr.
University of Arizona
1133 E North Campus Drive
Tucson, AZ 85721

C. Franks and B.J.J. Zelinski
Dept. of Materials Sci. and Engr.
University of Arizona
1235 E North Campus Drive
Tucson, AZ 85721

D.P. Birnie, III
Dept. of Ceramic and Materials Engr.
Rutgers, The State University of New Jersey
607 Taylor Rd.
Piscataway, NJ 08854

H. Smith and G. Smith
Pacific Northwest National Lab.
902 Battelle Boulevard
Richland, WA 99352

ABSTRACT
This work explores the use of an aqueous-based emulsion process to create an epoxy/rubber matrix for separating and encapsulating waste components from salt-laden, arsenic-contaminated, amorphous iron hydrate sludges. Such sludges are generated from conventional water purification precipitation/adsorption processes, used to convert aqueous brine streams to semi-solid waste streams, such as ion exchange/membrane separations, and from other precipitative heavy metal removal operations. In this study, epoxy resin and polystyrene butadiene (PSB) rubber emulsions are mixed together, combined with residual sludge, and cured and dried at 80°C to remove water. The microstructure of the resulting waste form is characterized by scanning electron microscopy (SEM), which confirms that the epoxy/PSB matrix surrounds and encapsulates the arsenic-laden amorphous iron hydrate phase, while allowing the salt to migrate to internal and external surfaces of the sample. Soluble salts leach from the sample at a rate given by diffusion coefficients of the order of 10^{-8} cm^2/s. Long-term leaching studies reveal no evidence of iron migration and, by inference, arsenic migration, and demonstrate that diffusivities of the unextracted salt yield leachability indices within regulations for non-hazardous landfill disposal.

INTRODUCTION
Efforts to implement new water quality standards, increase water reuse and reclamation, and reduce the cost of waste storage motivate the development of new processes for stabilizing wastewater residuals that minimize waste volume, water content and the long-term environmental risk from related by-products. Processes to remove toxic heavy metals from drinking water and other aqueous streams are not new, but a number of recent trends have both greatly increased the frequency of their use and drawn critical attention to the nature of their by-products.[1] The Environmental Protection Agency (EPA) has recently passed more strict standards for the allowable concentrations of many heavy metals in drinking water. The recently revised arsenic standard and the currently debated lead and copper rules are examples of this tightening of water quality standards.[2,3] As the readily available sources of new potable water diminish, there is an increasing emphasis on reuse and reclamation of water to augment supplies. Furthermore, as the available space for solid waste disposal becomes scarcer and the potential

negative impacts of such disposal are better realized, the scrutiny and cost of by-product disposal has greatly increased. As a consequence of these trends, there is a rapidly increasing volume of by-products requiring disposal at the same time as there is a need to minimize the volume, water content and long-term environmental risk of the by-products.

Since heavy metals are elemental, removal processes do not destroy them, but merely transfer them from the primary water stream to an alternative fluid or solid stream which constitutes the by-products (or as is often termed, residuals) of the process. Toxic heavy metal removal generates a metal-laden solid waste stream or, more commonly, a concentrated brine stream of the toxic components mixed with various nontoxic salts. Large-volume, troublesome examples include the ion exchange/membrane separation concentrated streams generated from the use of arsenic removal technologies by water treatment utilities and the aluminum and iron sludges emanating from precipitative, heavy metal removal operations utilized by semi-conductor, pulp and paper and metal plating industries. The emphasis of the research is on treating the sludges generated from conventional precipitation/adsorption technologies that are used to convert aqueous brine streams to semi-solid waste streams. Even though precipitation/adsorption greatly reduces the waste volume, the sludges generated are still prone to toxics release unless disposed in hazardous waste sites. The only competing technologies to the proposed polyceram technology for stabilization of such wastes are grout/cement based, which are inapplicable to high salt wastes (as salt interferes with the calcination reactions necessary for cement curing) and cannot remove the added waste volume composed of benign salts.[4,5] To the authors' knowledge, the combination of a stabilization technology for toxic waste and controlled release for nontoxic, soluble salts has not been successfully attempted before.

The work described here uses an arsenic-laden sludge as the residual of interest. Arsenic sludges were chosen because in 2001, drinking water standards were enacted that reduced the allowable levels of arsenic from 50 ppb to 10 ppb.[2] These new standards will affect thousands of small utilities and cause production of an estimated six million pounds of solid residuals every year.[2,3] Options expected to be used for arsenic removal from drinking water are: modified lime softening, modified coagulation/filtration, and ion exchange. These technologies will produce arsenic-bearing brine streams, which in turn will be treated to generate a final, arsenic-laden residual of amorphous ferric hydroxide (AFH) sludge. These residuals typically will not pass the Toxicity Characteristic Leaching Procedure (TCLP) specified by EPA to discriminate between hazardous and non-hazardous wastes and, consequently, must be disposed of in controlled, hazardous waste landfills.[6] Hazardous waste disposal costs are a factor of 5 to 10 higher than non-hazardous disposal costs. Thus, considering that implementation of the new arsenic rule will cost an estimated $180-206 million per year,[2] it will cost water providers approximately $9-19/lb more to dispose of a hazardous treatment residual, as opposed to a non-hazardous one. In order to place the sludge in a non-hazardous landfill, toxic components have to be stabilized so they do not leach significantly into the environment. At the same time, reducing the total volume of residuals will minimize landfill costs; hence there is an incentive to separate benign components from wastes prior to disposal. Consequently, this work focuses on an overall program objective to develop a separation/fixation technology for arsenic-laden AFH sludges that is simple, cost effective, safe, efficient, and durable. The development effort is based on application of polyceram waste-form technology as a means of meeting the proposed objectives.

The sludge used in this study was modeled after an AFH sludge that represents the solid residuals expected from an ion exchange (IX) operation used for removal of arsenic from drinking water. To treat water by IX, contaminated water containing arsenic passes through an

IX column filled with an anion exchange resin containing free chloride ions. As the water passes through, the chloride ions in the resin are exchanged with the arsenic ions in the water and the treated arsenic-free water exits the system. After about one thousand bed volumes of contaminated water have passed through the column, the resin is exhausted and must then be regenerated. At this point, four bed volumes of a regenerating solution containing 3% sodium chloride (NaCl) are passed through the column to return chloride ions to the resin and remove the arsenic ions. This process concentrates the contaminated water down to four bed volumes as opposed to one thousand bed volumes. The resulting brine, containing 2% NaCl and arsenic, then exits the column and enters a settling tank. Sodium hydroxide (NaOH) is added to the brine to precipitate ferrihydrite ($Fe(OH)_3$) to adsorb the arsenic. Enough ferric chloride ($FeCl_3$) is added to bring the iron to arsenic mass ratio to 10 to reduce the arsenic concentration below the drinking water standard of 10 ppb. The supernatant liquid in the settling tank contains a toxic-free brine solution of 2% NaCl which is extracted and recycled. The sludge that is deposited on the bottom of the settling tank consists of solid $Fe(OH)_3$, arsenic (both sorbed/coprecipitated with the AFH and as free species in the sludge-associated water), NaCl, and water. It is this sludge that requires the fixation of the iron and arsenic components while simultaneously separating the salt in order to reduce the volume to minimize the cost of disposal.

Polycerams, a hybrid of organic and inorganic components, have been known to be mechanically stable, chemically durable, to have low processing temperatures (under 100°C), and are easily fabricated, which make them feasible materials for fixating $Fe(OH)_3$. However, one disadvantage of polycerams involved the use of flammable and volatile solvents.[7] This disadvantage is mitigated in this work by the application of an aqueous processing route in which an emulsion was created by mixing two polymers throughout a continuous water phase.[8] When cured, the water evaporates and the emulsion undergoes a phase inversion causing the polymer to become the continuous phase with the salt and solid residuals distributed throughout. In previous work by our group, it was found that soluble salts were successfully encapsulated in the polymer matrix using this water-based route,[9] and the salt diffused at a slow rate yielding leachability indices well within the minimum requirements for land disposal. Microstructural characterization revealed that some portion of the dissolved salt was carried to the outer regions of the waste form as water evaporated during the curing process. This migration of salt can be further exploited as a separation and fixation technology that would separate the salt from the toxic components in a brine stream and fix those toxic components in a durable waste form matrix.

MATERIALS AND METHODS

A representative sludge was created to mimic the end residual of an IX operation for arsenic removal. All the chemicals used to make the sludge were reagent grade or better. Sodium hydrogenarsenate heptahydrate was first added to deionized water. Ferric chloride, $FeCl_3$, was then added and completely dissolved in solution. While the solution was being mixed, NaOH was added to precipitate $Fe(OH)_3$. Once the mixture was stirred to homogeneity, the batch was allowed to settle for 24 hours, during which time the arsenate adsorbed onto the $Fe(OH)_3$. After the 24-hour period, free liquid was decanted from the settled sludge layer and more NaCl was added to bring the chloride concentration up to that typical of an AFH sludge from IX. Sodium nitrate ($NaNO_3$) was also added because it typically coexists in the brine stream with the arsenate. The pH was adjusted to neutral with NaOH, the batch was mixed, and then allowed to settle to facilitate liquid decanting when needed.

Sludge of varying water content was used in sample fabrication to cover the range of sludge produced by industries. Changing the liquid content in the wet sludge affects the salt content of the dry sludge. At low liquid content (~70% w/w), the sludge is mud-like and, when dried, contains 33 wt% salt. At high liquid content (~95% w/w), the sludge is a fine particulate slurry and dries to contain 75 wt% salt (the remainder is the AFH phase). X-ray powder diffraction reveals that the AFH in the sludge is structurally amorphous. The final sludge composition (dry basis by weight) is NaCl: 29.6%; NaNO$_3$: 1.9%; Fe(OH)$_3$: 64.5%; HAsO$_4$: 4.0 %.

All waste forms were prepared in order to obtain equal parts, by weight, of polystyrene-butadiene (PSB) rubber and epoxy resin in the final cured sample. To fabricate the waste forms, PSB latex (Styronal ND 656, BASF) and the surfactant sorbitan monooleate (Span-80, Aldrich) were mixed for 10 minutes in a narrow, round-bottom mixing vessel to create an emulsion. While continually mixing, the arsenic-containing sludge was added. Then, epoxy resin (Epo-Kwick Resin, Buehler) was added drop-wise with a syringe while stirring. Next, a cross-linking agent, diethylenetriamine (DETA, Aldrich), was added to the mixture. The emulsion was mixed for 10 additional minutes to ensure uniformity and then cast in shallow aluminum dishes and placed in an oven to dry and cure at 80°C until no further weight loss was detected (approximately 3 days). Cured samples were flat "cookie" shapes with diameters of about 4 cm and ranging between 0.15 and 0.4 cm in thickness. In some cases, cylindrical monoliths of about 2 cm in diameter and 3.5 cm long were cast. Samples with various iron loadings, salt loading, and initial water content were fabricated.

Scanning Electron Microscopy (SEM) was used to analyze sample microstructure. Energy-Dispersive X-Ray Spectrometry (EDS) was used to obtain qualitative compositional analyses. An initial analysis of the polymeric matrix of which the waste forms are composed was performed. Microstructural features were revealed using the modified osmium tetroxide (OsO$_4$, Aldrich) staining method for rubber-toughened epoxies.[10] Salt leaching and extraction were conducted to calculate the diffusivity of salt through the polymeric matrix. Iron leaching was also performed. Arsenic leaching is assumed to be identical to the behavior of iron; however, leaching experiments will be conducted in the future to confirm this hypothesis.

To study the leaching behavior, portions of each sample were cut and suspended in a beaker of well-stirred water of known volume. A conductivity probe was placed in the beaker to measure the changes in the conductivity of the liquid as the salt leached out of the sample. The measurements were correlated to salt concentration and a mathematical model based on Fick's law of diffusion was fit to the data. The model equations are presented elsewhere.[9] Assuming that the samples leach to completion, the percent salt retention of the material is calculated using the following equation:

% Salt Retention = (Total salt in sample – Salt Concentration) / Total salt in sample × 100%   (1)

The model yields the effective diffusivity of the salt in the waste form, which is then used to calculate the leachability index for each sample.[9] Figure 1 shows an example of the model fit to actual experimental data. The model only starts correlating with the data after approximately one hour of initial extraction of salt from the outside of the sample. During the first part of the leaching experiment, solid salt close to the surface of the sample dissolves rapidly without participating in diffusion-limited transport. After this initial hour, all of the unincorporated salt has dissolved, and the remainder of the curve represents the diffusive mechanism of the leaching process.

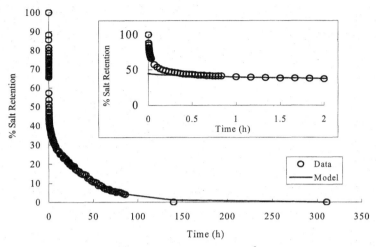

Figure 1. Example of experimental data fit to diffusion model[9] with an inset of the first two hours of leaching for sample 3.08 (Table I).

RESULTS AND DISCUSSION

An initial study of the sample microstructure of the polymeric matrix alone revealed that it is not homogeneous but in effect is composed of dispersed epoxy droplets in a continuous PSB phase. This was determined by the osmium tetroxide staining procedure, which provides contrast for SEM viewing, as the electron-dense OsO_4 binds preferentially to unsaturated bonds present in the PSB molecule.[10] The rubber components appear as brighter regions in a SEM micrograph and epoxy components appear darker (micrographs not shown due to space limitations). The epoxy droplets have a large size distribution ranging between 0.1 and 5 μm. Usually, addition of rubber to epoxy resins is associated with a dispersed rubber phase to reduce the stiffness of the epoxy matrix.[11] In our case, the relatively high rubber to epoxy ratio leads to a continuous rubber phase, whose hydrophobicity represents, in principle, an advantage for the encapsulation process. SEM analysis showed that the phase distribution of the polymeric matrix is not influenced by the presence of the arsenic-containing iron sludge and salt.

Table I. Sample numbers and overall compositions based on dry weight

Sample ID	Dry wt% Fe(OH)$_3$	Dry wt% HAsO$_4$	Dry wt% NaCl	Dry wt% NaNO$_3$
3.08	12.5	0.8	35.7	2.3
3.12	16.3	1.0	22.3	1.5
3.13	12.5	0.8	35.9	2.3
3.17	10.0	0.6	4.6	0.3
3.18	10.0	0.6	4.6	0.3

Table I lists the compositions of the waste form samples investigated in this study. Microstructural analysis by SEM reveals that AFH incorporation is similar in all waste forms but that salt features in the microstructure vary with the relative salt loading. For example, Figure 2a shows the microstructure of a waste form containing a low salt loading, sample 3.17. While Figure 2a shows only a portion of the sample, it is representative of the entire cross-section. Because the elements in the salt and AFH phase scatter electrons more strongly than the carbon and hydrogen atoms in the polymer, these phases appear light in the micrographs. The dark region is the polymeric matrix. As seen in the figure, the AFH phase forms dense, elongated particles with aspect ratios around 1/5 and lengths along the largest dimension in the range of 20-150 μm. The imaging done at a higher magnification and shown in Figure 2b shows that, at this low level of salt loading, the salt appears as cubic crystals about 5 μm in size embedded in the polymeric matrix. EDS analysis was conducted on the region containing the ferrihydrite particle circled in the magnified image of Figure 2a and shown in more detail in Figure 3a. Figures 3b-e show the elemental maps that correspond to this region. Lighter regions in these maps indicate the presence of the element indicated. The dark area surrounding the AFH particle is composed mainly of carbon as shown by the elemental carbon map (Figure 3b). Most of the light regions in the micrograph correspond to the iron containing phase, as shown in Figure 3d. The arsenic map (Figure 3c) shows that arsenic is strongly associated with the AFH phase. Finally, the maps indicate that the groups of crystals seen on either end of the AFH particle are composed of both sodium (Figure 3e) and chlorine (Figure 3f).

Figure 2. SEM micrographs of a sample with a low salt loading (sample 3.17) showing (a) a cross-section of cookie where the upper boundary of the image indicates the top of the cross-section and the lower boundary indicates the bottom; (b) the matrix of the cookie at high magnification; the image directed by the arrow is a magnified image of the region shown in the rectangular box in (a). Dry salt content of sample is 4.9%.

While the matrix phase does contain small salt crystals, EDS analyses of several samples indicate that the salt also concentrates as crystals in the cracks and interfacial regions of the AFH phase. The salt crystals seen on the ferrihydrite surface in Figure 3a originally lay inside cracks that ran parallel to the plane of EDS analysis. They were exposed for analysis when the sample was fractured for viewing.

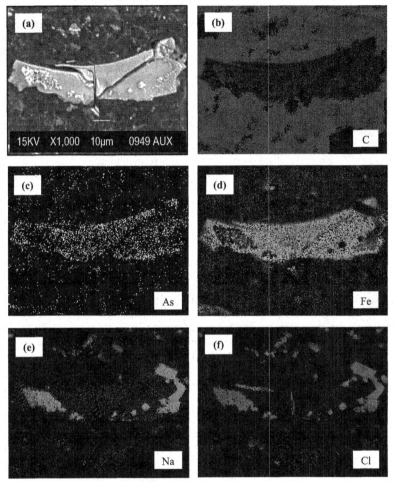

Figure 3. (a) SEM micrograph of sludge particle found in sample 3.17; maps of compositional elements: (b) carbon, (c) arsenic, (d) iron, (e) sodium, and (f) chlorine. Light areas indicate presence of specified element.

The microstructural features of a waste form containing a high salt loading are shown in Figure 4. The image in Figure 4a illustrates sample 3.08 at the middle of the monolith, which is representative of the majority of the sample. In contrast to sample 3.17, the polymeric matrix of sample 3.08 contains large amounts of salt in the form of agglomerated regions distributed uniformly throughout the waste form and high concentrations of salt at the outer edges (Figure 4b). In particular, the bottom of sample 3.08 includes a well-defined salt crust approximately 100 μm in thickness. Additionally, the matrix near the bottom contains a number of pores around 400 μm in diameter whose inner surfaces are coated with a thick layer of salt.

Figure 4. SEM micrographs showing microstructure of a sample with a high salt loading (sample 3.08) at (a) middle of monolith, and (b) bottom of monolith. Dry salt content of sample is 38%.

The high salt loading of sample 3.08 limited its structural integrity and caused it to be somewhat crumbly when cured, which may limit its applicability as a viable waste form. However, the fact that the waste form retains a large amount of salt in regions that are accessible by a very open network of pores suggests that the underlying process has great potential as a technology for extracting or separating the salt to minimize the amount of waste for disposal. A significant reduction in the quantity of waste would be achieved if this salt were separated or extracted from the waste form before disposal and storage.

To test the capabilities of the manufacturing processes for salt separation, salt extraction experiments were performed on a variety of samples containing varying amounts of AFH and salt. As seen in Table II, most of the salt (60 to 90%) in the sample was removed in the first hour of extraction. These results demonstrate that for a higher initial salt loading, a larger fraction of the salt is successfully removed.

Table II. Amount of salt dissolved after one hour of initial salt extraction for samples of various sludge loadings

Sample ID	Dry wt% $Fe(OH)_3$	Dry wt% Salt	% Salt Extracted
3.12C	16.3	23.8	86.1
3.13A	13.9	31.4	91.3
3.13D	13.9	31.4	85.7
3.17A	10.0	4.9	59.2
3.17E	10.0	4.9	61.0

To be an effective separation medium, the waste form must allow the rapid removal of salt while still retaining the arsenic-containing AFH phase. Separate iron leaching tests were performed, which indicated that no appreciable iron existed in the leachates after 300 hours of extraction. It is inferred that iron is therefore not extracted in 1 hour, and because arsenic adsorbs to the iron species in the waste form, it is also assumed that no arsenic is removed. This shows that the waste form can be used to separate the salts from toxic components from water treatment sludges. Upon contact with water, an important fraction of the salt will leach out of the waste form while keeping the AFH and associated arsenic still encapsulated.

The salt extraction experiments measure the behavior of the salt that is easily removed by immersion in water. When the salt leaching experiments are continued beyond the first hour of extraction, the rate of leaching of the salt that does not have ready access to internal or external surfaces of the waste form (e.g. the salt that is embedded in the polymeric matrix) can be determined. The diffusion rate of the salt in this period should be representative in magnitude of the rate at which arsenic or iron will diffuse from the waste form. It is important to recall that driving forces for diffusion of iron or arsenic species are drastically lower than those of salt because of their low solubility (iron) and their strong adsorption to AFH (arsenic).

Table III shows the diffusivities and their corresponding leachability indices for samples of different relative salt loadings, high and low. The minimum required leachability index for land disposal is 6. Both waste forms present leachability indices well within the required limit. The diffusion model used is amenable to calculating the diffusivity coefficient of the most mobile species in the waste form, i.e. the salt, and these results, compared to those shown in Table II, show that at various salt loadings, the majority of the salt can be separated using this waste form while indicating that the remaining salt, and thus the toxic, least mobile species, will be encapsulated well within leachibility limits for non-hazardous land disposal.

Table III. Diffusivities and leachability indices of unextracted salt in samples of different salt contents

Sample ID	Relative Salt Content	Diffusivity (cm^2/s)	Leachability Index
3.08	High	6.00×10^{-8}	7.22
3.17	Low	3.50×10^{-8}	7.46

CONCLUSIONS

The important step in making a waste form for the encapsulation of toxic residuals occurs at the phase inversion during the curing process. The evaporation of water causes the material to transform from a continuous water phase to a continuous polymeric phase. The evaporation of water also drives the salt to the outer surface of the polymer, while the arsenic-laden AFH particles remain in the host material. Subsequent water immersion successfully extracts a high percentage of salt from the host material by dissolution and transport through interconnected porosity, which thereby promotes a reduction in waste volume and ultimately reduces disposal costs. Most importantly, based on leaching studies of the unextracted salt, the diffusion coefficients of both a high and low salt loaded sample yielded leachability indices within regulations for non-hazardous landfill disposal. Because salt is the most mobile species, it is inferred that arsenic leaches from the host material at an even slower rate, making the waste forms amenable to unregulated land disposal options.

The environmentally-benign, water-based emulsion processing of epoxy/PSB polymeric hosts show great promise as a separation and fixation technology for treating brine streams from wastewater treatment facilities. Future work should address the explicit measurement of arsenic leaching rates, the increase in residual loading, and the waste form volume consolidation by post cure heat treatment of extracted host material. A sensitivity study determining the impact of processing variations should be performed before the waste form fabrication is scaled-up for marketing.

ACKNOWLEDGEMENTS

The authors wish to thank the United States Department of Energy, the National Science Foundation, and Pacific Northwest National Laboratory for providing the opportunity to participate in the Faculty and Student Teams program. The research team also thanks Rod Quinn and the Environmental Technology Directorate, the Advanced Process and Application Group, especially Evan Jones, Bradley R. Johnson, Gordon Xia, Michael Schweiger and Jim Davis, and the Process Science and Engineering Division at PNNL. Finally, the team thanks William Velez and the CSEMS for facilitating the organization of the FaST appointment at UA.

REFERENCES

[1]J. Roberson, "Complexities of the New Drinking Water Regulations," *Journal American Water Works Association*, **95** [3] 48-56 (2003).

[2]US EPA, "National Primary Drinking Water Regulations; Arsenic and Clarifications to Compliance and New Source Contaminants Monitoring, Final Rule," *Federal Register*, **66** 6976-7066 (2001).

[3]F. Pontius, "Update on USEPA's Drinking Water Regulations," *Journal American Water Works Association*, **95** [3] 57-68 (2003).

[4]T.M. Krishnamoorthy, S.N. Joshi, G.R. Doshi, and R.N. Nair, "Desorption Kinetics of Radionuclides Fixed in Cement Matrix," *Nuclear Technology*, **104** [3] 351-357 (1993).

[5]M. Leist, R.J. Casey, and D. Caridi, "The Management of Arsenic Wastes: Problems and Prospects," *Journal Hazardous Materials*, **76** [1] 125-138 (2000).

[6]G. Amy, M. Edwards, P. Brandhuber, L. McNeill, M. Benjamin, F. Vagliasindi, K. Carlson, and J. Chwirka, *Arsenic Treatability Options and Evaluation of Residuals Management Issues*, AWWA Research Foundation, Denver, CO, 2000.

[7]J.R. Conner, *Chemical Fixation and Solidification of Hazardous Wastes*, Van Nostrand Reinhold, New York, NY, 1990.

[8]L. Liang, H. Smith, R. Russell, G. Smith, and B.J.J. Zelinski, "Aqueous Based Polymeric Materials for Waste Form Applications"; pp. 359-368 in *Proceedings of the International Symposium on Environmental Issues and Waste Management Technologies in the Ceramic and Nuclear Industries VII*. Edited by G.L. Smith, S.K. Sundaram, and D.R. Spearing. Ceramic Transactions, The American Ceramic Society, Westerville, OH, 2002.

[9]R. Evans, A. Quach, D.P. Birnie, A.E. Sáez, W.P. Ela, B.J.J. Zelinski, G. Xia, and H. Smith, "Development of Polymeric Waste Forms for the Encapsulation of Toxic Wastes Using an Emulsion-Based Process," *U.S. Department of Energy Journal of Undergraduate Research*, **3** 56-63 (2003).

[10]C.K. Riew, R.W. Smith, "Modified Osmium Tetroxide Stain for the Microscopy of Rubber-Toughened Resins," *Journal Polymer Science Part A-1*, **9**, 2739-2744 (1971).

[11]R.A. Pearson, A.F. Yee, "Toughening Mechanisms in Elastomer-Modified Epoxies, Part 2, Microscopy Studies," *Journal Material Science*, **21**, 2475-2488 (1986).

EFFECT OF THERMAL TREATMENT CONDITIONS ON MICROSTRUCTURE AND COMPOSITION OF HIGH TEMPERATURE REACTOR FUEL KERNEL

François Charollais[1], Anne Duhart[1], Patrice Felines[1], Pierre Guillermier[2], Christophe Perrais[1]
[1]CEA Cadarache
DEN/CAD/DEC/SPUA/LCU
Bât 315 - BP1
13108 St Paul Lez Durance - France
[2]AREVA FRAMATOME-ANP
Fuel Sector
10, rue Juliette Récamier
69456 Lyon Cedex 06 - France

ABSTRACT

This paper describes aspects of an R&D project, conducted by both the Commissariat à l'Energie Atomique (CEA) and FRAMATOME-ANP, bearing on Fuel Fabrication with two main goals: (1) to restore and to improve coated and fissile kernels manufacturing knowledge including innovative technology perspectives and (2) to develop modern characterization methods able to satisfy nuclear industrial needs.

In this paper, we will focus on the first objective, i.e., fabrication of UO_2 fissile kernels. Due to the spherical specificity of HTR Fuel kernel, the less expensive fabrication route is a process based on the sol-gel principle and each country involved in a HTR program has developed its own recipe. Our sol-gel method, so-called Gel Supported Precipitation, generates U-based droplets whose diameter is calibrated by a vibrational dropping technique. Then, thermal treatments - drying, calcining and reducing-sintering stages - are required to produce dense UO_2 kernels. The influences of thermal treatments temperature, of the thermal treatment gas nature (i.e. oxygen potential of atmosphere) on the kernels' microstructure (grain size, porosity) and composition are studied by Thermal Gravimetric Analysis, X-Ray Diffraction, Optical and Scanning Electronic Microscopy and EDS characterization and discussed.

INTRODUCTION

The well-known characteristics of High Temperature Reactors (HTR) make this type of nuclear reactor a very serious candidate for the future, specially considering the concept of modular units (100-300MWe) adapted to actual worldwide needs as a supplement to large nuclear reactors (>1000MWe). Moreover, the high temperature gives a better efficiency and opens up possibilities of industrial use in non-energetic fields like hydrogen production and other chemical applications, e.g. desalination. The use of refractory coated spherical kernel fuel allows retention of every fission product, even in the most severe accidental conditions such as loss of primary coolant system leading to dispose of an advantageous inherent safety. In this context, the Commissariat à l'Energie Atomique and FRAMATOME-ANP are conducting an R&D project on HTR, in which one of the first steps is the recovery of coated particle fuel manufacturing know-how.

Present work focuses on the fabrication of the heart of the HTR fuel element, which is the fissile kernel. UO_2 kernels are produced using a sol-gel method called GSP process as Gel Supported Precipitation. Originally developed at an experimental scale by the Italians[1,2], the process has been further improved and validated at a higher production rate by HOBEG and KFA-Juelich in Germany[3,4,5]. Despite the German industrial proven experience, little key information is available in the open literature. Intensive work has been done at CEA, Cadarache, in order to understand and control each step of the HTR kernel fabrication process.

The present paper, after detailing the GSP process, deals with the influence of the calcining atmosphere (argon and air have been studied) on the microstructure and the composition of the U-based sintered kernels. Tailored microstructures of sintered UO_2 kernels in term of grain size and porosity are also presented.

EXPERIMENTAL PROCEDURE

Samples of U-based beads are prepared by a sol-gel method known as the gel supported precipitation (GSP) technique. The distinguishing principle of this method is that a water-soluble organic polymer is added to a heavy uranium metal solution. The polymer supports the particle spherical shape while ammonia diffuses into the gel bead and precipitates the uranium metal. A simplified scheme for the GSP process is shown in Figure 1.

An uranium feed solution, so-called broth solution, is prepared using uranyl nitrate with an additional organic polymer such as a polyvinyl alcohol (PVA) and an additional modifier agent such as tetrahydrofurfuryl alcohol (THFA). The drop formation is accomplished by flowing the feed solution through a vibrating nozzle into air or helium. The droplets form their spherical shape in air from the effect of surface tension, before passing through an ammonia gas layer, which hardened the surface into the final shape. The droplets are then collected in an aqueous concentrated ammonium hydroxide bath, where ammonia induces the precipitation of the uranyl nitrate in ammonium diuranate (ADU) and the gelation of the polymer. The gelled droplets are then aged for a suitable length of time to assure completion of the reaction of the ammonia with the uranyl nitrate to form the ADU. The desired overall reactions are shown in the following equations[6] :

$$(3 < pH < 7) \quad 2\ UO_2(NO_3)_2 + 4\ NH_4OH \rightarrow 2\ UO_2(OH)_2 + 4\ NH_4NO \quad (1)$$
$$(pH > 8) \quad 2\ UO_2(OH)_2 + 2\ NH_4OH \rightarrow (NH_4)_2U_2O_7 + 3\ H_2O \quad (2)$$
$$2\ UO_2(NO_3)_2 + 6\ NH_4OH \rightarrow (NH_4)_2U_2O_7 + 4\ NH_4NO_3 + 3\ H_2O \quad (3)$$

The excess ammonium nitrate produced during ADU precipitation is removed from the gelled droplets by washing in distilled water or in a diluted ammonium hydroxide solution. A second washing stage is realized using isopropanol to dehydrate the wet gel beads. Those two important washing steps prevent cracking on further thermal treatments. The thermal treatments, which are the drying, calcining and reduction-sintering stages, allow the elimination of all organic or nitrate products to produce dense UO_2 kernels. The influence of calcining atmosphere of air and argon (5 ppm O_2) has been investigated up to 600°C.

The crystalline phases of the beads, at the different steps of UO_2 kernel fabrication, are identified by powder X-ray diffraction on a diffractometer (XRD; D8Advance, Bruker) using CuKα radiation. The weight loss during the calcining and sintering stages is measured by thermogravimetric analysis (TGA, STA 409, Netzsch).

Sintered beads, mounted in an epoxy resin, are successively ground using 15 μm, then 9 μm silicon carbide grinding disks with water. Specimens are then polished with 1 μm diamond suspension on soft cloths. In order to reveal the grain boundaries, the polished bead sections are etched in a diluted sulphuric acid solution. Micrographs of polished sections and etched sections are acquired by image analysis using a black and white CCD camera linked to an optical microscope (PMG3, Olympus). The microstructure and chemical composition of the sintered beads are analyzed by scanning electron microscopy (SEM, XL30FEG, Philips) using energy-dispersive X-Ray spectroscopy (EDS, EDAX).

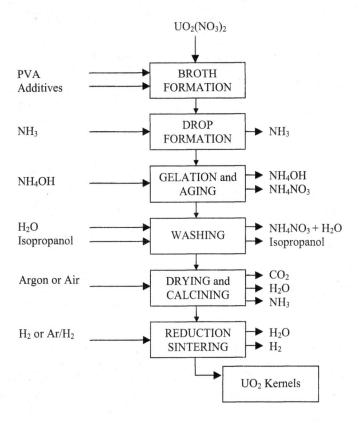

$UO_2(NO_3)_2$

PVA
Additives → BROTH FORMATION

NH$_3$ → DROP FORMATION → NH$_3$

NH$_4$OH → GELATION and AGING → NH$_4$OH / NH$_4$NO$_3$

H$_2$O
Isopropanol → WASHING → NH$_4$NO$_3$ + H$_2$O / Isopropanol

Argon or Air → DRYING and CALCINING → CO$_2$ / H$_2$O / NH$_3$

H$_2$ or Ar/H$_2$ → REDUCTION SINTERING → H$_2$O / H$_2$

UO$_2$ Kernels

Figure 1. Scheme for fabrication of UO$_2$ kernels by GSP process

RESULTS AND DISCUSSION : INFLUENCE OF CALCINING ATMOSPHERE ON MICROSTRUCTURE AND COMPOSITION OF U-BASED SINTERED KERNELS
X-Ray Diffraction

The evolution of X-Ray diffraction patterns as a result of the calcining stage of gelled beads (P1) in argon and in air and of the sintering step at 1600°C in a dry atmosphere of Ar/5%H$_2$ is shown in Figures 2 and 3 respectively. The JCPDS card numbers for all the compounds detected from XRD can be seen in Table I.

In agreement with GSP process chemistry[1,2,4], the crystalline phase of gelled beads (P1) is found to be $3UO_3(NH_3)_2.4H_2O$ ammonium uranium oxide hydrate, which is a hydrated form of ADU.

During the thermal calcining treatment in air, ADU is slightly decomposed; UO_3 phases are formed at a temperature of 500°C. At 600°C, a U_3O_8 phase appears mixed with UO_3. After sintering, uranium oxide is reduced to a stoichiometric UO_2.

As a result of calcining treatment in argon, a more reduced form of uranium oxide than that obtained in previous treatment in air is noticed by XRD. Two phases - U_3O_7 and U_4O_9 or a poorly crystallised UO_2 - are formed. After sintering, a mixture of uranium dioxide (UO_2) and uranium nitrides (UN and UN_2) appeared. Uranium nitride phases, resulting of a carbothermic reduction, could be formed only with presence of carbon and nitrogen in the calcined beads at high temperature[7,8].

Table I. JCPDS Cards of the crystalline compounds

Id. in Fig 2,3	Compound	JCPDS No.
A	Ammonium uranium oxide hydrate – $3UO_3(NH_3)_2.4H_2O$	43-366
B	Uranium oxide – UO_3	71-2124
C	Uranium oxide - $UO_{3.01}$	46-0947
D	Uranium oxide – U_3O_8	47-1493
E	Uraninite – UO_2	41-1422
F	Uranium oxide – U_3O_7	42-1215
G	Uranium oxide – U_4O_9	72-0125
H	Uranium nitride – UN	32-1397
I	Uranium nitride – UN_2	73-1713

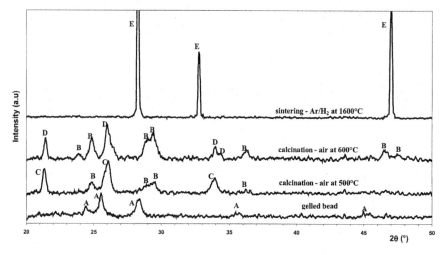

Figure 2. Evolution of XRD pattern of gelled bead during the calcination in air and sintering in Ar/H_2

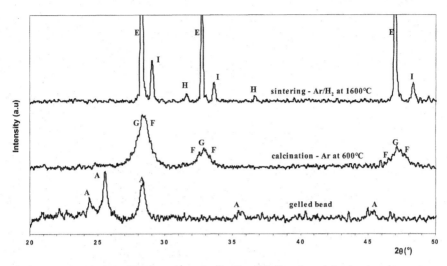

Figure 3. Evolution of XRD pattern of gelled bead during the calcination in argon
and sintering in Ar/H₂

Optical and SEM observations and EDS results

Optical micrographs of sintered beads ex-calcination under air and under argon are shown respectively in Figures 4a-4e and 5a,b. The microstructure of sintered beads which were calcined under air appears as a highly dense polycrystalline ceramic with very small porosity (cf. Figure 4a). We will notice that, depending on sintering parameters (temperature and chemical oxygen potential of sintering atmosphere), tailored microstructures (cf. Figures 4b-4e) in terms of grain size and pore size are possible to be optimised.

Whereas the sintered beads which were calcined under argon exhibit two kinds of highly porous microstructure:

- a first bead structure (cf. Figure 5a) with a dense shell of about 40 μm of thickness which seems to be a UO_2 monophase if we refer to the uniformity of grey levels of the optical image,

- a second bead structure (cf. Figure 5b) without any shell but exhibiting two phases, a white one located in the centre of the sintered beads and a grey one in the cortical zone.

SEM observations of this latter biphasic microstructure in backscattered electron mode (BSE mode) can be seen in Figure 6. On the image obtained in this BSE mode, (whose contrast is depending on the mean atomic numbers of the bead constituent elements) the presence of two phases is clearly detected. In order to evaluate the respective composition of those phases, EDS analyses have been conducted on spots as marked on Figure 6. The percentages of the main elements analysed are given in Table II. Those semi-quantitative results associated with XRD measurements allow us to consider the white zone (on BSE or optical images) as an uranium nitride phase, whereas the dark zone is constituted of uranium dioxide, UO_2.

Table II. EDS results

	Composition (atom%)		
	U	O	N
Spot A	29.2	21.6	49.2
Spot B	26.3	73.7	-
Spot C	25.6	74.4	-

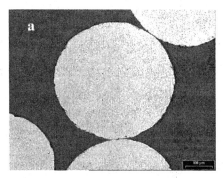

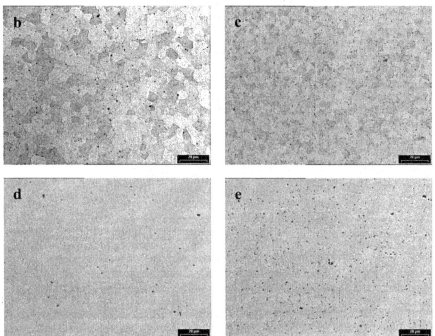

Figure 4. Optical micrographs of sintered UO_2 kernels after calcination in air: (a) general overview of UO_2 kernel; (b,c) etched cross-sections showing tailored grain size microstructure; (d,e) polished cross-sections showing tailored porosity microstructure

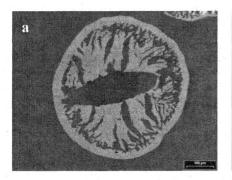 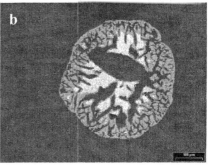

Figure 5. Optical micrographs of sintered U-based kernels after calcination in argon showing a highly porous structure (a) with a single phase of UO_2 (b) with a mixture of UO_2 (grey phase) and Uranium Nitride (white phase)

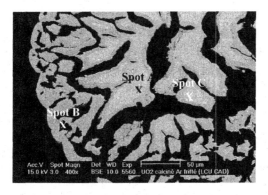

Figure 6. SEM micrograph in BSE mode of sintered U-based kernel after calcination in argon showing a mixture of UO_2 and UN_x

Thermogravimetric analysis and discussion

Calcining stage: In order to compare the effect of argon and air on the decomposition of gelled beads occurring during calcination, the weight loss and rate of weight loss have been measured by TGA up to 600°C (cf. Figure 7). It can be observed that the shape of the two weight loss curves is quite different and the final weight loss obtained in the case of the calcination under argon (18.7%) is significantly lower than that under air (21%), despite a more important reduction of the UO_x phase in the case of the calcination under argon than under air since a mixture of U_3O_7 and U_4O_9 is obtained in argon at 600°C whereas in air, XRD results show a mixture of UO_3 and U_3O_8. This can seem contradictory, thus we will first comment the both TGA curves and try to give an explanation of the phenomena occurring during both kind of calcinations, based on literature and on early internal works at CEA, Cadarache.

The main weight loss under air is achieved as soon as the temperature reaches about 400°C, above 450°C, a weight loss step is noticed. The curve of rate of weight loss, i.e. the kinetic of gelled beads decomposition, presents four peaks, respectively at 70°C, 170°C, 270°C and 340°C.

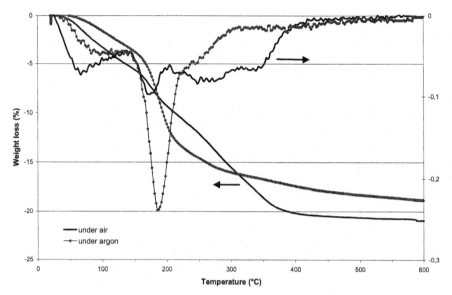

Figure 7. Weight loss and rate of weight loss curves of gelled beads
during calcining stage under air and argon

The weight loss obtained under argon decreases rapidly in two stages before 250°C reaching about 15%. Both steps are well exhibited by the rate curve; the first peak is observed at about 100°C and the second one of higher intensity is noticed at 180°C. Above 250°C, a weak but constant weight loss is detected up to 600°C, no asymptote being reached contrary to that observed for the calcination under air.

Before 200°C, it is known that the ammonium diuranate loses weakly and hardly adsorbed water[8] in two stages at temperatures equivalent to the two first peaks observed on our both experimental TGA curves, under air and under argon.

The peak of rate weight loss observed under air at 270°C is to be linked to the well known rapid decomposition[6,9] of the ammonium nitrate which is formed during the reaction between ammonia and uranyl nitrate (cf. Equation 3). The fourth stage of decomposition of gelled beads observed under air (with a rate peak at about 340°C) is assumed to be due to the decomposition of the polymer supporting the ammonium diuranate in a spherical shape. This assumption is in very good agreement with the results published by Weimin et al.[9] and with early non published CEA works at Cadarache. Concerning the calcination under argon and keeping in mind that the third and fourth stages of decomposition analysed under air are not observed under argon and that the main part of the weight loss (80%) measured in argon has happened before 250°C, we can assume that the argon atmosphere (which is a less oxidising gas than air) will induce a higher reduction of the UO_x phase of the ADU at low temperature, before decomposing the polymer and the nitrate. This phenomenon of accelerated reduction will be enhanced precisely by the presence of organic compounds which will impose locally severe reducing conditions. Finally, as the total weight loss under argon is lower than that under air, but with an uranium oxide phase with a O/U ratio smaller (cf. XRD analysis), we can conclude that the amount of oxygen brought by the argon atmosphere and the uranium oxide reduction is insufficient for a complete combustion of both polymer and nitrate. Thus, calcined beads under argon must contain residues of polymeric organic chains and nitrogenous compounds.

Sintering Stage: Figure 8 shows the weight loss of both types of calcined beads measured during the sintering in a Ar/5%H$_2$ atmosphere.

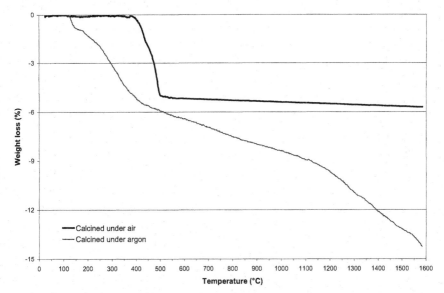

Figure 8. Weight loss curves of beads calcined under air and under argon
during sintering stage under Ar/H$_2$

The curve of the beads ex-air calcination decreases rapidly between 400 and 500°C. The weight loss associated with this decomposition is 5%, which is between the theoretical weight loss value corresponding to the reduction of UO$_3$ into UO$_2$ (5.6 %) and that of U$_3$O$_8$ into UO$_2$ (3.8 %). This is in good agreement with XRD analysis of beads calcined in air. Above 500°C, the shape of the weight loss is typical of normal UO$_2$ sintering in our TG apparatus, taking into account the buoyancy on the TG crucible.

The shape of the TGA curve of the beads ex-argon calcination is totally different. We can observe a first rapid decrease up to 400°C, then a slower constant decrease between 400°C and 1100°C, then an acceleration of weight loss up to 1600°C. On the basis of the XRD and MEB EDS results on sintered kernels, it is obvious that a carbothermic reduction occurred leading to the formation of UN and UN$_2$ phases (cf. Figures 3 and 6). This carboreduction must have happened between the UO$_x$ phase, the organic residue supplying the carbon element and the nitrogenous rests supplying the nitrogen element necessary for processing the nitriding reaction[7,8]. The particular porous structure observed on the sintered kernels ex-argon calcination (cf. Fig. 5a,b) is due to the slow thermal decomposition (without real combustion because of a lack of oxygen) of the excess of C and N-based residues during the sintering stage in an Ar/H$_2$ atmosphere, obstructing the expected densification of the kernels.

CONCLUSION

In the HTR fuel kernel fabrication process, the thermal treatments such as the calcining and the sintering stages are essential to eliminate organic or nitrate sub-products, to reduce the uranium oxide into UO_2 and finally to sinter the HTR kernel. Our work has shown that the oxygen chemical potential of the calcining step controls the behaviour of decomposition of the gelled beads and thus the final composition of the sintered kernel. Calcining atmosphere with low oxygen amount such as argon will induce after sintering in Ar/H_2 the formation of a mixture of UO_2 and uranium nitride in a highly porous microstructure because of a higher reduction of UO_x phase concomitant with an insufficient decomposition of carbon and nitrogenous compounds.

Calcining under air and sintering in Ar/H_2 atmosphere allow the obtaining of tailored microstructure of sintered UO_2 kernels.

In the common HTR fuel manufacture R&D programme carried out by CEA and FRAMATOME-ANP, each step of UO_2 kernel sol-gel fabrication has been revisited and understood with the help of sharp characterization techniques. Calcining and sintering stages appear as key-points in the manufacture process. By choosing adequate conditions (in terms of oxygen chemical potential in the oven's atmosphere, temperature, etc.), uranium compositions and microstructures could be precisely adjusted to be in accordance with the specification.

REFERENCES

[1] G. Brambilla, P. Gerontopoulos and D. Neri, "The SNAM Process for the Preparation of Ceramic Nuclear Fuel Microspheres: Laboratory Atudies", *Energia Nucleare*, **17** [4] 217-24 (1970).

[2] A.G. Facchini, "Wet Route Microsphere Production and Evaluation", IAEA Conference: Panel on Sol Gel Processes for Fuel Fabrication, Vienna, Austria, 21-24 May 1973.

[3] H. Huschka, M. Kadner, R. Förthmann and E. Zimmer, "Kernel Fabrication for Different Fuel Cycles in Germany", IAEA Conference: Panel on Sol Gel Processes for Fuel Fabrication, Vienna, Austria, 21-24 May 1973.

[4] M. Kadner and M. Baier, "Production of Fuel Kernels for High Temperature Reactor Fuel Elements", *Kerntechnik*, **18** [10] 413-20 (1976)

[5] W. Heit, H. Huschka and W. Rind, "Status of Qualification of HTR Fuel Element Spheres", *Nuclear Technology*, **69**, 44-54 (1985).

[6] C. Garrigou, "Etude de la Précipitation Ammoniacale du Diuranate d'Ammonium", Thèse Université Aix-Marseille III, 20 July 1995.

[7] G. Ledergerber, Z. Kopajtic, F. Ingold and R.W. Stratton, "Preparation of Uranium Nitride in the Form of Microspheres", *Journal of Nuclear Materials*, **188**, 22-35 (1992).

[8] S. Daumas, "Etude et Réalisation de Support-Matrices Inertes par le Procédé Sol-Gel pour l'Incinération des Actinides Mineurs", Thèse Université Aix-Marseille I, 4 July 1997.

[9] B. Weimin and W. Xuejun, "Preparation of Dense UO_2 Fuels Kernels by External Gelation Process (II)", *Atomic Energy Science and Technology*, **24** [4] 10-16 (1990).

Glass Waste Forms—
Modelling, Properties, and Testing

PREDICTING PHASE EQUILIBRIA OF SPINEL-FORMING CONSTITUENTS IN WASTE GLASS SYSTEMS

Theodore M. Besmann and Nagraj S. Kulkarni
Metals and Ceramics Division
Oak Ridge National Laboratory
Oak Ridge, TN 37831-6063

Karl E. Spear
44 Tidnish Head Road, RR#2
Amherst, NS B4H 3X9 Canada

John D. Vienna
Pacific Northwest National Laboratory
Richland, WA 99352

ABSTRACT

A modified associate species thermochemical model has been developed for the liquid/glass in nuclear waste glass systems, and provides a simple means for relatively accurately representing the thermochemistry of the liquid/glass phase. A modification of the methodology is required when two immiscible liquids are present, such that a positive interaction energy is included in the representation. The approach has been extended to include spinel-forming constituents together with the base glass system as well as development of a models for spinel phases.

INTRODUCTION

The production of nuclear materials for defense applications at several sites in the United States over almost six decades has resulted in the accumulation of a substantial quantity of radioactive waste. These materials are currently stored in a variety of forms including liquids, sludges, and solids. In addition, there are similar wastes that have resulted from the reprocessing of commercial spent fuel, although this has occurred to a much smaller extent. While the composition and characteristics of the various high-level wastes (HLW) differ, their behavior is similar in many respects. The focus of current U. S. Department of Energy efforts with regard to permanent disposal of these materials is that they will be incorporated in a stable, insoluble host solid (a glass or specific crystalline phase).

Components such as Fe, Cr, Zr, and Al have limited solubility in HLW glasses (1-3). These components precipitate as oxide minerals such as spinel, zircon, and nepheline once their solubility in glass is exceeded. Precipitated minerals may cause melter failure (4) and can alter the physical properties such as the leach resistance of the glass (2). To avoid these problems, current HLW glasses are formulated to assure oxide minerals do not precipitate in the melter (2,5,6). The solubility of these components can dictate HLW glass volume produced at the Savannah River Site (7) and West Valley and to be produced at Hanford (5, 8).

Thermochemical assessment of the phase equilibria and modeling of the liquid/glass phase can support optimization of glass formulations with regard to stability and waste loading. In order to provide a sufficient thermochemical understanding of the liquid and glass system used for sequestering HLW, an approach using the associate species technique was chosen (9, 10). It is attractive because it (a) accurately represents the thermodynamic behavior of very complex chemical systems over wide temperature and composition ranges, (b) accurately predicts the activities of components in metastable equilibrium glass phases, (c) allows logical estimation of unknown thermodynamic values with an accuracy much greater than that required for predicting useful engineering limits on thermodynamic activities in solutions, and (d) is relatively easy for non-specialists in thermochemistry to understand and use.

Ideal mixing of associate species accurately represent the solution energies in which end member components exhibit attractive forces. A modification to the associate species model, hence the term "modified" associate species model, is the incorporation of positive solution model constants to represent any positive interaction energies in a solution. With these it is possible to accurately represent reported immiscibility in solution phases (e.g., the liquid-liquid immiscibility common in many silica-containing systems). The results are simple, well-behaved equations for free energies that can be confidently extrapolated and interpolated into unstudied temperature and composition ranges. Thus, in support of the nuclear waste glass development effort, a model of the Na_2O-Al_2O_3-B_2O_3-SiO_2 was developed using the modified associate species approach and described elsewhere (9,10).

The work described here is focused on modeling spinel phases along with attendant liquid/glass and other crystalline phases in HLW systems. As noted above, previous efforts have successfully modeled the base glass system. Progress to date to include spinel-related constituents has resulted in modeling of the Fe-O, Mn-O, Al-Fe oxide, Cr-Fe oxide, and Al-Fe-Cr oxides. The basic data for the calculations are obtained from the 1996 version of the Scientific Group Thermodata Europe (SGTE) Pure Substance Database (12) and calculations are performed using the ChemSage (13) and FactSage (14) thermochemical software packages. Continuing efforts will seek to include other important elements in spinel phases, most notably Mn.

Fe-O SYSTEM

With the multivalent nature of iron and the importance of redox potential in many glass systems, the Fe-O system must be correctly modeled for its accurate inclusion in any liquid/glass system. This is less of an issue for HLW glass as efforts are made to fully oxidize species in the melter. Following the formalism described by Spear, et al. (9), liquid species' stoichiometry are chosen such that they contain 2 non-oxygen atoms per formula weight. The liquid/glass for Fe-O has been treated as a solution of Fe_2, Fe_2O_2, Fe_3O_4:2/3, and Fe_2O_3 species. The nomenclature for Fe_3O_4:2/3 indicates that the species has the Fe_3O_4 relative stoichiometry, although all values are multiplied by 2/3 in order to obtain 2 non-oxygen atoms per formula weight. The thermodynamic values for crystalline phases and liquid species were derived and given in Tables I-IV. These were based on the SGTE database (12), the procedures described by Spear et al. (9), and from fitting published phase equilibria,. The liquid-liquid immiscibility of the Fe-O system, however, required that the solution be described using positive (repulsive) energetic terms. A simple Redlich-Kister (15) model, for which the values were manually fit to reproduce the phase equilibria, is adequate. It has the general formalism

$$G_{ex} = xixj\Sigma\ (Ln\ (xi\text{-}xj)\ n\text{-}1)\quad\text{(J/mol)}\qquad\qquad(1)$$

where G_{ex} is the excess free energy of the solution, xi and xj are the mol fractions of species i and j, respectively. For the liquid phase the interacting species and interaction parameters are given in Table V.

Table I. Thermodynamic values for the crystalline phases based on the SGTE database and modified as necessary to develop associate species models. ($\Delta H_{f,298}$ is the 298K heat of formation, S_{298} is the 298K entropy, T is absolute temperature, T_{fus} is the melting temperature, and ΔH_{fus} is the heat of fusion.)

Crystalline Phase	$-\Delta H_{f,298}$(J/mol)	S_{298}(J/K-mol)	T_{fus}(K)	ΔH_{fus}(J/mol)
Mn	---	32.008	1517	12058.3
MnO	384928.	59.831	2058	54392.
Mn$_3$O$_4$	1386580.	153.971	1833	20920.
Mn$_2$O$_3$	956881.	110.458	---	decomposes
MnO$_2$	520071.	53.053	---	decomposes
Fe	---	27.280	1809	13807.2
FeO	279140.	43.2	1650	24058.
Fe$_3$O$_4$	1120000.	143.2	1870	138072.
Fe$_2$O$_3$	821500.	85.5	---	decomposes
Al$_2$O$_3$	1675692.	50.94	2327	111085.
Cr$_2$O$_3$	1150600.	81.1	2705	125000.
CrO$_2$	581576.	53.555	---	decomposes
Al$_2$FeO$_4$	1980870.	106.274	---	decomposes
Al$_2$Fe$_2$O$_6$	2405175.5	206.	---	decomposes
Cr$_2$FeO$_4$	1450760.	146.022	---	decomposes

Table II. Liquid species thermodynamic values based on the SGTE database and modified using the heat of fusion and fitting to the phase equilibria.

Liquid Species	$-\Delta H_{f,298}$(J/mol)	S_{298}(J/K-mol)
Mn$_2$	-20916.6	79.914
Mn$_2$O$_2$	658072.	172.517
Mn$_3$O$_4$:2/3	907440.	112.299
Mn$_2$O$_3$	955881.	110.458
Fe$_2$	-27614.	69.808
Fe$_2$O$_2$	475000.	134.5
Fe$_3$O$_4$:2/3	651000.	146.
Fe$_2$O$_3$	724000.	136.
Al$_2$O$_3$	581576.	53.555
Cr$_2$O$_3$	1015600.	127.3107
Al$_2$FeO$_4$:2/3	1309913.	70.84933
Cr$_2$FeO$_4$:2/3	955206.	97.348

Table III. Crystal phase SGTE-based heat capacity, transition temperature (T_{trans}), and heat of transition (ΔH°_{trans}). Heat Capacity Coefficients: $C_p = a + b \bullet T + c \bullet T^2 + d/T^2$ (J/K-mol)

Phase	a	b	c	d	T_{trans}(K)	ΔH°_{trans}(J/mol)
Mn	18.0891	0.0283786	-.928773E	-0548977.5	980	2225.9
	33.3004	0.435914E-02	0.0	0.0	1360	2121.3
	31.714701	0.8368E-02	0.0	0.0	1410	1878.6
	33.552799	0.82810E-02	0.0	0.0	1517	---
MnO.	46.484200	0.811696E-02	0.0	-368192.00	2115	---
Mn_3O_4	144.93401	0.04527090	0.0	-920480.	1445	---
Mn_2O_3	103.47000	0.0350619	0.0	-1351430.	1350	---
MnO_2	69.454399	0.010209	0.0	-1623390.	800	---
Fe	28.18	-7.32E-03	2.E-05	-290000.	1184	899.6
	28.	8.6E-03	0.0	0.0	1665	836.8
	24.64	9.9E-03	0.0	0.0	1809	---
FeO	48.	1.2E-02	-1.E-06	-200000.	1650	---
Fe_3O_4	153.55	5.E-02	0.0	0.0	700	0*
	175.	2.E-02	0.0	0.0	1184	0*
	165.	3.E-02	0.0	0.0	1870	---
Fe_2O_3	110.	5.E-02	0.0	-1700000.	700	0*
	138.	0.0	0.0	0.0	1050	0*
	130.	7.3E-03	-5.E-07	0.0	1980	---
Al_2O_3	117.49	1.038E-02	0.0	-3711000.	2327	---
Cr_2O_3	134.439	-.0126191	0.84377E-05	-2839800.	2705	---
CrO_2	48.534401	0.0118826	0.0	-1138050.	750	---
Al_2FeO_4	165.49	0.02238	-0.000001	-3911000.	1650	0*
	185.69	0.01038	0.0	-3711000.	2327	---
$Al_2Fe_2O_6$	227.49	0.06038	0.0	-5411000.	700	0*
	255.49	0.01038	0.0	-3711000.	1050	---
Cr_2FeO_4	182.439	-0.0006191	7.43775E-06	-3039800.	1650	0*
	202.639	-0.0126191	8.43775E-06	-2839800.	2705	---

*A value of zero for the enthalpy of transition indicates no transition occurs, but a new Cp equation is used for the next temperature range.

Table IV. Liquid species SGTE-based heat capacity, transition temperature (T_{trans}), and heat of transition (ΔH°_{trans}). Heat Capacity Coefficients: $C_p = a + b \bullet T + c \bullet T^2 + d/T^2$ (J/K-mol)

Species	a	b	c	d	T_{trans}(K)	ΔH°_{trans}(J/mol)
Mn$_2$	36.1782	0.0567572	-1.85755E-05	97955.	980	4451.8
	66.6008	0.00871828	0.0	0.0	1360	4242.6
	63.4294	0.016736	0.0	0.0	1410	3757.2
	67.1056	0.01656202	0.0	0.0	1517	fusion
	92.047996	0.0	0.0	0.0	2400	---
Mn$_2$O$_2$	92.9684	0.016233919	0.0	-736384.	2115	fusion
	121.335998	0.0	0.0	0.0	3000	---
Mn$_3$O$_4$:2/3	96.6227	0.0301806	0.0	-613653.3	1445	13947.
	140.0246667	0.0	0.0	0.0	1833	fusion
Mn$_2$O$_3$	103.47	0.0350619	0.0	-1351430.	1350	---
Fe$_2$	56.36	-1.464E-02	4.E-05	-580000.	1184	1799.2
	56.	1.72E-02	0.0	0.0	1665	1673.6
	49.28	1.98E-02	0.0	0.0	1811	fusion
	92.04	0.0	0.0	0.0	3000	---
Fe$_2$O$_2$	96.	2.4E-02	-2.E-06	-400000.	1650	fusion
	136.4	0.0	0.0	0.0	3000	---
Fe$_3$O$_4$:2/3	102.	3.33333E-02	0.0	0.0	700	0*
	116.667	1.33333E-02	0.0	0.0	1184	0*
	110.	2. E-02	0.0	0.0	1870	fusion
	133.888	0.0	0.0	0.0	3000	---
Fe$_2$O$_3$	110.	5. E-02	0.0	-1700000.	700	0*
	138.	0.0	0.0	0.0	1050	0*
	130.	7.3E-03	-5.E-07	0.0	1980	fusion
	148.	0.0	0.0	0.0	3000	---
Al$_2$O$_3$	117.49	1.038E-02	0.0	-3711000.	2327	0*
	192.464	0.0	0.0	0.0	3000	---
Cr$_2$O$_3$	134.439	-.0126191	0.84377E-05	-2839800.	2705	fusion
	170.	0.0	0.0	0.0	4500	---
Al$_2$FeO$_4$:2/3	110.3267	0.01492	-6.6667E-07	-2607333.	1650	0*
	123.7933	0.00692	0.0	-2474000.	2327	fusion
	173.776	0.0	0.0	0.0	3000	---
Cr$_2$FeO$_4$:2/3	121.626	-4.12733E-04	4.9585E-06	-2026533.	1650	0*
	135.0927	-0.008413	5.62517E-06	-1893200.	2705	fusion
	113.33333	0.0	0.0	0.0	4500	---

*A value of zero for the enthalpy of transition indicates no transition occurs, but a new Cp equation is used for the next temperature range.

Table V. G_{ex} coefficients (J/mol) (Eq. 1) for interacting species where T is absolute temperature.

Interacting Species	L_0	L_1	L_2
$Fe_2 - Fe_2O_2$	50,000	40,000	10,000
$Fe_2 - Fe_3O_4$	60,000	-	-
$Fe_3O_4:2/3 - Fe_2O_3$	25,000	-	-
$Mn_2 - Mn_2O_2$	100,000	50,000	-
$Mn_3O_4:2/3 - Mn_2O_3$	40,000	-	-
$Al_2O_3 - Fe_2O_3$	$90,000 - 40T$	-	-
$Cr_2O_3 - Fe_2O_3$	$-80,000 + 10T$	$-30,000 + 10T$	-

The published diagram of Fig. 1a compares well to the computed phase diagram seen in Fig. 1b. Unlike the simple, modified associate species model, the published diagram, which was also computed, utilized a compound energy model with ionic constituents for the crystalline phases and an ionic two-sublattice model (which required 8 polynomial expansions) for the liquid phases (16).

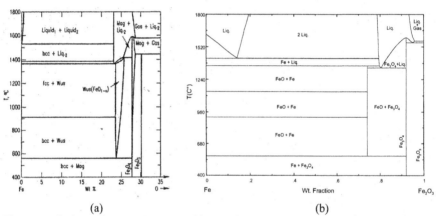

(a)   (b)

Fig. 1. Fe-O phase diagram (a) computed by Sundman (16) and (b) computed in this work. (wus=wustite; mag=magnetite; bcc= body-centered cubic phase; fcc=face-centered cubic phase)

Mn-O SYSTEM

The Mn-O system, which is analogous to the Fe-O system, was approached in a similar manner. The liquid/glass phase was treated as a solution of Mn_2, Mn_2O_2, $Mn_3O_4:2/3$, and Mn_2O_3 species having the thermodynamic values listed in Table I-IV. The system also exhibits liquid-liquid immiscibility. The necessary interaction energy terms for the Redlich-Kister (15) excess free energy that were derived are listed in Table V. The published phase diagram, which was determined in the same manner as the published Fe-O diagram (18), is seen in Fig. 2a and reasonably reproduces the computed diagram of this work seen in Fig. 2b.

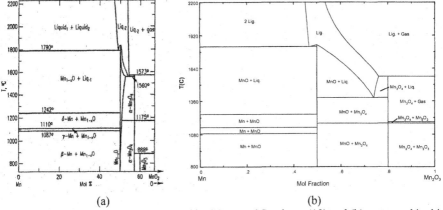

(a)   (b)

Fig. 2. Mn-O phase diagram (a) computed by Wang and Sundman (18) and (b) computed in this work.

Al-Fe OXIDE LIQUID/GLASS, SESQUIOXIDE, AND SPINEL

The alumina-iron oxide solution that represents the liquid/glass is made up of the species: Al_2O_3, Fe_2, Fe_2O_2, $Fe_3O_4{:}2/3$, Fe_2O_3, $FeAl_2O_4{:}2/3$. No additional interaction parameters were used beyond those for the unary systems and only one binary oxide associate was included.

The two-sublattice model (19) was used for the spinel phase, with an excess energy for the interactions between Fe^{3+} and Al^{3+} on the sublattice of

$$G_{xs} = x_{Fe3+} \cdot x_{Al3+} \cdot A \qquad \text{(j/mol)} \qquad (2)$$

where x is the concentration of the subscripted atom on the sublattice and A=20 kj/mol. In addition, adjustments were necessary to the 298 K heat of formation and entropy of the $AlFeO_3$ phase.

The sesquioxide phases, hematite and corundum, required the zeroth-order Redlich-Kister (15) interaction parameters (Eq. 1) shown in Table V to adequately represent the end-member solution phases.

The experimental diagram is seen in Figure 3a (20) and can be compared with the computed phase diagram of Figure 3b. The model does not reproduce all the details accurately, but given the sparsity of data and the reasonable agreement in temperature, it is a good fit.

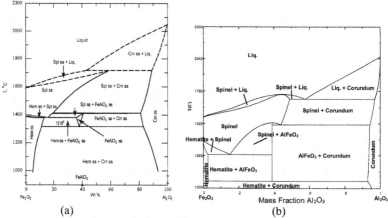

(a)   (b)

Fig. 3. Al_2O_3-Fe_2O_3 phase diagram (a) of Muan and Gee (20) and (b) computed in this work.

Cr-Fe OXIDE LIQUID/GLASS, SESQUIOXIDE, AND SPINEL

The chromia-iron oxide solution that represents the liquid/glass is treated very similarly to that of the alumina-iron oxide solution, made up of the species: Cr_2O_3, Fe_2, Fe_2O_2, Fe_3O_4:2/3, Fe_2O_3, $FeCr_2O_4$:2/3. Again, no additional interaction parameters were used beyond those for the unary systems and only one binary oxide associate was included.

The two-sublattice model (19) was used for the spinel phase, with an excess energy for the interactions between Fe^{3+} and Cr^{3+} on the sublattice of A=8 kj/mol (Eq. 2). The sesquioxide solution phase required zeroth- and first-order Redlich-Kister (15) (Eq. 1) interaction parameters (Table V) to adequately represent the end-member solution phases. The experimental diagram is seen in Fig. 4a (21) and can be compared with the computed phase diagram of Figure 4b. The model reproduces the essential phase equilibria, but disagrees significantly in the uncertain area of the liquidus.

Al_2O_3-Cr_2O_3-Fe_2O_3 PHASE EQUILIBRIA

The systems described above were combined to obtain a model for the alumina-chromia-iron oxide system and it includes all the liquid/glass species and interaction parameters, as well as all crystalline phases. No ternary associates were used in the model as well as no additional interaction parameters. An example phase diagram at 1500°C was computed and can be seen in Fig. 5 compared to an experimentally determined diagram (22). While the phase regions do not completely match, again the major features are correct and the results can be considered good for so simple an approach.

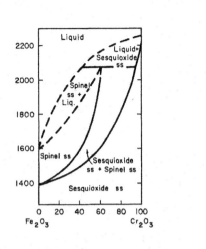

(a)

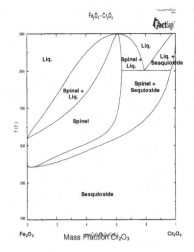

(b)

Fig. 4. Cr_2O_3-Fe_2O_3 phase diagram (a) of Muan and Somiya (21) (b) computed in this work.

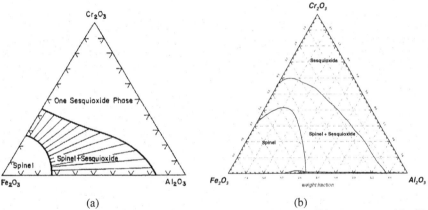

(a)   (b)

Fig. 5. Al_2O_3-Cr_2O_3-Fe_2O_3 phase diagram (a) of Muan and Somiya (22) (b) computed in this work.

CONCLUSIONS

The modified associate species approach for complex systems is simple, relatively accurate, and highly usable for describing liquidus surfaces, conditions for crystalline phase formation, and chemical activities of glass constituents. A base model for HLW glass systems that agrees reasonably with published phase diagrams has been developed. The very simple and usable modeling approach has successfully described important features of the Fe-O and Mn-O systems.

Efforts to similarly model the Cr_2O_3-Fe_2O_3 and Al_2O_3-Fe_2O_3 systems have been reasonably successful. The results have been combined to provide a comprehensive model of the alumina-chromia-iron oxide system.

ACKNOWLEDGMENTS
The authors thank T.R. Watkins and M.J. Lance for their useful comments. Research supported by the DOE Office of Biological and Environmental Research, U.S. Department of Energy, under Contract DE-AC05-00OR22725 with Oak Ridge National Laboratory, managed and operated by UT-Battelle, LLC.

REFERENCES
1. P. Hrma, G.F. Piepel, M.J. Schweiger, D.E. Smith, D.S. Kim, P.E. Redgate, J.D. Vienna, C.A. LoPresti, D.B. Simpson, D.K. Peeler, M.H. Langowski, "Property / Composition Relationships for Hanford High-Level Waste Glasses Melting at 1150°C"; PNL-10359, , Pacific Northwest Laboratory, Richland, WA, Vol. 1 and 2, 1994.

2. K.S. Kim, D.K. Peeler, and P. Hrma, "Effects of Crystallization on the Chemical Durability of Nuclear Waste Glasses," *Ceramic Transactions* **61** 177-85 (1995).

3. M. Mika, M.J. Schweiger, J.D. Vienna, P. Hrma, "Liquidus Temperature of Spinel Precipitating High-Level Waste Glasses," pp. 71-78 in *Scientific Basis for Nuclear Waste Management XX*, Edited by W.J. Gray, I.R. Triay, MRS Processing Vol. 465; Materials Research Society: Warrendale, PA, 1990;.

4. D.F. Bickford, A. Applewhite-Ramsey, C.M. Jantzen, K.G. Brown, "Control of Radioactive-Waste Glass Melters. 1. Preliminary General Limits At Savannah River," *Journal of the American Ceramic Society* **73**, 2896-2902 (1990).

5. P. Hrma, J. D. Vienna, and M. J. Schweiger, "Liquidus Temperature Limited Waste Loading Maximization for Vitrified HLW," *Ceramic Transactions,***72** 449-56 (1996).

6. J.D. Vienna, P. Hrma, M.J. Schweiger, D.E. Smith, D.S. Kim, "Low-Temperature High-Waste Loading Glass for Hanford DST/SST High-Level Waste Blend - Technical Note, " PNNL-11126, Pacific Northwest Laboratory, Richland, WA, 1995.

7. P. Hrma, J.D. Vienna, M. Mika, J.V. Crum, T.B. Edwards, "Liquidus Temperature Data for DWPF Glass," PNNL-11790, Pacific Northwest National Laboratory, Richland, WA, 1998.

8. S.L. Lambert, G.E. Stegen, J.D. Vienna, "Tank Waste Remediation System Phase I High-Level Waste Feed Processability Assessment Report," WHC-SD-WM-TI-768, Westinghouse Hanford Company, Richland, WA, 1996.

9. K.E. Spear, T.M. Besmann, and E.C. Beahm, "Thermochemical Modeling of Glass: Application to High-Level Nuclear Waste Glass," *MRS Bulletin,* **24** [4] 37-44 (1999).

10. T.M. Besmann and K.E. Spear, "Thermochemical Modeling of Oxide Glasses," *Journal of the American Ceramic Society,* **85** 2887-94 (2002).

12. A.T. Dinsdale, "SGTE Data for Pure Elements," *Calphad* **15** 317-425 (1991).

13. G. Eriksson, and K. Hack, "ChemSage - A Computer Program for the Calculation of Complex Chemical Equilibria," *Metallurgical Transactions B,* **21B** 1013-23 (1990).

14. C.W. Bale, P. Chartrand, S.A. Degterov, G. Eriksson, K. Hack, R. Ben Mahfoud, J. Melançon, A.D. Pelton, S. Petersen, "FactSage thermochemical software and databases" *CALPHAD,* **26** 189-228 (2002).

15. M. Hillert, *Phase Equilibria, Phase Diagrams, and Phase Transformations*, Cambridge University Press, NY, p. 462 (1998).

16. B. Sundman, "An Assessment of the Iron-Oxygen System," *Journal of Phase Equilibria*, **12** 127-40 (1991).

18. M.S. Wang, B. Sundman, "Thermodynamic Assessment of the Mn-O System," *Metallurgical Transactions B*, **23B** 821-31 (1992).

19. A. D. Pelton, H. Schmalzried, and J. Sticher, "Thermodynamics of Mn_3O_4-Co_3O_4, Fe_3O_4-Mn_3O_4, and Fe_3O_4-Co_3O_4 Spinels by Phase Diagram Analysis," *Ber. Nunsenges. Phys. Chem.*, **83** 241-52 (1979).

20. A. Muan and C.L. Gee, "Phase Equilibrium Studies in the System Iron Oxide-Alumina in Air and at One Atmosphere Oxygen Pressure," *Journal of the American Ceramic Society*, **39** 207-14 (1956).

21. A. Muan and S. Somiya, "Phase Relations in the System Iron Oxide-Cr_2O_3 in Air," *Journal of the American Ceramic Society*, **43** 204-9 (1960).

22. A. Muan and S. Somiya, "Phase Equilibrium Studies in the System Iron Oxide-Al_2O_3-Cr_2O_3," *Journal of the American Ceramic Society*, **42** 603-13 (1959).

LIQUIDUS TEMPERATURE AND ONE PERCENT CRYSTAL CONTENT MODELS FOR INITIAL HANFORD HLW GLASSES

J.D. Vienna[1], T.B. Edwards[2], J.V. Crum[1], D.S. Kim[1], and D.K. Peeler[2]
[1]Pacific Northwest National Laboratory, Richland, Washington
[2]Savannah River National Laboratory, Aiken, South Carolina

ABSTRACT

Preliminary models for liquidus temperature (T_L) and temperature at 1 vol% crystal (T_{01}) applicable to WTP HLW glasses in the spinel primary phase field were developed. A series of literature model forms were evaluated using consistent sets of data for model fitting and validation. For T_L, the ion potential and linear mixture models performed best, while for T_{01} the linear mixture model out performed all other model forms. T_L models were able to predict with smaller uncertainty. However, the lower T_{01} values (even with higher prediction uncertainties) were found to allow for a much broader processing envelope for WTP HLW glasses.

INTRODUCTION

Roughly 200,000 m^3 of high-level waste (HLW) are stored at the Hanford site. This waste will be separated into HLW and low-activity waste (LAW) fractions, and each fraction will be immobilized for final storage/disposal. The U.S. Department of Energy (DOE) Office of River Protection (ORP) is constructing a Waste Treatment and Immobilization Plant (WTP) that will be capable of separating the waste, vitrifying the entire HLW fraction of the waste, and vitrifying roughly 50% the LAW fraction.[1,2]

There are process and product performance constraints that have to be met by each of these glasses for vitrification to be successful. One constraint, of interest in this paper, deals with avoiding crystal accumulation in the HLW melter. If solid phases accumulate in the HLW glass melter, then the glass discharge chamber may plug or cause processing difficulties. Therefore, constraints are imposed on glass composition to avoid the deleterious effects of crystal accumulation. The traditionally imposed constraint limits the liquidus temperature (T_L) of the melt to be below the nominal melt temperature by some margin.[3,4] Zero tolerance to crystalline materials in the Hanford HLW glass melt is not practically achievable because of insoluble phases such as noble metal oxides. In addition, a T_L constraint may be over-conservative because a small fraction of unsettling crystals can be harmlessly discharged from the melter.[5,6] Therefore, two separate constraints were considered in this study: a T_L constraint and the temperature at which the equilibrium crystal fraction is 1 vol% (T_{01}) constraint.

The composition of the initial HLW feeds (from Tanks AY-101/C-106, AZ-101, and AZ-102) is such that resulting borosilicate waste glasses will be within the spinel primary phase field. The spinel formed from these glass melts is a solid solution of the general composition $[Fe,Ni,Mn,Zn][Fe,Cr]_2O_4$.[7-9] A number of models have been developed to predict the T_L of HLW glass melts in the spinel primary phase field:

1. Linear mixture model (LMM) (i.e., a first-order expansion of T_L data in composition) described in general by Cornell[10] and applied to waste glass T_L by Hrma et al:[11]

$$T_L = \sum_{i=1}^{N} T_i g_i^n \tag{1}$$

2. Ion potential model (IPM) described by Vienna et al.:[8]

$$T_L = \sum_{Ni,Cr,Mn} T_i x_i + \sum_{Alk,AlkE} (a + bP_i)x_i + \sum_{remaining} (c + dP_i)x_i \tag{2}$$

3. Solubility-product model (SPM) based on trevorite (SPMT) described by Plodinec:[12]

$$T_L = \frac{1000}{a + b\ln(g_{Fe_2O_3} g_{NiO})} \tag{3}$$

4. SPM based on nichromite (SPMN) described by Plodinec:[12]

$$T_L = \frac{1000}{a + b\ln(g_{Cr_2O_3} g_{NiO})} \tag{4}$$

5. Sub-lattice model (SLM) described by Gan and Pegg:[13]

$$\frac{1}{T_L} = a + b\left[\frac{g_{NiO}}{g_{NiO} + g_{MnO_2} + 0.03g_{Fe_2O_3}}\right]\left(\frac{g_{Cr_2O_3}}{g_{Cr_2O_3} + 0.97g_{Fe_2O_3}}\right)^2 + \sum_{i=1}^{N} D_i g_i \tag{5}$$

6. Defense Waste Processing Facility (DWPF) T_L model (DWPFM) described by Brown et al.:[14]

$$\frac{1}{T_L} = a\ln(M2) + b\ln(M1) + c\ln(MT) + d \tag{6}$$

where,

a	is a fit parameter
b	is a fit parameter
c	is a fit parameter
d	is a fit parameter
D_i	is the ith glass component coefficient in units of inverse temperature
g_i	is the mass percent of the ith glass component
g_i^n	is the normalized mass fraction of the ith component in glass
$M1$	is a distorted 6 to 8 coordinated site parameter given by $\sum \varphi_{i,j} z_i$
$M2$	is an octahedral site parameter given by given by $\sum \varphi_{i,j} z_i$
MT	is a tetrahedral site parameter given by given by $\sum \varphi_{i,j} z_i$
P_i	is the ion potential of the ith electropositive ion
$\varphi_{i,j}$	is the ith element coefficient for the jth pyroxene site
T_i	is the ith glass component coefficient in units of T_L
T_L	is liquidus temperature (°C for Equations 1 to 4 and K for Equations 5 and 6)
x_i	is the mole fraction of the ith electropositive element in glass
z_i	is the concentration of the ith electropositive element in moles per 100 g of glass.

The parameters for each of these model forms were fitted and validated using a consistent Hanford HLW T_L data set to compare their ability to predict the T_L of glasses to be produced from the initial tanks of HLW to be treated at Hanford. They were also fitted and validated to a subset of the data with T_{01} values rather than T_L values to determine if a T_{01} constraint could be applied in the same manner as current T_L constraints. It should be noted that these models have not been specifically designed to predict T_{01}.

The details of data collection and splitting and model fitting and validating are reported elsewhere.[15] This paper summarizes these results as they are pertinent to a comparison of T_L and T_{01} constraints, but focuses on comparing the effectiveness of model use to control glass composition at Hanford using these constraints.

MODEL FITTING AND VALIDATION RESULTS

T_L Data Set

The T_L of 163 glasses in the spinel primary phase field within the composition region of interest to WTP was obtained from experiment in this study (25) and literature (138). These data had T_L values ranging from 780 to 1350°C with a roughly normal distribution (centered on 1063°C). Three of these data points were removed from the data set because they showed an unexplainable difference between the measured T_L and an extrapolation of temperature versus crystal fraction to zero percent (leaving 160 data points). The details of these data are in Vienna et al.[15]

The data set was sorted by T_L value, and then every fifth data point was removed for validation. Since about one-fifth of the data is to be used for validation, selecting these data starting at the first, second, third, fourth, or fifth values in the database provides five possible splitting outcomes. Each was used to fit and validate each of the six models to give the best possible evaluations of model performance over this database. Table I summarizes the results of model fitting and validation (mean over all datasets). The IPM and LMM out perform the other models. Either of these two models may be considered a better model over this data set, depending on which comparison metric in Table I is considered.

Table I. Average T_L Model Fit and Validation Results

Parameter	LMM	IPM	SPMT	SPMN	SLM	DWPFM
R^2	0.90	0.86	0.16	0.21	0.89	0.65
R^2_{val}	0.83	0.83	0.14	0.18	0.79	0.59
Avg Prediction Uncertainty	67	72	188	183	78	126
Avg Prediction Uncertainty, val	68	72	188	183	80	126
UVESS[a]	1.17	1.11	13.8	12.8	2.00	4.68
p[b]	17	7	2	2	18	4

(a) Uncertainty-weighted validation-error sum of squares (UVESS) is an estimate of the overall ability to predict, normalized so that a value of 1 gives the "best" predictions while higher values give progressively worse predictions.
(b) p is the number of fit parameters used in a given model.

T_{01} Data Set

The T_{01} data set was more limited. A total of 45 T_{01} data points in the spinel primary phase field within the composition region of interest to WTP were obtained from experiment in this study (14) and literature (31). Four of these data points showed an unexplainable difference between the measured T_L and an extrapolation of temperature versus crystal fraction to zero percent crystal and so were removed from the data set (leaving 41 data points). The coverage of the composition region of interest to WTP with these 41 glasses was less than desired for development of models in 18 compositional dimensions. These data had T_{01} values ranging from 633 to 1180°C. The details of these data are in Vienna et al.[15]

The data set was sorted to obtain five groups of model fit and validation sets as was done for T_L modeling. Each data set was used for fitting and validation of each of the six models to give the best possible evaluations of model performance over this database. Table II summarizes the results of model fitting and validation. The LMM clearly out performs the other models. The relatively poor fit to the T_{01} data is largely attributed to the lack of data.

Table II. Average T_{01} Model Fit and Validation Results

Parameter	LMM	IPM	SPMT	SPMN	SLM	DWPFM
R^2	0.88	0.64	0.34	0.20	0.89	0.64
R^2_{val}	0.75	0.054	0.41	0.047	-3.8	-0.27
Avg Prediction Uncertainty	130	195	297	313	208	214
Avg Prediction Uncertainty, val	148	200	294	310	774	228
UVESS	1.00	5.95	7.16	9.29	671	43.2
p	11	7	2	2	18	4

Model Comparison

It is interesting to compare the data and models for T_L and T_{01}. Figure 1 shows the differences between T_L and T_{01} for the 41 available data points within the WTP composition region. T_L-T_{01} ranged from 46 to 417°C, suggesting a composition effect on the difference. To investigate the composition influence on T_L-T_{01}, the LMM coefficients for T_L and T_{01} were estimated using all of the available data sets (e.g., without removing data for validation). These coefficients are listed in Table III and compared in Figure 2.

For most components (i) the coefficients (T_i) are very similar between the models. Surprisingly, the coefficient for Cr_2O_3 is higher, relatively, for T_{01} than for T_L. This was unexpected; since Cr_2O_3 is consumed by spinel growth from these glasses before 1 vol% crystal is formed. The effect of other components is similar to those determined in other studies (see references [4, 7, 8, and 11] for example).

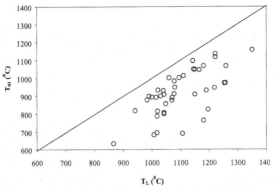

Figure 1. Comparison of T_{01} and T_L for Available Data

MODEL APPLICATION TO TYPICAL HANFORD HLW GLASSES

The implications of using either of these two constraints to limit crystal formation in the WTP HLW glass melter were evaluated to determine if further pursuit of a T_{01} model was warranted. T_L and T_{01} predictions for selected WTP HLW glasses were obtained using the LMMs with coefficients in Table III. Uncertainties (u) were estimated using simultaneous 90% confidence intervals. Figure 3 shows the predicted T_L and T_{01} for the current WTP baseline glasses for the first four HLW tanks and measured T_L for three glasses. The predicted T_L+u values are all above the 950°C limit and the predicted T_L values match the measured values within prediction uncertainty. With the exception of C-104 glass, the predicted T_{01}+u values are all below the 950°C limit. Although the T_{01} predictions have much higher prediction uncertainties, their values are low to make their predicted value plus uncertainty well below the 950°C limit. To see if this trend is valid for a broader set of possible WTP HLW glasses, predictions were made for a database of glasses that are to be made and characterized as part of an ongoing WTP glass property model-development task. These glasses were specifically designed to systematically cover the composition region of interest to WTP.[16] Figure 4 compares the predicted values for T_L and T_{01} for these glasses. Of the 102 glasses in this matrix, 95 would fail the constraint of T_L+u < 950°C while only 49 would fail the constraint of T_{01}+u < 950°C. Therefore, a tremendous benefit in increased available glass processing envelope could be achieved by using a T_{01}+u < 950°C constraint for WTP.

Table III. LMM Coefficients (T_i)

Comp.	T_L (T_i, °C)	T_{01} (T_i, °C)
Al_2O_3	2831.3	3391.7
B_2O_3	755.7	378.1
CdO	6240.6	
Cr_2O_3	25944.9	27121.9
F	5337.4	
Fe_2O_3	2759.1	3637.9
K_2O	-1211.1	
Li_2O	-2019.2	-2655.9
MgO	2233.8	
MnO	1862.0	2852.6
Na_2O	-827.1	-1786.5
NiO	9316.2	13169.6
P_2O_5	-3949.2	
SiO_2	862.7	393.8
SrO		-479.8
ThO_2	1766.9	
U_3O_8	2270.2	
ZrO_2	2122.2	4056.8
# of data	160	41
p	17	11
R^2	0.89	0.87
Adj R^2	0.88	0.83
RMSE (s)	32.2	53.5
Mean	1062	921

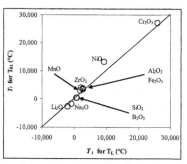

Figure 2. Comparison of LMM Coefficients for T_L and T_{01}

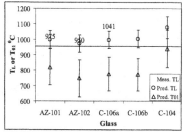

Figure 3. Comparison of Predicted T_L and T_{01} Values with Uncertainty

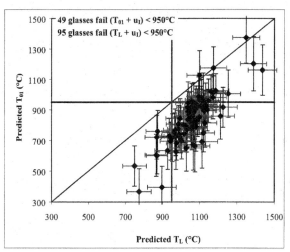

Figure 4. Comparison of T_L and T_{01} Predictions for WTP Matrix Glasses

CONCLUSIONS AND RECOMENDATIONS

Preliminary models for T_L and T_{01} applicable to WTP HLW glasses in the spinel primary phase field were fit. Several literature model forms were evaluated using consistent model fitting and validation datasets. For T_L, the IPM and LMM performed best, while for T_{01}, the LMM outperformed all other model forms. T_L models were able to predict with smaller uncertainty than T_{01} models. However, the lower T_{01} values (even with higher prediction uncertainties) will allow for a much broader processing envelope for WTP HLW glasses. It is recommended that additional T_{01} measurements are needed followed by further development of T_{01} models. The risk of operating the WTP HLW melter with T_{01} instead of T_L constraints should be weighed against the flexibility of the larger operating window.

ACKNOWLEDGMENTS

This study was partially funded by Bechtel National Inc, through U.S. Department of Energy (DOE) Contract DE-AC27-01RV14136. The authors thank Chris Musick (BNI) and Walt Tamosaitis (WGI) for support and guidance during this study. Greg Piepel and Scott Cooley (PNNL) designed the test matrix and developed LMMs for this study. Pacific Northwest National Laboratory is operated for the DOE by Battelle under Contract DE-AC06-76RL01830.

REFERENCES

[1]P.J. Certa, J.A. Reddick, J.O. Honeyman, and R.D. Wojtasek, *"River Protection Project System Plan"*, ORP-11242, Rev. 2, U.S. Department of Energy, Office of River Protection, Richland, WA (2003).

[2]DOE, *"Design, Construction, and Commissioning of the Hanford Tank Waste Treatment and Immobilization Plant"*, Contract Number: DE-AC27-01RV14136, as amended, U.S. Department of Energy, Office of River Protection, Richland, WA (2000).

[3]K.G. Brown, R.L. Postles, and T.B. Edwards, *"SME Acceptability Determination for DWPF Process Control"*, WSRC-TR-95-000364, Rev. 4, Westinghouse Savannah River Company, Aiken, SC (2002).

[4]P. Hrma, J.D. Vienna, and M.J. Schweiger, "Liquidus Temperature Limited Waste Loading Maximization for Vitrified HLW"; pp. 449-456 in *Ceramic Transactions 72*. American Ceramic Society. Westerville, OH, 1996.

[5]P. Schill, M. Trochta, J. Matyas, L. Nemec, and P. Hrma. "Mathematical Model of Spinel Settling in a Real Waste Glass Melter." In: *Waste Management '01*, University of Arizona, Tucson, AZ (2001).

[6]W.K. Kot, and I.L. Pegg, *"Glass Formulation and Testing With RPP-WTP HLW Simulants"*, VSL-01R2540-2, Vitreous State Laboratory, Washington, D.C. (2001).

[7]M. Mika, M.J. Schweiger, J.D. Vienna, and P. Hrma, "Liquidus Temperature of Spinel Precipitating High-Level Waste Glasses"; pp. 71-78 in *Scientific Basis for Nuclear Waste Management XX*. Materials Research Society. Pittsburgh, PA 1997.

[8]J.D. Vienna, P. Hrma, J.V. Crum, and M. Mika, "Liquidus Temperature-Composition Model for Multi-component Glasses in the Fe, Cr, Ni, and Mn Spinel Primary Phase Field", *Journal Non-crystalline Solids,* **292** (1-3):1-24 (2001).

[9]S. Annamalai, H. Gan, M. Chaudhuri, W.K. Kot, and I.L. Pegg, "Spinel Crystallization in HLW Glass Melts: Cation Exchange Systematics and the Role of Rh_2O_3 in Spinel Formation"; pp. 279-288 in *Ceramic Transactions 155*. American Ceramic Society. Westerville, OH 2004.

[10]J.A. Cornell, *Experiments with Mixtures*. 3rd ed., John Wiley & Sons, NY 2002.

[11]P. Hrma, G.F. Piepel, P.E. Redgate, D.E. Smith, M.J. Schweiger, J.D. Vienna, and D.-S. Kim, "Prediction of Processing Properties for Nuclear Waste Glasses"; pp. 505-513 in *Ceramic Transactions 61*. American Ceramic Society. Westerville, OH 1995.

[12]M.J. Plodinec,. "Solubility Approach for Modeling Waste Glass Liquidus", pp. 223–230 in *Scientific Basis for Nuclear Waste Management XXII*. Materials Research Society. Warrendale, PA 1999.

[13]H. Gan and I.L. Pegg, *"Development of Property-Composition Models for RPP-WTP HLW Glasses:*, VSL-01R3600-1, Vitreous State Laboratory, Washington, D.C. (2001).

[14]K.G. Brown, C.M. Jantzen, and G. Ritzhaupt, *"Relating Liquidus Temperature to Composition for Defense Waste Processing Facility (DWPF) Process Control"*, WSRC-TR-2001-00520, Westinghouse Savannah River Company, Aiken, SC (2001).

[15]J.D. Vienna, J.V. Crum, T.B. Edwards, D.E. Smith, J. Matyas, D.K. Peeler, G.F. Piepel, and S.K. Cooley, *"Liquidus Temperature Testing and Model Evaluation Results"*, PNWD-3369, WTP-RPT-085, Rev. 0, Battelle-Pacific Northwest Division, Richland, WA (2003).

[16]S.K. Cooley, G.F. Piepel, H. Gan, W.K. Kot, and I.L. Pegg, *"Augmentation Test Matrix to Support TCLP Model Development for RPP-WTP HLW Glasses"*, VSL-03S3780-1, Rev.1, Vitreous State Laboratory, The Catholic University of America, Washington, DC. (2003).

DEPENDENCY OF SULFATE SOLUBILITY ON MELT COMPOSITION AND MELT POLYMERIZATION

Carol M. Jantzen, Michael E. Smith, and David K. Peeler
Savannah River National Laboratory
Aiken, South Carolina 29808

ABSTRACT

Sulfate and sulfate salts are not very soluble in borosilicate waste glass. When sulfate is present in excess it can form water soluble secondary phases and/or a molten salt layer (gall) on the melt pool surface which is purported to cause steam explosions in slurry fed melters. Therefore, sulfate can impact glass durability while formation of a molten salt layer on the melt pool can impact processing. Sulfate solubility has been shown to be compositionally dependent in various studies, e.g. B_2O_3, Li_2O, CaO, MgO, Na_2O, and Fe_2O_3 were shown to increase sulfate solubility while Al_2O_3 and SiO_2 decreased sulfate solubility. This compositional dependency is shown to be related to the calculated melt viscosity at various temperatures and hence the melt polymerization.

INTRODUCTION

If the sulfate limit of a borosilicate glass is exceeded, the sulfate can form water soluble secondary phases and/or a molten salt layer (gall) on the melt pool surface. These sulfate salts, which are soluble, are often enriched in cesium and strontium, which can impact radionuclide release from the cooled glass if the salts are present as inclusions or a frozen gall layer [1]. The alkali and alkaline earth sulfate salts, in conjunction with alkali chlorides, can collect on the melt surface as a low melting (600-800°C), low density, and low viscosity melt phase. At moderate concentrations, the salts have a beneficial effect on melting rates [2, 3]. At excessively high feed concentrations, molten alkali sulfates float on the surface of the melt pool or become trapped as inclusions in the glass. The presence of this low viscosity (estimated to be ~1 centipoise at 1150°C) melt phase increases corrosion rates of the materials of construction (off-gas, refractories primarily at the melt line, and lid heaters due to splatter). The molten salt layer is purported to enhance the potential for steam explosions in waste glass melters that are slurry fed [4]. In addition, there is potential for undesirable current paths that could deplete energy delivered to the melter due to the electrical conductivity of the molten salt layer and the formation of corrosive off-gases [5].

In order to avoid the formation of sulfate inclusions and/or the formation of a molten sulfate rich phase on the melt pool in the Defense Waste Processing Facility (DWPF), a sulfate solubility limit has been imposed since DWPF startup in 1996. The sulfate limit is expressed as 0.59 wt% Na_2SO_4 which is equivalent to 0.4 wt% $SO_4^=$ in the vitrified waste form product. The $SO_4^=$ solubility limit in the glass represents the total sulfate that the glass can accommodate from both the liquid (Na_2SO_4) and solid ($CaSO_4$, $BaSO_4$, $PbSO_4$) fractions of High Level Waste (HLW) sludge and not form a layer or partial layer of molten salt on the melt pool. The complete absence of a molten salt layer on the melt pool is derived from the current DWPF safety basis which eliminates any potential for steam explosions.

Ferrous sulfamate, used as a reducing agent in the separation of plutonium from uranium, is the major source of sulfate in Savannah River Site (SRS) waste. The majority of the waste sulfate is water soluble and is removed from the HLW sludge solids during washing. Therefore, the wastes processed in DWPF since 1996 have had insignificant quantities of sulfates in them. However, the DWPF is preparing to vitrify Sludge Batch 3 (SB3) which may contain higher than normal sulfate levels. A large portion of this $SO_4^=$ is from ferrous sulfamate associated with the NpO_2^+ that will be added to SB3 directly from SRS separations after the sludge has already been prepared. The total amount of $SO_4^=$ in SB3 will be higher than the sulfate processed in any of the previous DWPF sludge batches and, when processed, may exceed the current DWPF limit for $SO_4^=$. Therefore, the limit for $SO_4^=$ was revisited in order to establish criteria for raising the limit without impacting safety.

BACKGROUND

Sulfate Saturation and Volatility

Sulfate solubility is difficult to determine because of supersaturation effects. Different researchers define sulfate solubility phenomenologically in terms of physical observations, e.g. vacuoles or inclusions in frozen glass generated in crucible or dynamic melter tests, complete or partial molten layer observations in melter tests, complete or partial frozen sulfate layers observed in quenched crucible tests. These phenomenological observations describe different "degrees of saturation" of the melt (or quenched melt).

Walker [4] was the first to describe the varying degrees of sulfate saturation in detail. Walker noted that in the presence of excess sulfate (when a heavy layer of gall was present) that the glass was "supersaturated" and Na_2SO_4 vacuoles formed in the glass. More sulfate was retained in the glass if it were in equilibrium with a layer of molten Na_2SO_4 than if the glass were in equilibrium with the gaseous SO_2 in the melter plenum, e.g. supersaturation could be induced by high p_{SO_2} and the formation of a layer of gall which inhibited SO_2 volatilization. Subsequently, the amount of saturation of the melt was determined to depend strongly on feed rate and reductant concentration in addition to sulfate concentration in the feed [6]. Faster feed rates were found to allow a molten salt layer to accumulate. Therefore, during this modeling effort, the phenomenological observations from different static crucible and dynamic melter tests were systematized based on the "degree of saturation" criteria given in Table I.

Saturation with respect to Na_2SO_4 is actually saturation with respect to a "mixed salt layer" since the salt contains chlorides, fluorides, chromates [7], and other sulfates, e.g. $CaSO_4$ [3,7,8].

Table I. Degree of Saturation Criteria

Degree of Saturation	Melt Pool/Glass Surface
Under	No gall present
At	No gall present
Over	Partial coverage of surface with gall
Super	Complete coverage of surface with gall and/or vacuoles or secondary phase observed in glass after cooling

Determination of sulfur saturation is complex because sulfur can volatilize by different reaction paths depending on the melt temperature, the sodium content, the REDuction/OXidation (REDOX) equilibrium, and the p_{SO_2} inside the melter plenum or crucible vapor space. The p_{SO_2} and the REDOX equilibrium combine to alter the type of sulfur species that vaporize and the amount of each species that vaporizes. For example, reducing REDOX conditions [9] and/or higher temperatures in a melter will force oxidized $SO_4^=$ species to the S^{2-} ion as $SO_2(g)$. If the system is open or has a low p_{SO_2} in the plenum or vapor space then the reaction shown in Equation 1 wants to progress to the right hand side (RHS) and liberate $SO_2(g)$. If the system is closed and there is a high p_{SO_2} in the plenum or vapor space then the equilibrium in Equation 1 is shifted to the left hand side (LHS) and the sulfate vaporization is inhibited. Likewise, oxidizing REDOX conditions [9] allow the $SO_4^=$ to decompose to the $SO_4^{2=}$ ion and vaporize as Na_2SO_4 which can condense in the melter off-gas line and be problematic [10,11]. If the system is open with respect to p_{SO_2} then Equation 2 proceeds to the RHS. This equilibrium to the RHS is accelerated if a melter is aggressively bubbled with oxygen or air [12]. However, if the p_{SO_2} is high in the plenum or vapor space then the release of Na_2SO_4 in Equation 2 is inhibited, and the equilibrium is forced to the LHS.

$$2Na_2SO4 \leftrightarrow 2Na_2O + 2SO_2\uparrow + O_2\uparrow \tag{1}$$
$$2Na_2SO_4 \leftrightarrow 2Na_2SO_4\uparrow \tag{2}$$

Moreover, higher temperatures, e.g. in the range of 1250-1400°C, and the addition of SiO_2 to a melt forces the decomposition of Na_2SO_4 to SO_2 per the Keppler reaction [13]

$$Na_2SO_4 + xSiO_2 \rightarrow Na_2O \bullet xSiO_2 + SO_2\uparrow + 0.5O_2 \tag{3}$$

The literature suggests that in conventional Joule heated melters (without bubbling) the sulfate volatility is between 40-70 wt% depending on REDOX, melt temperature, and melt viscosity. Sulfate volatility affords an extra margin of safety when setting a $SO_4^=$ solubility limit because no credit is taken for the volatility of $SO_4^=$ when the soluble and insoluble $SO_4^=$ concentration in the sludge is mathematically converted into the $SO_4^=$ glass solubility limit. The range of measured volatility based on various glasses, including those from nuclear waste glass studies, commercial glass studies, and even a mining waste study are very similar:

- ~75% of the Na_2SO_4 was vaporized in a pilot scale melter test at 1150°C with a high alumina containing glass having a viscosity of 160 poises [14]
- ~50% of the total sulfur (as S) was vaporized in a pilot scale melter demonstration [6]. Higher reductant content vaporized ~70% to the off-gas as SO_2 gas at $Fe^{+2}/\Sigma Fe$ REDOX ratios of 0.8, well above the $Fe^{+2}/\Sigma Fe$ limit of 0.33 to prevent nickel sulfide precipitation [15].
- ~ 45% of the total sulfate was vaporized in crucible tests when the mining waste was coupled with Frit 165 [16]
- 36-42% of the $SO_4^=$ is vaporized during routine commercial glass vitrification [14]
- ~40% of the $SO_4^=$ vaporized during Slurry-Fed Melt Rate Furnace (SMRF) testing reported in this study on SB3 feeds at a target REDOX of $Fe^{+2}/\Sigma Fe = 0.2$.
- ~55% of the Na_2SO_4 vaporized during pilot scale testing at 1150°C [3]

Previous Sulfate Solubility Modeling
A model for sulfate solubility was developed by Papadopoulos [17] in 1973 for soda-lime-silica melts. In this model, the number of bridging oxygen (O^0), free oxygen (O^{2-}), and non-bridging oxygen (O^-) are related to the dissociation of $SO_4^=$ by the equilibrium $O^0 + O^{2-} \leftrightarrow 2O^-$ and $SO_4^{2-} \leftrightarrow SO_3 + O^{2-}$. Combining the equilibrium constants (K_A and K_B) of these two

equations provides the relationship $[SO_4^{2-}] \propto \dfrac{P_{SO_3}[O^-]^2}{K_A K_B [O^0]}$ where P_{SO_3} is the partial pressure of

SO_3 in the melter atmosphere and $[SO_4^{2-}]$ is the sulfate solubility in the glass. Papadopoulos defined a linear relationship between the chemical composition parameter $[O^-]^2/[O^0]$, which is also known as the ratio of [non bridging oxygen]^2/[bridging oxygen] or $[NBO]^2/[BO]$, and $[SO_4^{2-}]$ retained in various commercial glasses when the melt temperature and the P_{SO_3} above the melt were constant. The dependency of the sulfate solubility on $[O^-]^2/[O^0]$ was confirmed by Ooura and Hanada [18] in 1998 for a series of alkali-silicate and alkaline earth-lime-silica glasses including Ba, Sr, Pb, Ca, Mg, and Zn species. The $[O^0]$ concentration was calculated as twice the SiO_2 mol% minus the alkali oxide (mol%) while the $[O^-]$ was calculated twice the R_2O (mol%) of the glass. Ooura and Hanada defined a linear relationship between the $[O^-]^2/[O^0]$ glass composition term and sulfate solubility. The sulfate solubility increased as the alkali content of the melt increased.

In 2001, Li et.al. [19] modified the Papadopoulos $[O^-]^2/[O^0]$ parameter to include many of the species found in simulated nuclear waste glasses, e.g. B_2O_3, Fe_2O_3, Al_2O_3. Li assumed that B, Fe, and Al were all network formers, e.g they formed $NaBO_3$, $NaFeO_2$, and $NaAlO_2$ structural groups, in order to predict sulfate solubility in waste glasses being considered for stabilization of Hanford Low Activity Waste (LAW). Use of the modified NBO term in the Li study caused the relationship between $[O^-]^2/[O^0]$ composition term and the sulfate solubility to be parabolic instead of linear. In addition, the P_{SO_3} of the melts studied was not controlled nor considered. Two distinct trends were observed when the data was modeled although the sulfate solubility did increase as the alkali content of the melt increased as found in previous studies [17,18].

Another empirical sulfate solubility model known as the "rule of five" was developed for Hanford LAW waste glasses [20]. This empirical model suggests that sulfate solubility decreases with increasing alkali content, a trend completely opposite from all previous studies [17,18,19]. This empirical model is based on the wt% of SO_3 ($w_{max}^{SO_3}$) and the wt% of Na_2O

(w_{Na_2O}) in the glass and is expressed as $w_{max}^{SO_3} = \dfrac{5}{w_{Na_2O}}$. The waste glasses modeled by Li [19] and

Pegg [20] contained only ≤ 2.5wt% Fe_2O_3 compared to ~12 wt% in DWPF waste glasses.

SULFATE SOLUBILITY MODELING AND VALIDATION DATABASES

A broad range of literature and experimental data on sulfate solubility was surveyed (see Table II) to construct a sulfate solubility versus glass composition database that included both LAW glasses with low Fe_2O_3 content and High Level Waste (HLW) waste glasses with high Fe_2O_3 contents. The data surveyed included both crucible and pilot scale melter tests on both SRS HLW glasses [1,3,4,7,8,14,21,22] Hanford LAW glasses [6,33,34,35], and European intermediate level nuclear waste glasses [23]. All of the glasses were classified as either at or under saturation with respect to sulfate if the melt pool surface in a pilot scale melter and or the solidified glass surface observed in a crucible had no sulfate deposits or gall present. Glasses were classified as supersaturated with respect to sulfur if there were heavy surface deposits and gall present and sulfur vacuoles and or secondary phases in the bulk glass. Glasses were classified as over saturated if some surface sulfate deposits were observed and there were no vacuoles or secondary sulfate phases in the bulk glass (see Table I). This classification partially addresses the varying P_{SO_3} experienced during experimentation due to different amounts of melt pool coverage by gall.

If the literature did not adequately describe the visual appearance of the melt pool, glass surface, and/or bulk glass [21], if the literature did not adequately measure the sulfate in the final glass [1,22], or if the literature study overwhelmed the melt system with reductant [6] at REDOX values unachievable in Joule heated melters, e.g. $Fe^{+2}/\Sigma Fe > 0.33$, then the data were excluded from the modeling database. All of the data given in Table II met the glass description criteria, the REDOX criteria, and sulfate measurement criteria. Note that only borosilicate glasses are included in the current evaluation. Details of all of the data surveyed are given in Reference24.

It is noteworthy that about half of the sulfate solubility data given in Table II is data from pilot scale melters. Most notably, between 1983 and 1984 three Engineering Scale Ceramic Melter (ESCM) pilot scale campaigns were performed at the Pacific Northwest National Laboratory (PNNL). Each melter campaign was 10 days and processed average composition SRS waste (Stage 1) mixed with Frit 165 [8]. The $SO_4^=$ in the feed was from both soluble Na_2SO_4 and insoluble $CaSO_4$. The data from the ESCM-3B campaign is the basis for the current DWPF sulfate glass limit of 0.4 wt% $SO_4^=$.

Additional sulfate solubility data was developed in a Slurry-Fed Melt Rate Furnace (SMRF) with simulated DWPF SB3. The first test (SMRF-124) was performed at 31 wt% waste loading while the second test (SMRF-125) was performed at 35 wt% waste loading. The target sulfate concentration in the feeds were 0.47 and 0.52 wt% $SO_4^=$ on a calcined oxide basis, respectively. The resulting measured sulfate in the glasses formed were 0.29 and 0.53 wt% $SO_4^=$, respectively. No sulfate was visually observed on the melt pool (see Table II).

The validation data was generated at SRS by Peeler et. al. [25] and is summarized in Table III. In these studies DWPF SB3 compositions were tested in crucibles and in the Slurry-fed Melt Rate Furnace (SMRF). The criteria was to define the maximum $SO_4^=$ in the glass with no visual observation of gall on the glass surface. In the crucible study both batch chemicals (bc) and precipitated sludge were used. There was no reductant in the batch chemical tests but the precipitated sludge was made at a target REDOX of $Fe^{+2}/\Sigma Fe$ of 0.2 (adjusted with a combination of formic and nitric acids). The crucibles were sealed with nepheline gel which inhibited vaporization of $SO_2(g)$ and imparted a high p_{SO_2} in the vapor space of the crucibles thereby forcing Equations 1 and 2 to the left and inhibiting $SO_2(g)$ vaporization. In the SMRF studies precipitated sludge made at a REDOX target of 0.2 was also used but the SMRF is open to $SO_2(g)$ vaporization.

SULFATE SOLBUILITY AS A FUNCTION OF GLASS POLYMERIZATION

The literature surveyed while compiling the databases maintained that many individual oxides had an impact on sulfate solubility in glass but that overall compositional dependency of sulfate solubility had here-to-fore not been determined. Melt melt de-polymerizers like B_2O_3 [1,34], Li_2O [26], CaO and MgO [34,27], Na_2O [4,18 , 28], and Fe_2O_3 [4] were all shown to increase sulfate solubility while Al_2O_3 [4,17] and SiO_2 [13,17], melt polymerizers, decreased sulfate solubility [4,28].

The glass species cited in the above studies are the predominant polymerization (bridging oxygens) and depolymerization (non-bridging oxygen) terms in the DWPF viscosity model [29] given by

$$\log \eta(poise) = -0.61 + \left(\frac{4472.45}{T(^{\circ}C)}\right) - (1.534 * NBO) \tag{4}$$

where NBO = $\dfrac{2 \ (Na_2O + K_2O + Cs_2O + Li_2O + Fe_2O_3 - Al_2O_3) + B_2O_3}{SiO_2}$

The DWPF viscosity model assumes that a pure SiO_2 glass is fully polymerized and that each mole of alkali oxide added creates two non-bridging oxygen bonds, e.g. depolymerizes the glass. Each mole of Al_2O_3 creates two bridging oxygen bonds (polymerizes the glass structure) by creating tetrahedral alumina groups that bond to the $NaAlO_2$ structural groups as in the Li et. al. [19] model. In Al_2O_3 and/or SiO_2 deficient glasses, Fe_2O_3 can take on a tetrahedral coordination and polymerize a glass by forming $NaFeO_2$ structural groups as assumed in the Li et. al. [19] model. However, if sufficient Al_2O_3 and SiO_2 are present in a glass, Fe_2O_3 is octahedral and creates two non-bridging oxygen bonds, e.g. it depolymerizes the glass matrix as assumed in the DWPF viscosity model (Equation 4). This is consistent with the work of Mysen [30] who demonstrated that in high iron magmas (iron silicate glasses) at levels of 10 wt% that Fe_2O_3 decreased the melt viscosity. He concluded that $NaFeO_2$ structural groups were not incorporated into the glass network to the same degree as $NaAlO_2$ structural groups [30]. Lastly, the DWPF viscosity model assumes that each mole of B_2O_3 creates one non-bridging oxygen bond. This is based on a data by Smets and Krol [31], and Konijnendijk [32] who demonstrated that for sodium silicate glasses with low B_2O_3 content that B_2O_3 enters the glass network as BO_4^- tetrahedra that contribute no NBO while at higher concentrations these tetrahedra are converted into planar BO_3^- groups that contribute one non-bridging oxygen atom. The latter is assumed in the DWPF viscosity model.

The DWPF viscosity is model is used in this study rather than the NBO term since the relation of viscosity to glass structure is temperature dependent and the DWPF viscosity model is a three dimensional spline fit which includes NBO, log viscosity, and temperature. This is significant for modeling sulfate solubility as a function of viscosity since glasses of varying melt temperatures were modeled. In addition, the sulfate solubility boundary was determined to be temperature dependent [33].

The compositions given in Table II were used to calculate the glass viscosity used in the sulfate solubility model as given in Equation 4. If an alternate NBO term is used that includes a CaO term and it is assumed that a mole of CaO creates two non-bridging oxygen bonds as does a mole of alkali oxide, then the model fit, in terms of R^2, is greatly improved. For brevity only the dependency of the sulfate solubility on the DWPF viscosity model is discussed. The usage of a CaO term is discussed in detail elsewhere [24].

Modeling of the sulfate solubility as a function of calculated viscosity from Equation 4 was performed. The glasses were grouped by sulfate saturation based on the definitions given in Table I. This provided a series of three parallel models, one at saturation, one at over saturation, and one at supersaturation as shown in

Figure 1 and given below:

$(SO_4^= wt\%)_{\text{"at saturation"}} = 1.299 - 0.5501 \log \text{viscosity}_{calc}$ (poise) $R^2 = 0.88$ (5)

$(SO_4^= wt\%)_{\text{"over satuation"}} = 1.7810 - 0.5650 \log \text{viscosity}_{calc}$ (poise) $R^2 = 0.76$ (6)

$(SO_4^= wt\%)_{\text{"supersatuation"}} = 2.016 - 0.4681 \log \text{viscosity}_{calc}$ (poise) $R^2 = 0.90$ (7)

Table II. Glass Compositions (Wt%), Sulfate Solubility (Wt%), and Calculated Viscosity (Poise) For Model Data

Glass ID/Degree of Saturation	TYPE	Temp (°C)	Ref	Al₂O₃	B₂O₃	CaO	Cr₂O₃	CuO	Fe₂O₃	K₂O	Li₂O	MgO	MnO	Na₂O	NiO	SiO₂	TiO₂	ZrO₂	Sum	SO₄⁼ Solub.	Log Visc
DYNAMIC MELTER TESTS AND SLURRY-FED MELT RATE (SMRF) TESTS																					
Walker Melter F131-Highest S/SUPER	HLW SRS	1150	4	3.68	10.58	1.18	0.00	0.00	16.91	0.00	4.10	1.44	3.66	13.33	1.97	41.69	0.72	0.00	99.26	1.21	1.08
Walker Melter F131 – High S/OVER	HLW SRS	1150	4	3.68	10.58	1.18	0.00	0.00	16.91	0.00	4.10	1.44	3.66	13.33	1.97	41.69	0.72	0.00	99.26	0.67	1.08
Walker Melter F131 – Lowest S /AT	HLW SRS	1150	4	3.68	10.58	1.18	0.00	0.00	16.91	0.00	4.10	1.44	3.66	13.33	1.97	41.69	0.72	0.00	99.26	0.47	1.08
Hull 1982 PNNL PSCM-in glass/AT	HLW SRS	1150	14	15.00	7.30	0.60	0.00	0.00	4.00	0.00	5.10	0.70	2.70	11.09	0.60	52.10	0.00	0.70	99.89	0.06	2.29
Hull 1982 PNNL PSCM-5/AT	HLW SRS	1150	3	5.10	7.20	1.70	0.00	0.00	12.70	0.00	5.00	0.70	2.90	10.34	0.80	52.80	0.00	0.70	99.94	0.25	1.83
SCM-2 Noble Metals Test/AT	HLW SRS	1150	7	4.08	10.34	1.33	0.26	0.16	11.90	3.19	3.14	1.21	2.87	12.75	1.07	46.47	0.02	0.00	98.80	0.38	1.48
Hull ESCM-3B/AT	HLW SRS	1150	8	5.15	7.28	1.24	0.10	0.12	10.77	0.10	5.10	0.73	2.43	11.31	0.64	52.84	0.00	0.73	98.55	0.36	1.80
Hull ESCM-3A/AT	HLW SRS	1150	8	5.18	7.33	1.25	0.10	0.12	10.84	0.00	5.13	0.73	2.45	11.19	0.65	53.18	0.00	0.73	98.89	0.19	1.82
Smith – SMRF-124/AT	SB3 F418	1150	24	6.24	6.07	1.20	0.180	0.08	12.35	0.24	5.11	1.02	2.24	13.25	0.44	53.00	0.02	0.19	101.60	0.29	1.72
Smith – SMRF-125/AT	SB3 F418	1150	24	7.81	6.77	1.35	0.20	0.08	13.75	0.38	4.43	1.30	2.21	14.30	0.37	46.80	0.00	0.24	99.96	0.534	1.54
OPEN CRUCIBLE TESTS																					
Walker TDS-3A/AT	HLW SRS	1150	4	3.16	10.58	1.16	0.00	0.00	15.69	0.00	4.10	1.44	3.68	14.36	1.93	43.03	0.72	0.00	99.85	0.60	1.08
Walker W-Al/OVER	HLW SRS	1150	4	16.28	10.58	0.32	0.00	0.00	4.55	0.00	4.10	1.44	3.04	15.00	0.67	43.16	0.72	0.00	99.85	0.70	1.89
Walker W-Fe/AT	HLW SRS	1150	4	0.44	10.58	1.29	0.00	0.00	19.11	0.00	4.10	1.44	1.05	15.23	3.26	42.63	0.72	0.00	99.85	1.00	0.79
Crichton-Tomozawa//SUPER	LAW	1100	33	12.00	5.00	4.00	0.04	0.00	0.01	0.33	0.00	0.00	0.00	20.00	0.00	56.78	0.00	0.01	98.15	0.71	2.66
LRM-1/OVER	LAW	1334	34	12.00	2.00	2.00	0.14	0.00	6.00	0.03	1.00	0.00	0.01	20.00	0.00	48.54	0.00	4.00	95.72	0.91	1.64
LRM-5412/OVER	LAW	1356	34	12.00	5.00	4.00	0.04	0.00	0.00	0.03	0.00	0.00	0.01	20.00	0.00	54.54	0.00	0.00	95.62	0.83	1.87
LRM-3/OVER	LAW	1440	34	12.00	6.00	0.00	0.04	0.00	0.00	0.03	0.00	0.00	0.01	20.00	0.00	51.54	0.00	6.00	95.62	0.71	1.61
LRM-4/SUPER	LAW	1140	34	10.00	6.00	6.00	0.04	0.00	0.00	0.03	0.50	0.00	0.01	20.00	0.00	43.04	0.00	4.00	95.62	1.116	1.93
LRM 6-5412/AT	LAW	1350	35	12.00	5.00	4.00	0.00	0.00	0.00	0.00	0.00	0.00	0.00	20.00	0.00	56.78	0.00	0.00	97.78	0.35	1.92
LRM 6-112SP/SUPER	LAW	1350	35	11.76	4.90	3.92	0.00	0.00	0.00	0.00	0.00	0.00	0.00	19.60	0.00	55.63	0.00	0.00	95.81	1.056	1.92
LRM 5-012SP/SUPER	LAW	1350	35	11.76	8.82	0.00	0.00	0.00	0.00	0.00	0.00	0.00	0.00	20.00	0.00	55.63	0.00	0.00	95.81	1.21	2.04
LRM 4-9012/AT	LAW	1350	35	12.00	9.00	0.00	0.00	0.00	0.00	0.00	0.00	0.00	0.00	20.00	0.00	56.78	0.00	0.00	97.78	0.35	1.83
LRM 4-012SP/OVER	LAW	1350	35	11.76	8.82	0.00	0.00	0.00	0.00	0.00	0.00	0.00	0.00	19.60	0.00	55.63	0.00	0.00	95.81	0.68	1.83
LRM 4-909SP/OVER	LAW	1350	35	8.82	8.82	0.00	0.00	0.00	0.00	0.00	0.00	0.00	0.00	19.60	0.00	58.57	0.00	0.00	95.81	0.82	1.78

Table III. Glass Compositions (Wt%), Sulfate Solubility (Wt%), and Calculated Viscosity (Poise) For Validation Data

Glass ID/Degree of Saturation	TYPE	Temp (°C)	Ref	Al_2O_3	B_2O_3	CaO	Cr_2O_3	CuO	Fe_2O_3	K_2O	Li_2O	MgO	MnO	Na_2O	NiO	SiO_2	TiO_2	ZrO_2	Sum	$SO_4^=$ Solub.	Log Visc
SLURRY-FED MELT RATE (SMRF) TESTS																					
SMRF-124/AT	SB3	1150	25	6.6	5.035	1.44	0.103	0.071	14.25	0.06	4.78	1.53	2.22	14	0.411	47.1	0.012	0.13	97.742	0.76	1.53
SEALED CRUCIBLE TESTS																					
s-bc-100-30/AT	SB3	1150	25	5.77	5.23	1.16	0.05	0.05	10.86	0	5.51	1.27	1.96	12.26	0.36	53.22	0	0.1	97.8	0.61	1.78
s-bc-100-33/AT	SB3	1150	25	5.69	5.39	1.25	0.07	0.06	12.5	0	5.44	1.38	2.15	12.69	0.41	51.08	0	0.11	98.22	0.7	1.66
s-bc-100-35/AT	SB3	1150	25	6.58	5.18	1.31	0.09	0.06	13.21	0	5.19	1.44	2.25	13.07	0.42	49.58	0	0.11	98.49	0.73	1.64
s-bc-100-37/AT	SB3	1150	25	6.7	5.11	1.39	0.08	0.07	14.09	0	5.07	1.52	2.36	13.46	0.44	47.92	0	0.12	98.33	0.77	1.56
s-bc-100-40/AT	SB3	1150	25	7.22	4.82	1.46	0.09	0.08	15.04	0	4.76	1.61	2.51	13.78	0.45	45.94	0	0.14	97.9	0.72	1.51
s-bc-50-30/AT	SB3	1150	25	5.34	5.48	1.09	0.06	0.07	11.47	0	5.56	1.35	2	12.03	0.41	56.08	0	0.11	101.05	0.49	1.84
s-bc-50-33/AT	SB3	1150	25	5.89	5.57	1.17	0.07	0.07	13.02	0	5.35	1.45	2.14	12.32	0.43	53.47	0	0.11	101.06	0.54	1.76
s-bc-50-35/AT	SB3	1150	25	6.6	5.05	1.27	0.09	0.07	13.41	0	5.11	1.49	2.24	12.79	0.46	52	0	0.12	100.7	0.59	1.74
s-bc-50-37/AT	SB3	1150	25	6.38	5.09	1.32	0.09	0.08	14.15	0	5.01	1.58	2.33	12.98	0.47	49.83	0	0.13	99.44	0.59	1.65
s-bc-50-40/AT	SB3	1150	25	7.03	4.67	1.48	0.09	0.09	15.65	0	4.74	1.7	2.58	13.61	0.48	47.53	0	0.14	99.79	0.64	1.57
s-bc-50-48/AT	SB3	1150	25	8.6	4.2	1.74	0.11	0.1	18.53	0	4.19	2.02	3.05	14.65	0.6	41.88	0	0.17	99.84	0.66	1.35
s-sp-100-30/AT	SB3	1150	25	5.63	5.67	1.06	0.08	0.05	12.35	0	5.38	1.27	1.76	11.48	0.42	54.77	0	0.22	100.14	0.49	1.84
s-sp-100-33/AT	SB3	1150	25	6.64	5.43	1.18	0.08	0.07	13.31	0	5.23	1.45	2	11.94	0.47	52.73	0	0.24	100.77	0.51	1.79
s-sp-100-35/AT	SB3	1150	25	6.8	5.26	1.22	0.09	0.05	14.23	0	4.99	1.47	2.05	12.2	0.49	51.49	0	0.26	100.6	0.54	1.76
s-sp-100-37/AT	SB3	1150	25	6.88	4.85	1.26	0.09	0.05	14.92	0	4.65	1.53	2.14	12.31	0.5	50.69	0	0.27	100.14	0.53	1.77
s-sp-100-40/AT	SB3	1150	25	7.59	4.9	1.37	0.1	0.06	16.31	0	4.72	1.61	2.28	12.68	0.55	48.59	0	0.3	101.06	0.47	1.66
s-sp-100-42/AT	SB3	1150	25	7.94	4.82	1.47	0.11	0.06	17.07	0	4.68	1.78	2.5	13.13	0.59	45.96	0	0.31	100.42	0.66	1.55
s-sp-100-44/AT	SB3	1150	25	8.13	4.88	1.51	0.11	0.06	17.77	0	4.64	1.81	2.55	13.13	0.61	44.99	0	0.32	100.51	0.72	1.50
s-sp-50-30/AT	SB3	1150	25	6.3	5.4	0.98	0.08	0.07	12.07	0	5.3	1.23	1.68	11.55	0.4	54.9	0	0.22	100.18	0.38	1.88
s-sp-50-33/AT	SB3	1150	25	6.05	5.24	1.12	0.09	0.08	13.3	0	5.18	1.43	1.91	11.79	0.47	52.74	0	0.25	99.65	0.49	1.79
s-sp-50-35/AT	SB3	1150	25	6.34	5.06	1.15	0.09	0.08	13.99	0	5.09	1.44	1.99	11.92	0.47	51.55	0	0.26	99.43	0.46	1.76
s-sp-50-37/AT	SB3	1150	25	6.78	4.89	1.26	0.09	0.08	14.78	0	4.89	1.54	2.15	12.31	0.48	49.39	0	0.27	98.91	0.48	1.70
s-sp-50-40/AT	SB3	1150	25	7.39	4.8	1.37	0.1	0.09	15.92	0	4.75	1.63	2.3	12.73	0.52	48.26	0	0.29	100.15	0.5	1.65
s-sp-50-45/AT	SB3	1150	25	8.38	4.57	1.52	0.11	0.1	17.82	0	4.43	1.84	2.59	13.35	0.6	44.31	0	0.33	99.95	0.56	1.51

The "at saturation" model is based on 12 data points most of which are pilot scale melter tests and SMRF tests, only one crucible study is included. The melter runs span well beyond current DWPF viscosity limits of 20 and 110 poise since melter testing in the early 1980's was performed to define the DWPF viscosity operational limits. The "supersaturation model" is based on 5 data points and also includes melter and crucible studies that span well beyond the DWPF viscosity limits of 20 and 110 poise. The "over saturation" model (Equation 6) is more difficult to define since it includes variable sulfate layer coverage and is developed primarily on Hanford glass compositions (5 data points) and only 2 DWPF glass compositions. Both melter and crucible data are included. Thus the "over saturation" model (Equation 6) has the lowest R^2 goodness of fit.

It can be argued that Equation 5 is very conservative in that it is an "at saturation" to just under saturated sulfate solubility limit in glass that is applied to the feed composition. Hence, the glass solubility limit does not account for sulfate volatility in a melter which has been shown to be ~40% for melts of DWPF as well as for melts of different composition, viscosity and REDOX.

Therefore, a recommendation was made that a melter operational limit be set at the upper 95% confidence (U95) of the individual predicted $SO_4^=$ values representing "at saturation." The U95 limit remains extremely conservative compared to both the over saturated and the supersaturated sulfate in glass solubility models while allowing more $SO_4^=$ to be processed in the DWPF melter given the safety basis constraint of no visible sulfate layer or partial sulfate layer [24].

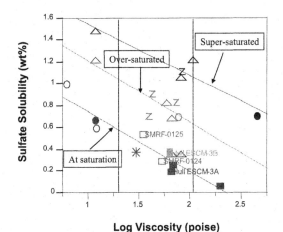

Figure 1. Model data used to define three separate sulfate saturation models as a function of melt polymerization and temperature, e.g. melt viscosity (poise). The DWPF viscosity limits of 20 poise (1.3 log viscosity) and 110 poise (2.04 log viscosity) are shown for reference.

The usage of Equation 5 as a conservative "at saturation" sulfate model was verified with the SB3 specific SMRF data which represented an open system, e.g. open to volatilization, and with sealed crucible data [25]. In the closed crucible tests the data given in Table III was adjusted for

40% vaporization to simulate the volatilization experienced in all the open system tests modeled in this study (see Table II). While no volatilization was observed of the tests run with batch chemicals (samples in Table III that include the letters bc) there was some minor volatilization of the sulfur in the tests run with SRAT product (sample identifications in Table III include the letters sp). Recognizing that application of the 40% volatilization factor for the sealed crucible tests is an approximation, this does demonstrate that the SMRF and crucible data for SB3 validates the "at saturation" sulfate solubility correlation (Figure 2). If the closed crucible data is not adjusted for volatilization then it validates the U95 of Equation 5 verifying the operational limit recommended to avoid sulfate accumulation in a waste glass melter [24].

The sulfate solubility-melt viscosity model presented in this study is linear like the Papadopoulos [17] model. A regression of the NBO term from Equation 4 with a CaO and an MgO term to a modified Papadopoulos $[O^-]^2/[O^0]$ term to account for the effects of Al_2O_3, Fe_2O_3 and B_2O_3 was linear $(NBO)=0.18+2.12*[O^-]^2/[O^0]$ with a correlation coefficient of 0.95. This correlation was based on 62 data points which included the data from Table II and Table III and the original 13 data points from Papadopoulos. The empirical Hanford sulfate model known as the "rule of five" [20] performed poorly with all the data modeled, e.g. R^2 values of 0.16, 0.36, and 0.56 for the at saturation, over saturation and supersaturation data presented in this study.

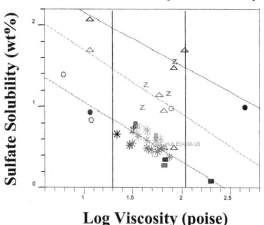

Log Viscosity (poise)

Figure 2. Sulfate solubility models with validation data for the "as-saturated" correlation (Equation 5) overlain. Note that the sealed crucible sulfate solubilities were adjusted for the lack of volatilization as noted in the text.

CONCLUSIONS

Sulfate solubility in waste glasses can be modeled as a function of melt polymerization and temperature through the compositionally and temperature dependent DWPF melt viscosity model. This modeling approach was validated in this study and shown to be linear like the Papadopoulos [17] and Ooura and Hanada [18] melt polymerization vs. sulfate models rather than parabolic like the Li et. al. model [19]. Sulfate solubility increases with additional alkali in the models presented in this study which include data from glasses developed for HLW and

LAW wastes in agreement with previous studies [17,18,19] but in disagreement with the empirical Hanford "rule of five" sulfate solubility model.

REFERENCES

1. M.J. Plodinec and J.R. Wiley, "Evaluation of Glass as a Matrix for Solidifying Savannah River Plant Waste: Properties of Glasses Containing Li_2O," U.S. DOE Report DP-1498, E.I. duPont deNemours & Co., SRL, Aiken, SC 29808 (February 1979).
2. D.F. Bickford, P. Hrma, B.W. Bowan and P.K. Smith, "Control of Radioactive Waste Glass Melters: II. Foaming and Melt Rate Limits," *Journal of the American Ceramic Society,* **73** [10] 2903-15 (1990).
3. H.L. Hull, "Trip Report Battelle-PNL Slurry-Fed Melter Test May 17-21, 1982," U.S. DOE Report DPST-82-718, E.I. duPont deNemours & Co., SRL, Aiken, SC 29808 (November 1982).
4. D.D. Walker, "Sulfate Solubility in DWPF Glass," U.S. DOE Report DPST-86-546, E.I. duPont deNemours & Co., SRL, Aiken, SC 29808 (October 1986).
5. D.F. Bickford, A. Applewhite-Ramsey, C.M. Jantzen, and K.G. Brown, "Control of Radioactive Waste Glass Melters: I, Preliminary General Limits at Savannah River," *Journal of the American Ceramic Society,* **73** [10] 2896-2902 (1990).
6. J.G. Darab, D.D. Graham, B.D. MacIssac, R.L. Russell, H.D. Smith, and J.D. Vienna, "Sulfur Partitioning During Vitrification of INEEL Sodium Bearing Waste: Status Report," U.S. DOE Report PNNL-13588, PNNL, Richland, WA (July 2001).
7. P.G. Walker, "Melter Performance During the Second SCM-2 Precipitate Hydrolysis Test," U.S. DOE Report DPST-85-574, E.I. duPont deNemours & Co., SRL, Aiken, SC 29808 (August 1985).
8. H.L. Hull, "Trip Report Battelle-PNL Slurry-Fed Melter Tests June, August, October, 1983," U.S. DOE Report DPST-84-518, E.I. duPont deNemours & Co., SRL, Aiken, SC 29808 (May 1984).
9. H.D. Schreiber, S.J. Kozak, P.G. Leonhard, and K.K. McManus, "Sulfur Chemistry in a Borosilicate Melt, Part I. REDOX Equilibria and Solubility," *Glastech. Ber.* **60** [12], 389-398 (1987)
10. C.M. Jantzen, "Characterization of Off-Gas System Pluggages, Significance for DWPF and Suggested Remediation," U.S. DOE Report WSRC-TR-90-205, Westinghouse Savannah River Company, Aiken, SC 29808 (1991).
11. C.M. Jantzen, "Glass Melter Off-gas System Pluggages: Cause, Significance, and Remediation," pp. 621-630 in *Proceedings of 5th International Symposium on Ceramics in Nuclear Waste Management,* Edited by G.G. Wicks et. al. American Ceramic Society, Westerville, OH, 1991.
12. P. Hrma, J.D. Vienna, W.C. Buchmiller and J.S. Ricklefs, "Sulfate Retention during Waste Glass Melting," pp. 93-99 in *Environmental Issues and Waste Management Technology IX.* Edited by J.D. Vienna and D.R. Spearing. American Ceramic Society, Westerville, OH, 2004.
13. M.B. Volf, pp. 524-543 in *Chemical Approach to Glass,* Elsevier Publishing Company, New York, 1984.
14. H.L. Hull, "Trip Report Battelle-PNL Slurry-Fed Melter Test June 14-23, 1982," U.S. DOE Report DPST-82-1008, E.I. duPont deNemours & Co., SRL, Aiken, SC 29808 (November 1982).
15. C.M. Jantzen, J.R. Zamecnik, D.C. Koopman, C.C. Herman, and J.B. Pickett, "Electron Equivalents Model for Controlling REDuction/OXidation (REDOX) Equilibrium During High Level Waste (HLW) Vitrification," U.S. DOE Report WSRC-TR-2003-00126, Westinghouse Savannah River Company, Aiken, SC 29808 (May 2003).
16. C.M. Jantzen, R.F. Schumacher, and J.B. Pickett, "Mining Industry Waste Remediated for Recycle by Vitrification," pp. 65-74 in *Environmental Issues and Waste Management. Technologies VI.* Edited by D.R. Spearing, et. al. American Ceramic Society. Westerville, OH, 2001.
17. K. Papadopoulos, "The Solubility of SO_3 in Soda-Lime-Silica Melts," *Phys. Chem. Glasses,* **14**[3], pp., 60-65 (1973).

18. M. Ooura and T. Hanada, "Compositional Dependence of Solubility of Sulphate in Silicate Glasses," *Glass Technology*, **39** [2], 68-73 (1998).

19. H. Li, P. Hrma, and J.D. Vienna, "Sulfate Retention and Segregation in Simulated Radioactive Waste Borosilicate Glasses," pp. 237-245 in *Environmental Issues and Waste Management Technologies VI.* Edited by D.R. Spearing, et. al. American Ceramic Society. Westerville, OH, 2001.

20. I. Pegg, H. Gan, I.S. Muller, D.A. McKeown, and K.S. Matlack, "Summary of Preliminary Results on Enhanced Sulfate Incorporation During Vitrification of LAW Feeds," VSL-00R3630-1, Revision 1, Vitreous State Laboratory, The Catholic University of America, Washington, DC (2000).

21. J.A. Kelley, "Evaluation of Glass as a Matrix for Solidification of Savannah River Plant Waste," U.S. DOE Report DP-1382, E.I. duPont deNemours & Co., SRL, Aiken, SC 29808 (May 1975).

22. H. Hull, "Trip Report Battelle-Pacific Northwest Laboratory Slurry-Fed Melter Test, January 25-29, 1982," U.S. DOE Report DPST-82-387, E.I. duPont deNemours & Co., SRL, Aiken, SC 29808 (March 1982).

23. F.A. Lifanov, A.P. Kobelev, S.A. Dmitriev, M.I. Ojovan, A.E. Savkin, and I.A. Sobolev, "Vitrification of Intermediate Level Liquid Radioactive Waste," pp. 241-244 in *Environmental Remediation. and Environmental. Management Issues.* ASME, New York, 1993.

24. C.M. Jantzen and M.E. Smith, "Revision of the DWPF Sulfate Solubility Limit," U.S. DOE Report WSRC-TR-2003-00126, Westinghouse Savannah River Co., Aiken, SC 29808 (May 2003).

25. D.K. Peeler, C.C. Herman, M.E. Smith, T.H. Lorier, D.R. Best, T.B. Edwards, and M.A. Baich, "An Assessment of the Sulfate Solubility Limit of the Frit 418 – Sludge Batch 2/3 System", U.S. DOE Report WSRC-TR-2004-00081, Westinghouse Savannah River Company, Aiken, SC (January 2004).

26. I. Pegg, H. Gan, I.S. Muller, D.A. McKeown, and K.S. Matlack, "Summary of Preliminary Results on Enhanced Sulfate Incorporation During Vitrification of LAW Feeds," VSL-00R3630-1, Vitreous State Laboratory, The Catholic University of American, Washington, DC, (2001).

27. H.D. Schreiber, M.E. Stokes, and C.W. Schreiber, "VMI Final Report - Sulfur in Waste Melts: Enhancing the Capacity of Silicate Melts for Sulfate," VMI Research Laboratories Prepared for GTS Duratek, Inc. and BNFL, Inc. (August 2000).

28. M.H. Langowski, "The Incorporation of P, S, Cr, F, Cl, I, Mn, Ti, U and Bi into Simulated Nuclear Waste Glasses: Literature Study," U.S. DOE Report T3C-95-111, Pacific Northwest National Laboratory, Richland, WA (1994).

29. C.M. Jantzen, "Relationship of Glass Composition to Glass Viscosity, Resistivity, Liquidus Temperature, and Durability: First Principles Process-Product Models for Vitrification of Nuclear Waste," pp. 37-51 in *Proceedings of the 5th International Symposium on Ceramics in Nuclear Waste Management.* Edited by G.G. Wicks, D.F. Bickford, and R. Bunnell. American Ceramic Society, Westerville, OH, 1991.

30. B.O. Mysen, D. Virgo, C.M. Scarfe, and D.J. Cronin, "Viscosity and Structure of Iron- and Aluminum-Bearing Calcium Silicate Melts at 1 Atm.," *Am. Mineraologist*, **70**, 487-498 (1985).

31. B.M.J. Smets and D.M. Krol, "Group III Ions in Sodium Silicate Glass. Part 1. X-ray Photoelectron Spectroscopy Study," *Phys. Chem. Glasses*, **25** [5], 113-118 (1984).

32. W.L. Konijnendijk, "Structural Differences Between Borosilicate and Aluminosilicate Glasses Studied by Raman Scattering," *Glastechn. Ber.* **48** [10], 216-218 (1975).

33. S.N. Crichton, T.J. Barieri and M. Tomozawa, "Solubility Limits for Troublesome Components in a Simulated Lowe Level Nuclear Waste Glass," *Ceramic Transactions V.61*, 283-290 (1995).

34. X. Feng, P. Hrma, M.J. Schweiger, and H. Li, "Sulfur and Phosphorous Solubilities in Phase II Vendor Glass," Spectrum 96 (August 18-23, 1996), Seattle, WA 555-562 (1996).

35. H. Li, J.G. Darab, D.W. Matson, P.A. Smith, P. Hrma, Y. Chen, and J. Liu, "Effect of Minor Components on Vitrification of Low-Level Simulated Nuclear Waste Glasses," *Scientific Basis for Nuclear Waste Management* XIX, Vol. 412, Materials Research Society, Pittsburgh, PA, 1995.

EVALUATION OF GLASS FROM THE DWPF MELTER

A.D. Cozzi and N.E. Bibler
Westinghouse Savannah River Company
Aiken, SC 29808

ABSTRACT

The Defense Waste Processing Facility (DWPF) melter has operated for over eight years with more than six years of radioactive operations. For each sludge batch of waste processed a sample of the radioactive glass is analyzed. In conjunction with the pour stream sampling of Sludge Batch 2, a sample of the glass in contact with the pour spout insert was also collected for analysis. The samples were evaluated for chemical composition, crystal content and redox. This paper was prepared in connection with work done under Contract No. DE-AC09-96SR18500 with the U.S. Department of Energy.

INTRODUCTION

Two glass samples from the Defense Waste Processing Facility (DWPF) were characterized for chemical composition, crystal content and redox. The two glasses consisted of a pour stream sample taken while filling canister S01753 during processing of sludge batch 2 (SB2) and a sample from an Inconel™ pour spout insert recovered during insert removal/replacement and consisted of material that had spalled off the insert during cooling. To provide some comparative data, the analysis of SME batch 224[*] (processed shortly before or during the sampling of the pour stream) and the Savannah River Technology Center (SRTC) Tank 40 qualification sample are included.

SAMPLE ANALYSIS
Visual Observation

The two samples were placed in the SRTC Shielded Cells, removed from their primary containers, and photographed, Figure 1. The pour stream sample was contained in a platinum sampling boat and appeared dark with a reflective surface. The insert sample consisted of small thin flakes that were matte and dark

---

[*] The Slurry Mix Evaporator (SME) is the vessel where the frit is added to the modified sludge prior to being transferred to the melter feed tank.

gray to black with textured surfaces that had a gritty appearance. The pour stream sample was removed from the boat using an extractor provided by DWPF-Engineering to contain the glass during impact. The pour stream sample was 40.7 grams and the insert sample was 20.2 grams.

Figure 1. A) Dark and reflective pour stream sample, 40.7 grams and B) Dark gray and matte insert sample, 20.2 grams.

Chemical Composition

Samples were prepared for chemical analysis by pulverizing a portion of the glass in an agate vial with agate balls. The pulverized sample was sieved using a 100-mesh (149 μm) sieve. The –100-mesh sample was used for the dissolutions. The glass samples were dissolved by two methods[†] to account for all of the elements of interest. To provide a gross representation of the expected composition of the pour stream sample, the analysis of SME batch 224 was converted to oxides. Table I shows the composition of the two DWPF samples as well as the composition of the SRTC Tank 40 qualification sample and the measured composition of the vitrified SME batch 224. As expected, the composition of the pour stream sample resembles those of the SRTC Tank 40 qualification sample and SME batch 224. The composition of the insert sample was different from the other samples in that it was deficient in aluminum, boron, calcium, lithium, sodium, uranium, and silica. The insert sample was enriched in chromium, iron, and nickel with respect to the other samples. The low sum of oxides for the insert sample is a result of incomplete dissolution of the sample. The dissolution procedures are tailored for glass analysis and, while aggressive, is not designed for the dissolution of all ceramic materials. Table II presents the ratio of the major components of the pour stream sample to the other compositions from Table I. It can be discerned from Table II that the pour stream sample was close in composition to both the SRTC Tank 40 qualification and the SME batch 224 compositions (a ratio of 1 would indicate identical compositions

---

[†] ADS-2502 – Sodium Peroxide/Sodium Hydroxide Dissolutions of Sludge and Glass for Elemental and Ion Analysis.
ADS-2227 – Acid Dissolution of Glass and Sludge for Elemental Analysis.

for an analyte). These results were expected given that the SRTC Tank 40 qualification sample is intended to be representative of Sludge Batch 2 (Tank 40). Applying the DWPF process to a sample of Sludge Batch 2 produced the SRTC Tank 40 qualification sample. The glass from SME batch 224 was vitrified and analyzed in the DWPF analytical laboratory during the processing of Sludge Batch 2. It is also apparent that the composition of the insert sample is significantly different from that of the pour stream (and SRTC TK 40 and SME batch 224) sample. The ratio of the major non-spinel forming components (Al, B, Ca, Li, Na, Si, U) is approximately 0.5, indicating that the sample is 50% glass and 50% other materials (Probably from the Inconel™ pour spout insert. The components are the same, nickel, chromium and iron, but the ratios do not precisely match).

Table I. Compositions of Pour Stream, Insert and Qualification Glasses (in wt%; NM-not measured).

	Pour Stream	Insert	SRTC TK 40[1]	SME batch 224
Al_2O_3	4.22	2.12	4.27	4.36
B_2O_3	7.31	3.30	8.21	7.37
CaO	1.39	0.62	1.30	1.25
Cr_2O_3	0.06	21.44	0.33[a]	0.09
CuO	0.07	0.06	0.09	0.03
Fe_2O_3	12.29	21.05	11.80	12.08
La_2O_3	0.02	0.01	0.16	0.00
Li_2O	3.29	1.70	3.51	3.21
MgO	2.35	2.18	2.49	2.27
MnO	2.14	3.09	1.38	1.42
Na_2O	11.38	5.45	11.9	10.19
NiO	0.54	7.03	0.60	0.55
SiO_2	48.73	23.05	53.1	49.42
TiO_2	0.05	0.11	0.08	0.05
U_3O_8	3.57	1.99	2.98	3.43
ZnO	0.09	0.18	NM	NM
ZrO_2	0.09	0.04	0.13	0.08
Sum	98.59	94.00	102	95.80

[a]Sample prepared in stainless steel grinder for sludge batch 2 qualification.

Table II. Ratio of Major Components of the Pour Stream Sample to the Insert Sample.

	Insert / Pour Stream	SRTC TK 40[1]/ Pour Stream	SME batch 224/ Pour Stream
Al_2O_3	0.50	1.0	1.0
B_2O_3	0.45	1.1	1.0
CaO	0.45	0.94	0.90
Cr_2O_3	360	5.5[a]	1.5
Fe_2O_3	1.7	0.96	0.98
Li_2O	0.52	1.1	0.98
MgO	0.93	1.1	0.97
MnO	1.4	0.64	0.66
Na_2O	0.48	1.1	0.90
NiO	13	1.1	1.0
SiO_2	0.47	1.1	1.0
U_3O_8	0.56	0.83	0.96

[a]Sample prepared in stainless steel grinder for sludge batch 2 qualification.

Analysis for Noble Metals, Am-241 and Selected U-235 Fission Products in the Two Samples

The solutions that resulted from acid dissolution of the two samples were analyzed by Inductively Coupled Mass Spectroscopy (ICP-MS) for noble metals and gamma emitters to gain more detailed information about the composition of the samples. Isotopes selected were the gamma emitters detected, noble metals resulting from neutron fission of U-235, and a sampling of other U-235 fission products. Concentrations in weight percent along with the respective concentrations measured in the SRTC Tank 40 qualification sample are given in Table III. The ratios of the concentrations in respective glasses are given the last two columns of the Table III. The isotopes Co-60, Cs-137, Eu-154, Eu-155, and Am-241 were measured by gamma counting. All others were measured by ICP-MS.

The concentrations measured in the pour stream sample for most of the isotopes were nearly equal to their respective concentrations in the SRTC Tank 40 sample, Table III. Agreement is expected as the Tank 40 Qualification sample originated from the same material as the feed for the pour stream sample. This indicates that mixing in Tank 40 was sufficient to get a representative sample for the SRTC qualification demonstration For the gamma emitters, the agreement was within 20% or better. For some of the isotopes analyzed by ICP-MS the agreement was not as good. For the noble metals, isotopes of Ru and Pd were 40 to 50% higher in the pour stream than the SRTC Tank 40 sample. To the contrary, the noble metal Rh-103 had a concentration 64% less in the pour stream sample than that in the SRTC Tank 40 sample. These differences can be attributed to analytical error associated with the low concentrations of these isotopes in the glass. The measured concentrations in the samples were in some cases close to the sensitivity of the ICP-MS method; thus their relative error could be large (in some cases 30-50%).

Most of the concentrations of the radioisotopes measured in the insert sample were less than those measured in the pour stream sample. The Insert/Pour Stream column of Table III indicates that the ratio of differences was 0.4 to 0.6 as indicated by the major components in the sample that do not typically participate in spinel formation (see Table II). However, six isotopes had higher concentrations in the insert sample compared to the pour stream samples. These were Co-60 measured by gamma counting and the noble metals measured by ICP-MS. As shown in Table III, the concentration of Rh was significantly higher in the insert sample compared to the pour stream. The reason for these higher concentrations in the insert is not immediately apparent.

Table III. Comparison of Some Isotopic Concentrations (wt.%) of the SRTC Tank 40, Pour Stream, and Insert Glasses.

Isotope	SRTC Tank 40[1]	Pour Stream	Insert	Insert/ Pour Stream[a]	Pour Stream/ Tank 40[b]
Co-60	1.50E-07	1.54E-07	5.05E-07	3.3	1.0
Tc-99	2.28E-04	2.75E-04	9.68E-05	0.35	1.2
Ru-101	2.13E-03	2.97E-03	7.02E-02	24	1.4
Ru-102	1.99E-03	2.89E-03	6.69E-02	23	1.5
Rh-103	1.71E-03	6.22E-04	1.78E-01	286	0.36
Ru-104	1.41E-03	1.91E-03	4.48E-02	24	1.4
Pd-105	1.34E-04	2.07E-04	3.35E-04	1.6	1.5
Cd-112	9.83E-03	1.06E-02	6.51E-03	0.61	1.1
Cs-137	8.93E-05	1.03E-04	4.55E-05	0.44	1.2
La-139	5.55 E-03	6.54E-03	3.72E-03	0.57	1.2
Nd-143	5.96 E-03	6.10E-03	3.29E-03	0.54	1.0
Eu-154	8.47E-07	8.45E-07	3.94E-07	0.47	1.0
Eu-155	2.36E-07	2.52E-07	1.45E-07	0.57	1.0
Am-241	2.74E-04	2.85E-04	1.17E-04	0.41	1.0

[a]Ratio should be ~0.47 by dilution of the glass components by insert material
[b] Ratio should be ~1.0 as the Tank 40 sample should be representative of the pour stream.

REDOX Analysis

To prepare samples for redox analysis, portions of the pour stream and insert materials were pulverized in an agate vial with agate balls. The Environmental Assessment (EA) glass was prepared alongside the pour stream and insert samples as a control. EA glass is reported to have a redox ratio (Fe^{2+}/Fe^{tot}) of ~ 0.18^2. The dissolution of the samples was conducted so as to maintain the redox of the iron in the glass[3]. Not all of the dissolutions could be performed on the same day as they were measured. Therefore, some of the samples were prepared a day in advance. Previous research has indicated that the redox state of the dissolved material is stable when diluted with boric acid solution[3]. Using the method in Reference 3, dissolution of all three of the insert samples was incomplete. Incomplete dissolution is not a problem in a homogeneous glass because the redox value is a ratio. However, if there is selective dissolution of phases that contain disproportionate amounts of Fe^{2+} and Fe^{3+}, the results become meaningless. Table IV shows the results of the redox analysis. The measured redox ratio of the EA glass was greater than expected. For the amount of EA glass used (20 - 32 mg) the typical absorbance value at 562 nm for Fe^{tot} is 0.5 - 0.6. The average measured redox of the pour stream sample was 0.21. SME batch 224 had a predicted redox of 0.19. The relatively low absorbance of Fe^{tot} in the insert sample that had significantly more iron ($Fe_{insert}/Fe_{pour\ stream}$ =1.7), which indicates that there were undissolved iron compounds in the insert samples. One explanation is the formation of iron-rich spinel in the insert sample. Spinels are not as readily dissolved as glasses and may not have been completely taken into solution.

Table IV. Redox of Pour Stream Sample (PSS) and Insert Sample (IS).

Sample	Fe^{2+}/Fe^{tot}	Average
EA-1	0.31^a	
EA-2	0.37^a	
EA-3	0.23^b	0.30
PSS-1	0.23^a	
PSS-2	0.20^b	
PSS-3	0.19^b	0.21
IS-1^c	0.24^b	
IS-2^c	0.19^b	
IS-3^c	0.27^b	

^{a}prepared one day before analysis
^{b}prepared day of analysis
^{c}Sample only partially dissolved

Contained Scanning Electron Microscopy
    For contained scanning electron microscopy with energy dispersive spectroscopy (CSEM/EDS), the samples ranged from eight to twelve milligrams to minimize the interference of radiation with the detector and personnel exposure. The small size of the samples limits the representative nature of the analysis. The CSEM analysis of the pour stream sample revealed uniformity

across the entire sample. The insert sample viewed at 1000x appeared to have more surface texture than a typical glass sample. However, there were no ambiguous inclusions using SEI mode, Figure 2, although when the image viewed using the backscatter electron imaging (BSI) mode, several distinct features became apparent, Figure 3.

Figure 2. Micrograph of the insert sample, 1000x.

Figure 3. Micrograph of the image in Figure 3 using the BSI mode, 1000x.

EDS analysis of the insert sample revealed that there are at least three distinct compositional regions. One region, as indicated by spot "B" in Figure 3, is predominantly chromium and iron. The darker region of the photo, spot "D", has the components of a typical DWPF glass. Spot "C", not shown in Figure 3 (from a separate micrograph), is almost entirely chromium. Figure 4 is the EDS spectra for spots "B", "C", and "D". In the evaluation of the two insert samples, the majority of the material was represented by one of the three compositional regions

"B", "C", or "D". A fourth region, represented by a twenty-micron diameter region, indicates the presence of a sodium chromium sulfate. This has been detected in previous analysis of pour spout regions[4].

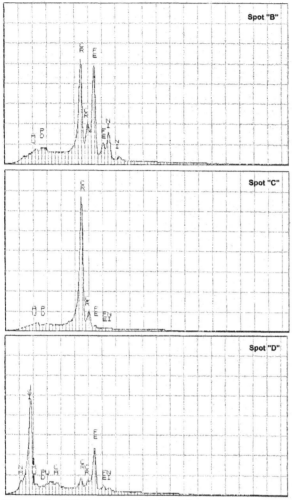

Figure 4. EDS spectra from the insert sample showing the chromium and iron rich region (spot "B"), the predominantly chromium region (spot "C", and the glass region (spot "D").

Contained X-ray Diffraction Analysis (CXRD)
In agreement with the SEM results, the XRD pattern of the pour stream sample was typical of a borosilicate glass and free of any indicators of crystalline matter.

The XRD analysis of the insert sample indicated the presence of three distinct phases. Along with the amorphous hump associated with a glassy phase, a spinel phase and a chromium oxide phase (eskolaite[‡]) were identified. The spinel phase resembles trevorite[§] with chromium partially substituting for iron and iron partially substituting for nickel. This reasoning is based on the EDS spectrum of spot "B". Figure 5 shows the XRD patterns of both the pour stream sample and the insert sample to demonstrate the differences between the two samples. With only 0.06 wt % Cr_2O_3 in the pour stream, the Inconel™ insert is most likely the primary source of the chromium for both the eskolaite and the trevorite.

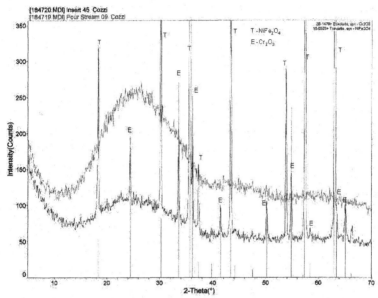

Figure 5. Overlaid x-ray diffraction patterns of the pour stream and insert samples.

CONCLUSIONS

Pour Stream Sample

Visual observation of the pour stream sample it sample to be typical of a DWPF-type glass (opaque and reflective). Compositional analysis by ICP-ES demonstrated a correlation among the pour stream sample and both the SME batch 224 and the SRTC Tank 40 qualification sample. Agreement of concentrations between the pour stream sample and the SRTC Tank 40 qualification sample was also observed for Am-241 and most of the measured U-235 fission products. However, agreement between many of the noble metal

---

[‡] Eskolaite. International Centre for Diffraction Data (ICDD) card 38-1479 Cr_2O_3
[§] Trevorite ICDD card 10-0325 $NiFe_2O_4$

concentrations was poor. The average measured redox of 0.21 in the pour stream sample matches well with the predicted redox value of 0.19 for SME batch 224. CSEM analysis revealed an amorphous sample with no indication of inclusions or crystalline material. The spectrum from the CXRD analysis reinforced the amorphous nature of the sample.

Pour Spout Insert Sample

The sample was received as small, thin dark gray flakes with a matte, grainy finish. Compositional analysis of the insert sample revealed significantly greater quantities of transition elements (Cr, Fe, Mn, Ni, and Zn) and reduced amounts of other components (Al, B, Li, Na, Si and several radioactive isotopes) as compared to the pour stream sample. Samples for redox analysis were not dissolved completely and were therefore not meaningful. CSEM analysis identified four distinct compositional regions ("glass", transition metals, high chromium, sulfate salt). CXRD analysis confirms the presence of an amorphous phase, a spinel phase, and a chrome oxide phase. CSEM indicated that the sulfate content was minimal and would not be discernible by CXRD. It can be hypothesized from the compositional ratios and the CXRD results that the insert sample is comprised of approximately 50% glass and 50% crystalline material. At the temperature of the insert in the pour spout, approximately 1100°C, Inconel™ rapidly oxidizes to form a protective chrome oxide layer. Under typical operation, glass moving over the insert will not significantly react with the Inconel™. However, glass that has spattered onto the insert, out of the path of the pour stream, is given substantial opportunity to incorporate not only the oxide coating, but also a portion of the underlying Inconel™. The amount of material (pour spout insert and spattered glass) involved should not affect normal melter operations.

REFERENCES

[1] T.L. Fellinger, N.E. Bibler, J.M. Pareizs, A.D. Cozzi, and C.L. Crawford, WSRC-RP-2001-01016, November 2001.
[2] C.M. Jantzen, N.E. Bibler, D.C. Beam, C.L. Crawford, and M.A. Pickett, "Characterization of the Defense Waste Processing Facility (DWPF) Environmental Assessment (EA) Glass Standard Reference Material (U)", WSRC-TR-92-346, Rev. 1. June 1994.
[3] C.J. Coleman, E.W. Baumann, and N.E. Bibler, "Colorimetric Determination of Fe^{2+}/Fe^{3+} Ratio in Radioactive Glasses (U)," WSRC-MS-91-375. April 1992.
[4] C.M. Jantzen and D.P. Lambert, "Inspection and Analysis of the Integrated DWPF Melter System After Seven Years of Continuous Operation," WSRC-MS-99-00336, August 1999.

REDOX ACTIVITY OF RHENIUM IN SILICATE GLASSES

Henry D. Schreiber, Charlotte W. Schreiber, and Amy M. Swink
Department of Chemistry
Virginia Military Institute
Lexington, Virginia 24450

ABSTRACT

An experimental study positioned the redox couples of rhenium within an electromotive force series for a reference soda-lime-silicate glass melt. The reduction potentials of the rhenium redox couples (Re^{7+}-Re^{4+} and Re^{4+}-Re^0) were estimated by measuring rhenium's interaction with Mn^{3+}-Mn^{2+} as well as with V^{5+}-V^{4+}-V^{3+}. The study determined that the Re^{7+}-Re^{4+} redox couple was operational only under very oxidizing conditions, and the Re^{4+}-Re^0 redox couple only under very reducing conditions. These extreme conditions should not be readily encountered in the processing of nuclear waste into a glass wasteform, because the oxidizing conditions needed to stabilize Re^{7+} would also result in melt foaming while the reducing conditions needed to stabilize Re^0 would precipitate other metals and metal sulfides. In fact, the redox constraints defined by the Fe^{2+}/Fe^{3+} ratio for suitable processing of nuclear waste should preclude the presence of Re^{7+} and Re^0 in the final glass wasteform. Thus, the dominant redox state of rhenium stabilized in the glass melt under processing conditions is Re^{4+}. Insofar as the redox chemistry of rhenium mimics that of technetium, technetium should also be incorporated as Tc^{4+} within the glass used for nuclear waste immobilization.

INTRODUCTION

In order to manage a significant portion of its stored nuclear waste, the United States plans to immobilize that waste by vitrifying a mixture of glass frit and waste, before storing the resulting glass wasteform in a repository. The waste becomes an integral component of the glass matrix so that the glass wasteform serves as a first level of protection against the dispersion of that waste. The borosilicate glass matrix is able to dissolve a wide range of nuclear waste, often including exotic elements not encountered in commercial glass-making.

Some of the nuclear waste to be vitrified contains relatively high concentrations of highly radioactive technetium [1]. Because technetium has no non-radioactive isotopes, it is difficult to perform laboratory experiments to characterize its properties during its processing into a glass wasteform. One strategy has been to employ an element in technetium's chemical family, because such an element would be expected to have analogous chemical properties. Thus, rhenium has been used as a surrogate for technetium in many experimental formulations [1]. However, much of the chemistry of rhenium (and thus, technetium) in glass-forming melts still remains to be determined.

One property of consequence in processing nuclear waste into a glass wasteform is the redox chemistry of each element in the waste. Inappropriate redox states in the melter may result in processing problems such as foaming, precipitation of metals and sulfides, corrosion of electrodes and refractories, and volatilization of radioactive isotopes [2]. In fact, guidelines on the redox state of the system, usually expressed as the Fe^{2+}/Fe^{3+} redox ratio in the resulting glass, have been established to circumvent many of the potential problems [2]. The basis for these guidelines is rooted within an electromotive force series, that is a series of relative reduction potentials of redox couples, which has been developed for a reference borosilicate composition to be used for nuclear waste immobilization [3]. Such a series can be used to predict redox reactions within the melt and, accordingly, the elemental redox states of concern. In order to understand the redox activity of rhenium, and consequently technetium, the first step would be to place the redox couples of rhenium within the electromotive force series for this reference composition.

Table I compiles the electromotive force series for redox couples in a reference soda-lime-silicate composition [4]. The ordering of redox couples in this base composition is analogous to that in an alkali borosilicate composition relevant for nuclear waste immobilization [5]. Figure 1 is a graphical representation of this same electromotive force series [4]. This figure shows how the imposed oxygen fugacity (expressed as -log fO_2, so that the figure plots from oxidizing to reducing conditions as one goes from the graph's bottom to top) affects the ratio of the reduced to oxidized redox ion concentrations (R) of each multivalent element. The experimental data for the multivalent elements generate straight lines with slopes equal to 4/n, where n is the number of electrons transferred by the redox couple, and x-intercepts equal to the relative reduction potentials (E') shown in Table I. All redox couples to the right of a given couple in Figure 1 are easier to reduce (more of that element being in its reduced state at a given oxygen fugacity) and therefore will act as oxidizing agents to that couple. For example, the Ce^{4+}-Ce^{3+} redox couple is to the right of the Fe^{3+}-Fe^{2+} redox couple; thus, cerium will oxidize Fe^{2+} to Fe^{3+} within the melt.

A previous study [6] has shown that the most stable redox states of rhenium in glass are Re^{7+}, Re^{4+}, and Re^0 and those of technetium are similarly Tc^{7+}, Tc^{4+}, and Tc^0. These are also the same redox forms of rhenium (and technetium) that are most stable in aqueous systems [1].

Table I. The electromotive force series of redox couples in the reference system[#]

Redox Couple		E'		Redox Couple		E'
Au	Au^{3+}-Au^0	>3.6		Te	Te^{4+}-Te^0	-1.2
Ir	Ir^{3+}-Ir^0	+3.4		V	V^{5+}-V^{4+}	-1.3
Pt	Pt^{4+}-Pt^0	+3.1		Sb*	Sb^{3+}-Sb^0	-1.4
Pd	Pd^{2+}-Pd^0	+2.9		Bi	Bi^{3+}-Bi^0	-1.6
Ni	Ni^{3+}-Ni^{2+}	+2.3		U*	U^{5+}-U^{4+}	-1.6
Rh	Rh^{3+}-Rh^0	+2.0		As*	As^{3+}-As^0	-2.1
Co	Co^{3+}-Co^{2+}	+2.0		Cu*	Cu^+-Cu^0	-2.7
Mn	Mn^{3+}-Mn^{2+}	+1.2		Cr*	Cr^{3+}-Cr^{2+}	-3.1
Ag	Ag^+-Ag^0	+1.1		V*	V^{4+}-V^{3+}	-3.4
Se	Se^{6+}-Se^{4+}	+0.8		Eu	Eu^{3+}-Eu^{2+}	-3.4
Se*	Se^{4+}-Se^0	+0.8		Ti	Ti^{4+}-Ti^{3+}	-4.0
Ce	Ce^{4+}-Ce^{3+}	+0.5		Ni*	Ni^{2+}-Ni^0	-4.7
Cr	Cr^{6+}-Cr^{3+}	+0.4		Se**	Se^0-Se^{2-}	-5.1
Sb	Sb^{5+}-Sb^{3+}	+0.3		Fe*	Fe^{2+}-Fe^0	-5.7
Re	Re^{7+}-Re^{4+}	0.0		Sn	Sn^{4+}-Sn^{2+}	-5.8
Cu	Cu^{2+}-Cu^+	-0.2		Co*	Co^{2+}-Co^0	-5.8
U	U^{6+}-U^{5+}	-0.9		Zn	Zn^{2+}-Zn^0	-8.1
Fe	Fe^{3+}-Fe^{2+}	-1.0		Re*	Re^{4+}-Re^0	-11.0
As	As^{5+}-As^{3+}	-1.1		S	S^{6+}-S^{2-}	-18.6

[#]reference system = soda-lime-silicate composition, 1400°C, 1 wt% M [4]; estimated error in E' = ±0.3 units; Re and Re* values from this study.

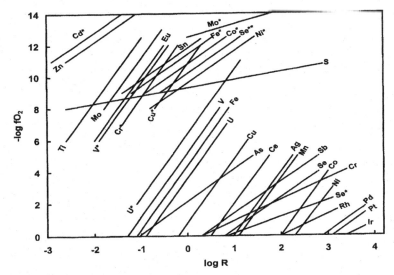

Figure 1. The electromotive force series of redox couples in soda-lime-silicate glass at 1400°C [4]. Identification of a couple is keyed to the symbol in Table I. R is the concentration ratio of reduced to oxidized species; fO_2 is oxygen fugacity.

OBJECTIVES AND APPROACH

The goal of this study is to estimate the relative reduction potentials of Re^{7+}-Re^{4+} and Re^{4+}-Re^{0} in a reference glass composition, that is to position these redox couples within the electromotive force series of multivalent elements in Table I and Figure 1. These results can then be extrapolated to the analogous technetium redox couples and to predict the prevailing redox state during nuclear waste processing.

One approach to ascertain the reduction potential of a redox couple is to determine what redox couples can be oxidized (or reduced) by that multivalent species under specified reference conditions, as has been used previously [7]. This study will monitor the interaction of rhenium with manganese (Mn^{3+}-Mn^{2+}) as an element sensitive to change only under oxidizing conditions, and of rhenium with vanadium (V^{5+}-V^{4+}-V^{3+}) as a system whose redox equilibria span the entire range of redox conditions.

EXPERIMENTAL METHODS

The reference soda-lime-silicate composition is 17 mole% Na_2O, 11 mole% CaO, and 72 mole% SiO_2 [4]. It was prepared by mixing appropriate amounts of ultrapure Na_2CO_3, CaO, and SiO_2; melting at 1400°C for about 24 hours; quenching to a glass; and powdering to a homogeneous frit. This soda-lime-silicate glass was employed as a base composition instead of the SRL-131 (Savannah River Laboratory frit# 131) alkali borosilicate composition because of possible redox interference by titanium (as Ti^{3+}) inherent to SRL-131 under reducing conditions. Both the soda-lime-silicate and alkali borosilicate compositions have been used interchangeably in the development of the ordering of redox couples in the desired electromotive force series [7].

A composition of the soda-lime-silicate glass with 0.50 wt% V was made by mechanically mixing V_2O_5 with the reference frit, followed by melting at 1400°C for 24 hours, rapidly quenching, and powdering. In analogous fashion, compositions containing 1.00 wt% Mn, 0.50 wt% V and 1.00 wt% Re, 0.50 wt% V and 2.00 wt% Re, and 1.00 wt% Mn and 2.00 wt% Re were prepared. Manganese was added to the frit as MnO, and rhenium as $KReO_4$.

Individual 0.5-g samples were synthesized in Pt capsules suspended within the hot zones of furnaces controlled at 1400°C, and whose atmospheres were regulated by the gas flow of pure O_2, air, pure CO_2, and CO_2/CO mixtures. Samples prepared under the very reducing C/CO buffer conditions were synthesized in graphite containers in a pure CO atmosphere. All samples were equilibrated for 24 hours, which was determined to be more than sufficient time to achieve redox equilibrium for this sample configuration. Samples were then vitrified by removal from the furnace in order to quench the redox ion concentrations.

Specimens of about 4 mm diameter were sectioned in order to manufacture finely polished glass slabs of about 1 mm thickness. Visible absorption spectra from 800 nm to 300 nm were obtained on a Shimadzu 3100 spectrophotometer. Manganese-containing samples were measured for its characteristic Mn^{3+} absorption at about 495 nm, and the vanadium-containing samples were monitored for characteristic V^{5+} (315 nm), V^{4+} (425, 625 nm), and V^{3+} (450, 680 nm) peaks [8]. Changes in the manganese (Mn^{3+}) and vanadium (V^{4+}, V^{3+}) absorbances upon the addition of rhenium were used to quantify how rhenium affected the redox ion concentration of manganese and vanadium, that is whether rhenium acted as an oxidizing or reducing agent.

RESULTS

Figure 2 displays the visible absorption spectrum of Mn^{3+} in the reference glass containing 1.0 wt% Mn with and without Re additions, when synthesized at pure oxygen (-log fO_2 = 0). The Mn^{3+} absorption decreases by about 30% upon addition of rhenium to the glass-forming melt, indicative that the following reaction is most likely operational.

$$Re^{4+} + 3\,Mn^{3+} \rightarrow 3\,Mn^{2+} + Re^{7+} \tag{1}$$

That is, rhenium (as the Re^{7+}-Re^{4+} couple) reacts as a reducing agent with respect to Mn^{3+}. The interaction becomes less effective at air (-log fO_2 = 0.7) and essentially nonexistent at pure CO_2 (-log fO_2 = 2.9) at 1400°C. This dependence on oxygen fugacity is not surprising as this redox potential of rhenium, a three-electron redox couple, is much more dependent on the imposed oxygen fugacity than the one-electron manganese redox couple. The Re^{7+}-Re^{4+} redox couple accordingly should be placed to the left of the Mn^{3+}-Mn^{2+} redox couple in Figure 1, perhaps near and parallel to the line represented by the three-electron Cr^{6+}-Cr^{3+} redox couple.

Under the same oxidizing atmospheric conditions, the prevailing vanadium redox couple is V^{5+}-V^{4+} (probably as VO_3^--VO^{2+}). V^{4+} is only a small percentage of the total vanadium concentration, probably less than 10% of the total vanadium, under these conditions. Rhenium induces no change in the vanadium absorption spectra. This is indicative that the Re^{7+}-Re^{4+} redox couple is not too far to the right of the line representing the V^{5+}-V^{4+} redox couple (that is, their E' values are not dramatically different) in Figure 1. The probable relative reduction potential E' of Re^{7+}-Re^{4+} in the reference glass-forming melt is estimated to be 0.0 ± 0.3 units.

Under intermediate redox conditions (-log fO_2 of about 5 through 8), the vanadium absorption spectra were identical with or without rhenium additives, indicative that rhenium was not redox active under such conditions. The rhenium was most likely present as the Re^{4+} ion in the reference melt under such conditions.

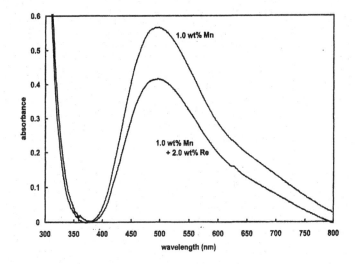

Figure 2. Visible absorption spectra of the soda-lime-silicate glass containing 1.0 wt% Mn with and without 2.0 wt% Re at 1400°C and pure oxygen. The intensity of the absorption peak is proportional to Mn^{3+} concentration.

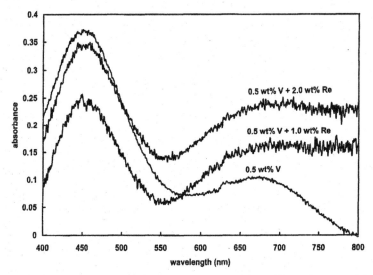

Figure 3. Visible absorption spectra of soda-lime-silicate glass containing 0.5 wt% V and varying concentration of Re at 1400°C and at the C/CO buffer.

Under relatively reducing conditions (below $-\log fO_2 = 9$), the rhenium-containing glasses appeared gray, due to finely dispersed rhenium metal (Re^0) precipitating from the melt. These reducing conditions are similar to where the divalent species of nickel, iron, and cobalt also precipitate as metals from the glass-forming melt. In addition, Figure 3 shows the vanadium absorption spectra at C/CO buffer conditions ($-\log fO_2 = 12$) without rhenium, with 1.0 wt% Re, and with 2.0 wt% Re. Subsequent difference spectra clearly show that the V^{3+} peaks increase at the expense of the V^{4+} peaks upon addition of rhenium, and that the magnitude of the change is proportional to the amount of added rhenium. In essence, the rhenium redox couple (Re^{4+}-Re^0) acts as a reducing agent with respect to the vanadium redox couple (V^{4+}-V^{3+}), as represented by the following reaction under these reducing conditions.

$$Re^0 + 4\ V^{4+} \rightarrow 4\ V^{3+} + Re^{4+} \tag{2}$$

These observations would place the Re^{4+}-Re^0 redox couple generally to the left of the V^{4+}-V^{3+} redox couple in Figure 1, probably between the line representing Zn^{2+}-Zn^0 and that of Fe^{2+}-Fe^0. This rhenium redox couple is probably similar to these other M^{2+}-M^0 redox couples, except that it exhibits more profound oxygen fugacity dependence based on its four-electron redox couple. An estimated relative reduction potential E' would be about -11 units in this reference system.

DISCUSSION

Estimates of the placement of the Re^{7+}-Re^{4+} and the Re^{4+}-Re^0 redox couples within the electromotive force series are shown in Figure 4 (as well as Table I), based on the interaction of rhenium with the vanadium and manganese redox couples. The range of rhenium's redox activity is very limited. The Re^{7+}-Re^{4+} redox equilibrium is operational only under oxidizing conditions, and that of Re^{4+}-Re^0 under very reducing conditions. In addition, Re^{7+} probably exists in the glass as ReO_4^- entities [1], much like other high valence species of chromium and vanadium exist as CrO_4^{2-} and VO_4^- respectively. One way to represent the rhenium redox equilibria in glass-forming melts is shown by the following equations under oxidizing and reducing conditions respectively.

$$4\ Re^{7+} + 6\ O^{2-} \rightarrow 4\ Re^{4+} + 3\ O_2 \tag{3}$$

$$Re^{4+} + 2\ O^{2-} \rightarrow Re^0 + O_2 \tag{4}$$

The placement of the rhenium redox couples within the electromotive force series for the reference glass is also consistent with their position in an aqueous series [9].

The extreme oxidizing and reducing conditions needed to stabilize Re^{7+} and Re^0 respectively should not be encountered in the processing of nuclear waste into a glass wasteform, the former because such conditions result in melt foaming and the latter because such conditions result in metal and metal sulfide precipitation. In fact, over the wide range of redox conditions defined by Fe^{2+}/Fe^{3+} suitable for processing and vitrifying nuclear waste, the prevailing redox state of rhenium is Re^{4+}. However, Re^{7+} may temporarily be present in the melter as the aqueous system is being mixed with the glass frit at high temperatures, resulting in volatilization of the rhenium as the higher valence species [1, 10].

Analogously, the redox chemistry of technetium should also be dominated by the Tc^{4+} redox ion in the reference glass-forming melt. Although Tc^{7+} might be expected under very oxidizing and Tc^0 possible under very reducing conditions (by comparing the aqueous reduction potentials of the rhenium and technetium redox couples), such conditions would never occur unless severe problems such as foaming or metal precipitation occurred in the melter. This would mean that the enhanced volatility of the Re^{7+} and Tc^{7+} probably occurs as a kinetic consequence and not as a thermodynamic consequence during processing in the melter [11]. On the other hand, potential concern during processing may result from co-precipitation of Tc (as TcO_2) with the highly stable RuO_2 from the melt [12]. In summary, this study confirms the general predictions of a previous work [1] concerning the redox properties of rhenium and technetium during processing of nuclear waste into a vitrified wasteform.

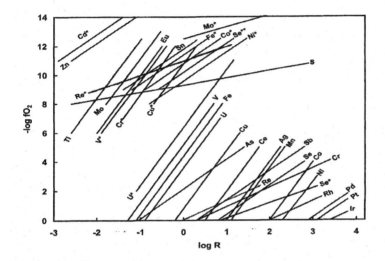

Figure 4. The electromotive force series of redox couples (including the rhenium redox couples) in soda-lime-silicate glass at 1400°C. Identification of a couple is keyed to the symbol in Table I. R is the concentration ratio of reduced to oxidized species; fO_2 is oxygen fugacity.

CONCLUSIONS

Rhenium, as Re^{7+}-Re^{4+}, reduces Mn^{3+} to Mn^{2+} in the reference glass-forming melt under oxidizing conditions. Rhenium, as Re^{4+}-Re^0, also reduces V^{4+} to V^{3+} in glass-forming melts under reducing conditions. Thus, the redox activity of rhenium is limited to the two extremes of redox conditions. Accordingly, the redox chemistry of rhenium in glass-forming melts is dominated by the redox inactive Re^{4+} species under most processing conditions. By analogy, the redox chemistry of technetium is predicted to be controlled by Tc^{4+} during the vitrification of nuclear waste. Both rhenium and technetium will be incorporated as tetravalent species, perhaps co-precipitating as ReO_2 and TcO_2 with RuO_2 in the final glass wasteform, in glass processed according to current redox guidelines.

REFERENCES

[1] J.G. Darab and P.A. Smith, "Chemistry of Technetium and Rhenium during Low-Level Radioactive Waste Vitrification", *Chemistry of Materials*, **8**, 1004-1021 (1996).

[2] H.D. Schreiber and A.L. Hockman, "Redox Chemistry in Candidate Glasses for Nuclear Waste Immobilization", *Journal of the American Ceramic Society*, **70**, 591-594 (1987).

[3] H.D. Schreiber, G.B. Balazs, B.E. Carpenter, J.E. Kirkley, L.M. Minnix, and P.L. Jamison, "An Electromotive Force Series in a Borosilicate Glass-forming Melt", *Communications of the American Ceramic Society*, **67**, C106-C108 (1984).

[4] H.D. Schreiber, B.T. Long, M.C. Dixon, and C.W. Schreiber, "The Redox Equilibrium of Tin in Soda-Lime-Silicate Glass", *Ceramic Transactions*, **109**, 357-365 (2000).

[5] H.D. Schreiber, N. Wilk, and C.W. Schreiber, "A Comprehensive Electromotive Force Series in Soda-Lime-Silicate Glass", *Journal of Non-Crystalline Solids*, **253**, 68-75 (1999).

[6] E. Freude, W. Lutze, C. Rüssel, and H. A. Schaeffer, "Investigation of the Redox Behavior of Technetium in Borosilicate Glass Melts by Voltammetry", *Materials Research Society Proceedings* (Scientific Basis for Nuclear Waste Management XII), **127**, 199-204 (1989).

[7] H.D. Schreiber, M.E. Stokes, and C.W. Schreiber, "Electrochemical Series of Polyvalent Elements in Sodium Disilicate Glasses and Melts", *Physics and Chemistry of Glasses*, **43C**, 355-357 (2002).

[8] M. Leister, D. Ehrt, G. von der Gönna, C. Rüssel, and F. W. Breitbarth, "Redox States and Coordination of Vanadium in Sodium Silicates Melted at high Temperatures", *Physics and Chemistry of Glasses*, **40** [6] 319-325 (1999).

[9] H.D. Schreiber and M.T. Coolbaugh, "Solvations of Redox Ions in Glass-Forming Silicate Melts", *Journal of Non-Crystalline Solids*, **181**, 225-230 (1995).

[10] H. Migge, "Thermochemical Comparison of the Systems Re-O and Tc-O", *Materials Research Society Proceedings* (Scientific Basis for Nuclear Waste Management XII), **127**, 205-213 (1989).

[11] H. Migge, "Simultaneous Evaporation of Cs and Tc during Vitrification – A Thermochemical Approach", *Materials Research Society Proceedings* (Scientific Basis for Nuclear waste Management XIII), **176**, 411-417 (1990).

[12] H.D. Schreiber, F.A. Settle, P.L. Jamison, J.P. Eckenrode, and G.W. Headley, "Ruthenium in Glass-Forming Borosilicate Melts", *Journal of Less-Common Metals*, **115**, 145-154 (1986).

ANALYSIS OF DEFENSE WASTE PROCESSING FACILITY PRODUCTS WITH LASER INDUCED BREAKDOWN SPECTROSCOPY

Fang-Yu Yueh, Hongbo Zheng, Jagdish P. Singh and William G. Ramsey
Diagnostic Instrumentation and Analysis Laboratory
Mississippi State University
205 Research Boulevard
Starkville, MS 39759-7404

ABSTRACT

Laser Induced Breakdown Spectroscopy (LIBS) is a powerful analytical tool. It is suitable for quick and on-line elemental analysis of any phase of material and has proved its importance in obtaining analytical atomic emission spectra directly from solid, liquid, and gaseous samples. LIBS is under evaluation for direct analysis of dried sludge or glass in radiation shielded cell facilities to increase the analytical throughput and reduce waste generation in radiological analytical facilities. In this paper, we have demonstrated LIBS' capability for this type of application. Some initial results on glass batch and glass are presented.

INTRODUCTION

Direct analysis of dried sludge or glass in shielded cell facilities will significantly increase the analytical throughput and reduce waste generation in radiological analytical facilities. LIBS uses a high pulse energy laser beam to produces a micro plasma to vaporize, dissociate, excite or ionize species on material surfaces. [1-3] The study of the atomic emission from the micro plasma provides information about the composition of the material. It is experimentally simple and can provide fast response with almost no sample preparation. Also the laser light and emitted signal can be delivered via optical fiber so it is useful for hazardous environment. LIBS has been applied to solid, [4-9] liquid, [10-12] and gaseous samples. [13-16] The capability of LIBS (remote, on-line, real-time) meets the needs for on-line analysis of simulated sludge and slurry products. However, the laser plasma involves ablation, atomization and excitation, and the process is complex and difficult to reproduce. Therefore, LIBS has poorer detection limits and reproducibility compared to alternate analytical techniques such as Inductively Couple Plasma (ICP) and Atomic Absorption (AA).

To improve LIBS' *Analytical Figure of Merits* for this application, in this work we use a high resolution broadband spectrometer (cover 200 – 800 nm) for simultaneously detecting all the elements of interest. The high- resolution broadband detection will allow us to select the most suitable analyte lines (sensitivity and interference-free line) for different matrix. The initial results of LIBS measurements of glass batch and glass with this detection system are discussed.

EXPERIMENTAL

The details of the experimental setup of the LIBS system are described in Reference 5. In brief, a frequency-doubled Nd:YAG laser beam is focused on the sample using a 30cm focal length fused silica convex lens (Figure 1). The focusing lens also collects the plasma emission, which is fed to an Echelle spectrometer through a UV grade optical fiber. The spectrograph is fitted with a gated 1024x1024 intensified charge coupled device (CCD). Both gate delay and gate width are controlled by a pulse generator, which is synchronized with the laser. The data

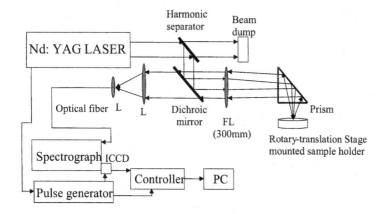

Figure 1. The schematic of LIBS system.

sampling time is set to 6s for good signal-to-noise ratio data. The pellet samples are mounted on a rotating platform, which can rotate with a rotational speed adjustable from 2.6 minutes per rotation to 8.6 minutes per rotation. The parameter such as laser energy, detection delay time and width and lens-to-samples distance were carefully studied to determine the optimum experimental parameters for the LIBS measurement.

In our previous study[5], we found that the LIBS measurement from a powder sample gave poorer precision. This is because of the uncertainty in the position of the laser focal spot due to shock-wave accompanying the laser breakdown. On the other hand the composition of powder samples can be accurately monitored by making pellets. The relative standard deviation (RSD) of the LIBS data obtained using pellets is about 10 times lower than that of the data acquired from powder samples. In this study, we analyze the glass batch only in pellet form. Four glass batches were prepared for this study. The glass batches are prepared by carefully weighing and thoroughly mixing the boron glass power with different percent of surrogate. 0.8 ml polyvinyl alcohol (2 wt% in distilled water) is added to the mixture as binder. 5g of this powder mixture is pressed into pellet by putting it in a 25mm bore die which is subjected to about 24MPa pressure. The pellets are about 5mm thick with 25mm diameter. All the glass batch samples were also analyzed chemically to get the exact concentration of the elements in particular samples (see Table I). To make glass sample for the LIBS calibration, we have prepared a stock mixture containing all the major component of glass (See Table II). Different concentration of oxides can be added to the stock mixture and heated in a furnace to produce glass with different concentration of metal for the calibration.

RESULTS

Glass batch measurements

The LIBS spectra of the glass batch were recorded at different focusing position, gate delay and laser energy to obtain the experimental condition for obtaining a good signal-to-noise ratio LIBS spectrum. A typical LIBS spectra of a glass batch pellet from a Echelle spectrometer is shown in Figure 2. The elements identified from the spectra are Si, Al, B, Ca, Mg, Mn, Na, K, Cr, Cu, Fe, Ti, Sr, Zr, Zn, Pb, and Ni. The quantitative spectral analysis involves relating the spectral line intensity of an element in the plasma to the concentration of that element in the target. The intensity of the desired atomic lines corresponding to the above elements was measured by integrating the peak area with an automatic spectral baseline correction. We have compared the analysis results using the peak area analysis with the peak height analysis. The peak height analysis provide more accurate results for the emission lines that have slight spectral interference while peak area analysis give better measurement precision for the interference free atomic lines. The four samples used in this study contain different elemental composition. Therefore, the data obtained can be used to create calibration data for the different elements as well as to evaluate the quantitative figure of merit of this system. Since the LIBS signal can be

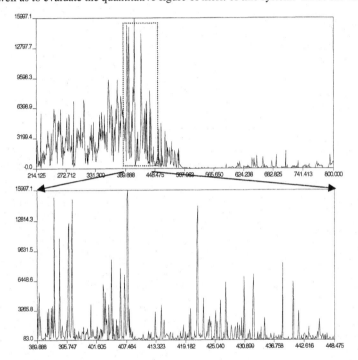

Figure 2. Typical LIBS spectrum from a Echelle spectrometer.

affected by a fluctuation in the experimental parameters, to reduce the effects of these variables, the intensity ratio provides a more reliable calibration in some cases. We have plotted the intensity ratio of the emission line of an element with the intensity of Si 288.158 nm line in a different sample against its concentration ratio to form a calibration curve. For all the elements identified, we obtained a linear calibration curve for concentration ratio with Si with a correlation coefficient very close to unity, except for Boron (B). Figure 3 shows some typical calibration curves for a glass batch pellet. The data in this study has a precision of 5% or better. Since the concentration ratio of the B-to-Si are very similar in these samples, we are unable to obtain reliable calibration from the intensity ratio of B to Si. However, we found that we can obtain a reasonable calibration using the intensity ratio of B-to-Fe using Fe 371.99 nm as the reference line. Figure 4 shows the calibration data for B. The nonlinear calibration for B/Fe is because the B line reabsorb the light at high concentration. A self-absorption model will be tested later to correct the nolinear calibration data. A comparison of LIBS analysis based on the calibration data of pellet with ICP analysis is shown in Table III. It is clear that the RA (= $(X_{LIBS}-X_{ICP})/X_{ICP}$) for most elements is better than 5%.

Table I. Concentration (wt%) of the major elements in glass batch

Glass Batch*	Al	Fe	Si	B	Ni	Zn	Mg	Sr	Zr
#1	6.18	11.39	18.88	2.20	0.78	2.41	7.435	6.22	9.68
#2	6.80	12.52	17.04	2.04	0.86	2.01	8.162	6.83	10.65
#3	8.04	14.80	13.35	1.73	1.02	1.20	9.649	8.08	12.59
#4	8.66	15.94	11.51	1.57	1.09	0.80	10.393	8.69	13.55

*Glass batch consists of surrogate and the glass powder.

Table II. Glass Batch and Glass Composition

Stock mixture used to make the test glass sample			Glass Analysis (DIAL Analytical Lab)		
Chemical	Weight %	Elemental Ratio to Si	Chemical	Weight %	Elemental ratio to Si
Al_2O_3	3.69	0.086	Al_2O_3	4.43	0.082
H_3BO_3	9.90	0.15	B_2O_3	8.47	0.092
$CaCO_3$	1.99	0.063	CaO	2.63	0.065
K_2CO_3	0.058	0.0015	K_2O	0.06	0.002
$MgCO_3$	3.65	0.046	MgO	0.98	0.021
MnO	1.34	0.045	MnO	1.12	0.030
Na_2CO_3	28.7	0.55	Na_2O	16.94	0.44
NiO	0.57	0.02	NiO	0.75	0.02
SiO_2	48.65	1.	SiO_2	61.37	1.
TiO_2	0.7	0.018	TiO_2	0.83	0.017
ZnO	0.26	0.0092	ZNO	0.4	0.011
ZrO_2	0.48	0.016	ZrO_2	0.52	0.013
Fe_2O_3	0	0	Fe_2O_3	0.38	0.009

Glass Measurement

The initial glass measurements were performed with the glass made from the stock mixture (see Table II). The best experimental parameters for glass measurement were also determined. We found that a lower laser energy (15mJ) and shorter delay time (1 μs) than with pellet samples (25mJ and 3 μs gate delay) can be used to achieve the same detection sensitivity. The measurement with the glass sample was repeated and the repeatability of the glass measurement was obtained for Mn/Si, Mg/SI, Al/Si, Ni/Si, Zn/Si, and B/Si. The RSTD of all these measurements are lower than 5% and Ti/Si has a RSTD lower than 6%.

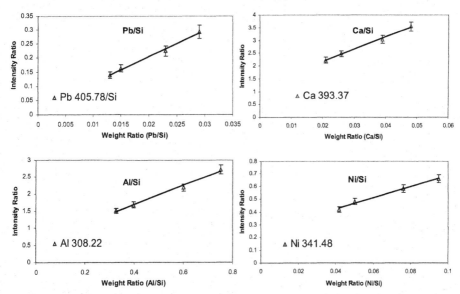

Figure 3.Typical calibration data obtained from the glass batch (in pellet)

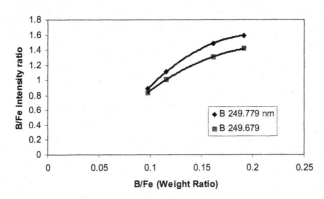

Figure 4. Calibration for B in glass pellet.

Table III. Comparison of LIBS and ICP Analysis Results of Pellets

| | ICP analysis | | LIBS Analysis | | |
X	X/Si	Wavelength(nm)	X/Si	RSTD	RA
Mn	0.239	257.61	0.251	0.12	0.052
Mg	0.005	280.27	0.0051	0.13	0.015
Al	0.398	308.22	0.395	0.02	0.0076
Ti	0.0053	499.11	0.0051	0.082	0.029
Fe	0.733	104.58	0.751	0.032	0.024
Ca	0.026	393.37	0.026	0.049	0.015
Ni	0.05	341.48	0.053	0.039	0.065
Sr	0.2	496.23	0.21	0.071	0.046
Pb	0.015	405.78	0.157	0.12	0.046
Ba	0.0042	455.4	0.00402	0.055	0.043
Cr	0.0024	425.43	0.0022	0.094	0.083
Cu	0.0023	427.4	0.0025	0.008	0.069
Zn	0.118	328.23	0.124	0.06	0.055
Zr	0.312	360.12	0.306	0.078	0.020
B	0.118	249.68	0.117	0.08	0.0096
La	0.0265	433.37	0.0272	0.12	0.029

* Reference line: Si 288.16 nm

CONCLUSION

This work demonstrates that LIBS is an analytical method which will allow a rapid on-line determination of the amount of waste and frit to be combined to produce a waste acceptance product specifications (WAPS) wasteform. A broadband Echelle spectrometer was used in this work to monitor all the major elements in the UV-VIS spectral region. LIBS spectra of glass batch were recorded to determine the best data analysis scheme. To perform quantitative measurements, we have also recorded LIBS spectra of the simulated samples with different elemental concentrations to plot the calibration curve for various elements. The results of LIBS analysis with the simulated samples were compared with ICP analysis to determine the accuracy of the measurement. We found that LIBS can provide an accuracy of 3-5% for elements with concentration >1% and 5-10% or better for minor elements in solid samples. The precision for glass batch (in pellet) and glass are better than 5% for most major elements (wt% great than 1%).

ACKNOWLEDGEMENT

Authors would like to thank Sharon Marra and David Peeler, SRNL, for discussions and Tracy Miller for help in experiment. This work is funded by U.S. DOE Cooperative agreement DE-FC26-98FT40395.

REFERENCES

[1]D.A. Cremers and L.J.Radziemski, "Laser Plasmas for chemical analysis" in *Laser Spectroscopy and Its Applications*, Chapter 5. Edited by L.J.Radziemski, R.W. Solarz and J.A. Paisner. Marcel Dekker, New York, 1987.

[2]F.-Y. Yueh, J. P. Singh and H. Zhang, "Laser-induced breakdown spectroscopy-elemental analysis", in *Encyclopedia of Analytical Chemistry,* Edited by R.A.Meyers. Wiley, New York, 2000.

[3]I. Schechter, "Laser induced plasma spectroscopy, a review of recent advances," Rev. Anal. Chem. **16,** 173-298 (1997).

[4] J-I, Yun, R. Klenze, and J.-I. Kim, "Laser-Induced Breakdown Spectroscopy for the On-Line Multielement Analysis of Highly Radioactive Glass Melt. Part I: Characterization and Evaluation of the Method", *Applied Spectroscopy* **56**, 437-448 (2002).

[5]B. Lal, H. Zheng, F.-Y. Yueh and J. P. Singh, "Parametric Study of Pellets for Elemental Analysis with Laser Induced Breakdown Spectroscopy", Appl. Opt. , 2004 (Accepted).

[6]G. Zikratov, R. Vasudev, F.Y. Yueh, J.P. Singh, and J.C. Mara., "Laser-induced Breakdown Spectroscopy of Hafnium doped Vitrified Glass," *Glass Technology*, **40**, 84-88.

[7]C. F Su, S. Feng, J.P. Singh, F.Y. Yueh, J.T. Rigby, D.L. Monts and R.L. Cook, "Glass Composition Measurement Using Laser-induced Breakdown Spectroscopy," *Glass Technology*, **41**, 16-21 (2000).

[8]A.K. Rai, H. Zhang, F.Y. Yueh, J.P. Singh and A. Weisburg, "Parametric Study of a Fiber Optic Laser Induced Breakdown Spectroscopy Probe for Analysis of Aluminum Alloy." *Spectrochemica Acta Part B:Atomic Spectroscopy,* **56**, 2371-2383 (2001).

[9]A.I. Whitehouse, J. Young, I.M. Botheroyd, S. Lawson, C.P. Evans and J. Wright, " Remote Material Analysis of Nuclear Power Station Steam Generator Tubes by Laser-Induced Breakdown Spectroscopy," *Spectrochimica Acta Part B* **56**, 821-830 (2001).

[10]F.Y., Yueh, R.C., Sharma, J.P. Singh, and H., Zhang, "Evaluation of the Potential of Laser-Induced Breakdown Spectroscopy for Detection of Trace Element in Liquid", *Journal Air & Waste Management Association* **52**, 1307-1315 (2002).

[11]A. Kumar, F. Y. Yueh, T. Miller, and J. P. Singh, "Detection of Trace Elements in Liquid using Meinhard Nebulizer by Laser-induced Breakdown Spectroscopy", *Applied Optics*, **42**, 6040-6046 (2003).

[12]G. Arca, A. Ciucci, V. Palleschi, S. Rastelli, and E. Tognoni, "Trace Element Analysis in water by the laser-induced breakdown spectroscopy technique," *Journal Applied Spectroscopy* **51** [8], 1102-1105 (1997).

[13]H. Zhang, F. Y. Yueh and J. P. Singh, "Laser Induced Breakdown Spectrometry as a Multi-Metal Continuous Emission Monitor," *Applied Optics* **38** [9] 1459-1466 (1999).

[14]H. Zhang, F. Y. Yueh and J. P. Singh, "Performance Evaluation of Laser Induced Breakdown Spectroscopy as a Multi-metal Continuous Emission Monitor", *Journal Air & Waste Management Association,* **51** 174 (2001).

[15]D. W. Hahn and M.M. Lunden, "Detection and Analysis of Particles by Laser Induced Breakdown Spectrometry", *Aerosol Science Technology*, **33** 30-48 (2002).

[16]D. W. Hahn, W. L. Flower, and K. R. Henken, "Discrete Particle Detection and Metal Emission Monitoring using Laser Induced Breakdown Spectrometry," *Applied Spectroscopy*, **51** 1836-1845 (1997).

THE STRUCTURAL CHEMISTRY OF MOLYBDENUM IN MODEL HIGH LEVEL NUCLEAR WASTE GLASSES, INVESTIGATED BY MO K-EDGE X-RAY ABSORPTION SPECTROSCOPY

Neil C. Hyatt,* Rick J. Short, Russell J. Hand and William E. Lee
Immobilisation Science Laboratory,
Department of Engineering Materials,
The University of Sheffield,
Mappin Street,
Sheffield, S1 3JD. UK.

Francis Livens
Centre for Radiochemistry Research,
Department of Chemistry,
The University of Manchester,
Oxford Road,
Manchester, M13 9PL. UK.

John M. Charnock and R.L. Bilsborrow
CLRC Daresbury Laboratory,
Warrington,
WA4 4AD. UK.

ABSTRACT

The immobilisation of molybdenum in model UK high level nuclear waste glasses was investigated by X-ray Absorption Spectroscopy (XAS). Molybdenum K-edge XAS data were acquired from several inactive simulant high level nuclear waste glasses, including specimens of Magnox and Magnox / UO_2 waste glasses prepared during commissioning of the BNFL Sellafield Waste Vitrification Plant. These data demonstrate that molybdenum is immobilised in the Mo (VI) oxidation state as the tetrahedral MoO_4^{2-} species, with a Mo-O contact distance of *ca* 1.76 Å. The MoO_4^{2-} species are not immobilised within the polymeric borosilicate network, instead they are likely to be located in extra-framework cavities, together with network modifier cations. Molybdenum K-edge XAS data acquired from "yellow phase" material, removed from a sample of simulant Magnox waste glass, show that this substance incorporates Mo (VI), also present as tetrahedral MoO_4^{2-} species. The presence of isolated MoO_4^{2-} tetrahedra within the glass matrix provides an explanation for the initial rapid release of molybdenum from simulant HLW waste glasses in static dissolution experiments and the strong correlation of the initial Mo leach rate with that of Na and B.

INTRODUCTION

In the UK and France the highly radioactive fission products arising from the reprocessing of nuclear fuel are currently immobilised in an alkali borosilicate glass matrix. The fission products are classified as High Level nuclear Waste (HLW) due to their heat generating capacity. Alkali borosilicate glasses are the matrix of choice for the immobilisation of waste fission products since they show considerable flexibility with respect to chemical constitution, permitting vitrification of waste streams of variable composition [1, 2]. In addition, alkali borosilicate glasses may be fabricated at reasonable temperatures, *ca* 1060°C, and show superior chemical durability with respect to other vitreous wasteforms (*e.g.* phosphate based glasses) [1, 2]. In Europe, vitrification of high level nuclear waste is currently undertaken at the Sellafield Waste Vitrification Plant (WVP), operated by British Nuclear Fuels plc, and the La Hague vitrification plants, R7 and T7, operated by COGEMA.

The incorporation of isotopes of molybdenum is of particular interest in the vitrification of HLW. This element is known to have a low solubility ($\leq$ 1 wt% MoO_3) in borosilicate glass compositions and its presence in excess results in the phase separation of a complex yellow molten salt during vitrification [1]. The formation of so-called "yellow phase" material during the vitrification process is undesirable since this leads to accelerated corrosion of the (inconel) melter. Furthermore, this material, which is reported to contain other fission products (including ^{137}Cs), is soluble in aqueous solution providing a potential route for the facile dispersal of radionuclides [1].

In order to understand the role of molybdenum in the fabrication of HLW glasses we have undertaken a Mo K-edge X-ray Absorption Spectroscopy (XAS) study of a range of molybdenum bearing glasses, as listed in Table I. Molybdenum bearing alkali borosilicate glasses were prepared, as simple model analogues of the chemically complex simulant waste glasses, under mildly oxidizing or reducing conditions by sparging the glass melt with compressed air (Sample 1) or N_2 / H_2 (95 / 5 %, Sample 2), respectively. Samples of inactive simulant HLW glass produced during commissioning of the Sellafield Waste Vitrification Plant were provided by BNFL plc, two compositions were examined corresponding to vitrified HLW arising from the reprocessing of Magnox nuclear fuel (Sample 3) and a blend of HLW arising from the reprocessing of Magnox and UO_2 fuel (Sample 4); these glasses were formulated to achieve a HLW loading of *ca* 25 wt% in the glass matrix, on an oxides basis. Laboratory simulant glasses designed to approximate vitrified HLW derived from electro-dissolution of spent UO_2 fuel elements, as a future alternative to the current Purex reprocessing technology, were also studied. These glasses were formulated to achieve a HLW loading of *ca* 20 wt% in the glass matrix, on an oxides basis, and were fabricated under both mildly oxidizing and reducing conditions by sparging the glass melt with compressed air (Sample 5) or N_2 / H_2 (95 / 5 %, Sample 6). In addition, a specimen of "yellow phase" material recovered from a sample of full scale Magnox waste glass was analysed (Sample 7). The Mo K-edge XAS data show that in materials fabricated under both mildly oxidising and reducing conditions Mo (VI) is present as the tetrahedral MoO_4^{2-} species. Electron spin resonance data show that Mo (III) species are stabilised within the model and simulant HLW glasses fabricated under reducing conditions; however, Mo (VI) remains the dominant species.

EXPERIMENTAL
Sample preparation
CaMoO$_4$ and SrMoO$_4$: Samples of $CaMoO_4$ and $SrMoO_4$ were prepared by solid state reaction of stoichiometric quantities of MoO_3 and $CaCO_3$ or $SrCO_3$ at 800°C for 48h, in air. These materials were confirmed to be single phase by X-ray powder diffraction.

Laboratory simulants (Samples 1, 2, 5 and 6): Stoichiometric quantities of metal oxides or carbonates were intimately mixed, according to the desired batch composition, and heated in an alumina crucible, at 1 °C min^{-1}, to 1000 °C; the crucible was then transferred to a glass melting furnace at 1150 °C and allowed to equilibrate at this temperature for 1h. Subsequently, a high density alumina tube was inserted into the molten glass allowing compressed air or N_2 / H_2 (95 / 5 %) to be bubbled through the melt at a rate of *ca* 1 dm^3 min^{-1}. The gas sparge allows the partial oxygen pressure in the melt to be varied and aids the formation of a homogeneous glass by continuously mixing the melt. The melts were sparged for 4.5 h at 1150 °C and then cast into rectangular blocks, annealed at 550 °C for 1 h and cooled to room temperature at 1 °C min^{-1} in air. Ag was used as a surrogate for the noble metals Pd and Ru in Samples 5 and 6. X-ray powder diffraction revealed the presence of metallic silver in Sample 6, whereas Samples 1, 2 and 5 were not found to contain any significant amount of crystalline material.

Table I: Compositions of molybdenum bearing alkali borosilicate glasses studied by XAS

Component (wt %)	Sample					
	1	2*	3	4	5	6*
SiO_2	61.18	61.18	46.10	46.28	49.44	49.44
B_2O_3	21.68	21.68	15.90	15.9	17.52	17.52
Na_2O	10.89	10.89	8.29	8.59	8.80	8.80
Li_2O	5.25	5.25	4.07	3.92	4.24	4.24
Ag_2O	-	-	-	-	1.41	1.41
BaO	-	-	0.50	0.24	0.94	0.94
CeO_2	-	-	0.84	1.86	1.33	1.33
Cs_2O	-	-	1.11	1.60	1.31	1.31
Gd_2O_3	-	-	-	2.92	2.77	2.77
La_2O_3	-	-	0.48	0.87	0.71	0.71
MoO_3	1.00	1.00	1.62	2.21	2.39	2.39
Nd_2O_3	-	-	1.44	2.77	3.00	3.00
Rb_2O	-	-	-	-	0.16	0.16
RuO_2	-	-	0.70	1.03	1.49	1.49
Sm_2O_3	-	-	0.22	0.44	0.44	0.44
SrO	-	-	0.30	0.55	0.41	0.41
TeO_2	-	-	-	0.31	0.12	0.12
TiO_2	-	-	0.01	0.06	0.53	0.53
Y_2O_3	-	-	0.10	0.36	0.25	0.25
ZrO_2	-	-	1.45	2.78	2.20	2.20
Fe_2O_3	-	-	3.00	1.10	0.33	0.33
NiO	-	-	0.37	0.21	0.09	0.09
Cr_2O_3	-	-	0.58	0.23	0.10	0.10
Al_2O_3	-	-	6.58	1.59	-	-
CaO	-	-	0.01	0.03	-	-
MgO	-	-	5.74	1.41	-	-
K_2O	-	-	0.01	0.15	-	-
P_2O_5	-	-	0.26	0.11	-	-
HfO_2	-	-	0.02	0.06	-	-
Pr_6O_{11}	-	-	0.44	0.85	-	-
Total	100.00	100.00	100.14	98.43	-	100.00

Samples 1 and 2 are simple "model" glasses; Samples 3 and 4 are specimens of full scale glasses designed to simulate vitrified Magnox / UO_2 and Magnox HLW, respectively; Samples 5 and 6 are laboratory glasses designed to simulate vitrified HLW arising from electro-dissolution of UO_2 fuel. * Denotes glasses prepared using a N_2 / H_2 sparge.

Full scale simulants (Samples 3 and 4): Full scale inactive simulant HLW glasses were prepared during commissioning trials at the Waste Vitrification Plant, Sellafield, UK. In outline, a charge of sodium lithium borosilicate glass frit was added to the inconel melter (volume ~0.12 m^3) and melted at *ca* 1080 °C with a compressed air sparge. Further frit, together with simulant high level nuclear waste material, was added to the melt over a period of hours to achieve the desired composition. The melt was allowed to homogenise over a period of 1h following the final addition of the simulant waste and frit, then discharged into a stainless steel canister (each canister holds ~400 kg of waste glass). Samples (~200 g) were randomly selected for study from the recovered material. X-ray powder diffraction revealed the presence of crystalline RuO_2 in the simulant Magnox and Magnox / UO_2 waste glasses (Samples 3 and 4, respectively).

"Yellow phase" material (Sample 7): An inclusion of "yellow phase" material, ~0.1 g, was removed from a sample of full scale simulant Magnox waste glass by scraping with a spatula. This material was found to be composed of $LiCsMoO_4$ and other unidentified phases, by powder X-ray diffraction.

X-Ray Absorption Spectroscopy

Mo K-edge X-ray absorption spectra were collected on Station 16.5 at the Synchrotron Radiation Source, Daresbury, UK. The storage ring operates at 2 GeV with a typical beam current of 150 mA. A double crystal Si (220) monochromator was used, detuned to 50% of maximum intensity, for harmonic rejection. The Mo K absorption edge was calibrated by measuring the K-edge from a Mo foil at E_0 = 19999.0 eV. Spectra were recorded at ambient temperature in transmission mode using ion chambers filled with a mixture of Ar (698 mbar) / balance He (incident beam, I_o) and Kr (368 mbar) / balance He (transmitted beam, I_t). The samples, in the form of fine powders, diluted where necessary with BN (in a *ca* 1:10 volume ratio), were packed into aluminium sample holders. At least two data sets were acquired per sample, over the energy range 19850 – 20850 eV. The spectral data were summed, calibrated and background subtracted using the programs EXCALIB, EXSPLINE and EXBACK. The background subtracted EXAFS data were analyzed using EXCURV98 [3], employing Rehr-Albers theory [4] and using single scattering. Phase shifts were derived from *ab initio* calculations using Hedin-Lundqvist potentials and von Barth ground states [5]. Theoretical fits were obtained by adding shells of backscattering atoms around the central absorber atom (Mo) and refining the Fermi energy, E_f, the absorber-scatterer distances, r, and the Debye – Waller factors, $2\sigma^2$, to minimize the sum of the squares of the residuals between the calculated and experimental EXAFS. The estimated precision in the refined absorber – scatterer contact distances is $\pm$ 0.02 Å, the relative precision in the refined Debye – Waller factors is estimated to be $\pm$ 25 %. Shell occupancies, N_{occ}, were fixed at the integral values which gave the best fit as indicated by the EXAFS goodness of fit parameter, or R-factor, defined in reference 6.

RESULTS AND DISCUSSION

Mo K-edge XAS data were acquired from samples of $CaMoO_4$ and $SrMoO_4$ to provide a set of reference data for comparison with that acquired from the model and simulant HLW glasses. $CaMoO_4$ and $SrMoO_4$ crystallise in the Scheelite ($CaWO_4$) structure, with isolated MoO_4^{2-} tetrahedra [7]. The measured position of the Mo K-absorption edges in $CaMoO_4$ and $SrMoO_4$, E_0 = 20007.3 eV, is consistent with the presence of Mo (VI). The background subtracted EXAFS data from both samples were fitted with two co-ordination shells: an inner shell of four oxygen atoms at *ca* 1.75 Å and an outer shell of 8 Ca or Sr atoms at *ca* 4.0 Å. The crystal structures of $CaMoO_4$ and $SrMoO_4$ are characterised by four equivalent Mo-O bond lengths and two distinct Mo-Ca/Sr contact distances to eight Ca/Sr near neighbours. For comparison, the Mo-O bond length and Mo-Ca/Sr contact distances for $CaMoO_4$ are, respectively, 4 x 1.757 Å, and 4 x 3.695 Å plus 4 x 3.402 Å; for $SrMoO_4$ the equivalent distances are 4 x 1.867 Å, and 4 x 3.814 Å plus 4 x 3.559 Å [7].

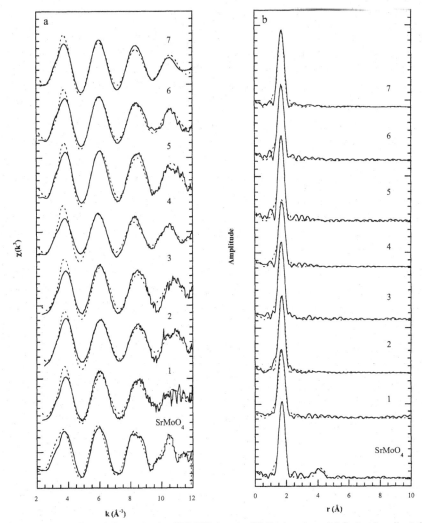

Figure 1: a) Experimental Mo K-edge EXAFS data (solid line) and model fit (broken line) for SrMoO$_4$, glass and "yellow phase" samples; b) corresponding radial distribution functions (see Introduction and Table I for sample identification).

The Mo-O contact distance for CaMoO$_4$ is in excellent agreement with the available crystallographic data for this compound. However, the Mo-O contact distance in SrMoO$_4$ is significantly longer than that determined by crystallographic methods. In both compounds the Mo-Ca and Mo-Sr contact distances are slightly larger than the average contact distance determined by crystallographic methods. The refined parameters are given in Table II and the background subtracted k^3-weighted EXAFS spectrum and associated Fourier transform for SrMoO$_4$ are shown in Figure 2.

The position of the Mo K-absorption edge determined for the glass and "yellow phase" samples was not significantly different from that determined for the $CaMoO_4$ and $SrMoO_4$ reference materials, within the experimental precision of ± 0.5 eV, see Table III. The Mo K-edge X-ray Absorption Near Edge Structure (XANES) of the glass and "yellow phase" samples are also similar to those of the $CaMoO_4$ and $SrMoO_4$ reference material, as demonstrated by Figure 2. These observations indicated that in these materials the oxidation state and speciation of molybdenum were essentially identical, $viz.$ Mo (VI) present as isolated MoO_4^{2-} species.

The background subtracted EXAFS from all glass samples and the "yellow phase" material were adequately fitted with a single co-ordination shell of four oxygen atoms at a distance of ca 1.76 Å from the central molybdenum absorber atom. The refined parameters are given in Table III and the background subtracted k^3-weighted EXAFS spectra and associated Fourier transforms are shown in Figure 1. The average Mo-O contact distance of 1.76 Å is consistent with the Mo-O bond length observed in isolated tetrahedral MoO_4^{2-} species; for example, the five different MoO_4 tetrahedra in $La_2(MoO_4)_3$ and $Ce_6(MoO_4)_8(Mo_2O_7)$ have average Mo-O bond lengths of 1.78 Å and 1.76 Å, respectively [8]. These data are also consistent with Mo K-edge XAS studies of molybdenum bearing $Na_2O-K_2O-SiO_2$ glasses in which an average Mo-O contact distance of 1.77 Å was determined [9].

Table II: Refined parameters for fit to XAS data for $CaMoO_4$ and $SrMoO_4$

Sample	E_0 (eV)	Shell	Type	N_{occ}	r (Å)	$2\sigma^2$ (Å^2)	$2\sigma_{Mo}^2$ (Å^2)	R
$CaMoO_4$	20007.3	1	O	4	1.76	0.007	0.010	24.49
		2	Ca	8	4.16	0.028		
$SrMoO_4$	20007.3	1	O	4	1.74	0.009	0.010	22.51
		2	Sr	8	4.08	0.018		

Table III: Refined parameters for fit to EXAFS data for glass and "yellow phase" samples, (see Introduction and Table I for sample identification).

Sample	E_0 (eV)	Shell	Type	N_{occ}	r (Å)	$2\sigma_{Mo}^2$ (Å^2)	$2\sigma_O^2$ (Å^2)	R
1	20007.5	1	O	4	1.74	0.010	0.013	36.7
2	20007.5	1	O	4	1.74	0.010	0.012	29.7
3	20007.3	1	O	4	1.78	0.010	0.012	30.2
4	20007.5	1	O	4	1.74	0.010	0.010	36.1
5	20007.5	1	O	4	1.76	0.010	0.009	19.0
6	20007.4	1	O	4	1.78	0.010	0.010	34.0
7	20007.4	1	O	4	1.77	0.010	0.005	26.2

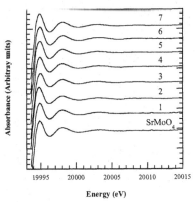

Figure 2: Mo K-edge XANES for SrMoO₄, glass and "yellow phase" samples, (see Introduction and Table I for sample identification).

The EXAFS and XANES data presented here show that Mo (VI) is present as the MoO_4^{2-} species in both simple model alkali borosilicate glasses and the more chemically complex simulant HLW glasses. Since only one co-ordination shell was required to adequately fit the EXAFS data and the corresponding Fourier transforms show no significant peaks (corresponding to potential next nearest neighbour Si atoms) at contact distances greater than 1.76 Å we infer that the MoO_4^{2-} species do not form part of the polymeric network in these materials. Instead, these MoO_4^{2-} species are likely to be encapsulated, together with network modifier cations, within cavities defined by the borosilicate glass network. This conclusion is in agreement with the model proposed by Calas et al., based on molecular dynamics simulations and a Mo K-edge XAS study of the French SON68 simulant HLW glass [10].

The presence of isolated MoO_4^{2-} species within cavities defined by the polymeric glass network raises pertinent questions with respect to leaching of molybdenum from (simulant) HLW glasses. In the general model for glass corrosion in aqueous solution, network hydrolysis precedes network dissolution and the release of alkali metals, followed by the precipitation of secondary alteration products as leached species reach solubility limited concentrations. Thus, on the basis of a model of isolated MoO_4^{2-} species encapsulated within cavities formed by the polymeric glass network, the rate of release of Mo, Na and B (a tracer for network dissolution) would be expected to be closely correlated. Indeed, static leach test data from simulant Magnox waste glasses indicate the initial leach rates of Na, B and Mo to be essentially equivalent [11].

X-band Electron Spin Resonance (ESR) spectra acquired at room temperature from the glasses fabricated under reducing conditions (Samples 2 and 6) show a sharp resonance centered at $g = 1.9060(1)$, see Figure 3. This is consistent with the ESR data acquired from reduced molybdenum bearing silicate glasses studied by Camara et al. and Horneber et al., and is attributed to Mo (III) species trapped within the alkali borosilicate glass matrix [12, 13]. In contrast, the ESR spectra acquired from glasses fabricated under mildly oxidising conditions were silent (Sample 1) or were characterised by a broad signal associated with paramagnetic transition metal species (Samples 3, 4 and 5), see Figure 3b. These data confirm that reducing conditions are indeed effective in stabilising Mo (III) species within these glasses, however, the position of the Mo K-absorption edge in the XAS spectra of these glasses, 20007 eV, demonstrates that Mo (VI) is the dominant species. Spin counts acquired using a $CuSO_4.5H_2O$ standard indicated that approximately 1.6 % of the total molybdenum in Sample 2 is present as reduced Mo (III) species.

It has been suggested that reducing the oxidation state of molybdenum may increase the solubility of this element in HLW waste glasses [1], thereby inhibiting the formation of the "yellow phase" molten salt during vitrification thus reducing the rate of melter corrosion. The data presented here confirm that reduced Mo (III) species may be stabilised in such glasses, although at low concentration. It is anticipated, however, that a reducing environment may result in the precipitation of noble metals from the glass, as indicated by the observation of metallic silver in Sample 6, prepared under reducing conditions. This is undesirable since noble metal precipitates are known to form a viscous "heel" at the bottom of the melter which may lead to blockage of the drain during discharge. Therefore, there would appear to be little incentive to fabricate HLW glasses under a reducing atmosphere in an effort to increase the incorporation of molybdenum.

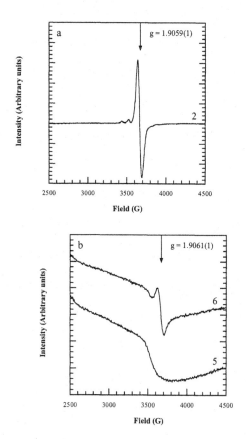

Figure 3: Electron spin resonance data from a) Sample 2 and b) Samples 5 and 6, showing a sharp resonance attributed to Mo (III) species in glasses fabricated under reducing conditions (Samples 2 and 5; see Introduction and Table I for sample identification).

CONCLUSIONS

Analysis of Mo K-edge XAS data has shown that Mo (VI) is present as isolated, tetrahedral, MoO_4^{2-} species in alkali borosilicate glasses and recovered "yellow phase" material. Electron Spin Resonance data confirm that reducing conditions allow the stabilisation of Mo (III) species within the glass melt, although at low concentration and Mo (VI) remains the dominant species. The presence of isolated MoO_4^{2-} tetrahedra within the glass matrix provides an explanation for the initial rapid release of molybdenum from simulant HLW waste glasses in static dissolution experiments and the strong correlation of the initial Mo leach rate with that of Na and B.

ACKNOWLEDGEMENTS

We thank Mr. Ian Watts (Department of Engineering Materials, The University of Sheffield) for assistance in melting of laboratory simulant glasses and Dr. Paul Anderson and Mr. Terry Green (School of Chemistry, University of Birmingham) for the acquisition of ESR data. We acknowledge Mr. Charlie Scales, Dr. Ivan Owens and Dr. Ewan Maddrell (BNFL plc) for useful discussions and provision of the Magnox and Magnox / UO_2 blended simulant glasses. We are grateful to BNFL plc and EPSRC for funding this work.

REFERENCES

[1] W. Lutze and R.C. Ewing, "Nuclear Wasteforms for the Future", Elsevier (1998).

[2] I.W. Donald, B.L. Metcalfe and R.N.J. Taylor, "The Immobilization of High Level Radioactive Wastes using Ceramics and Glasses", *Journal Materials Science*, **32** 5851 (1997).

[3] N. Binsted, Daresbury Laboratory EXCURV98 Program, (1998).

[4] J.J. Rehr and R.C. Albers, "Scattering-Matrix Formulation of Curved-Wave Multiple-Scattering Theory: Application to X-ray-Absorption Fine Structure", *Phys. Rev. B*, **41** 8139 (1990).

[5] L. Hedin and S. Lundqvist, *Solid State Phys.*, **23** 1 (1969).

[6] N. Binsted, R.W. Strange, and S.S. Hasnain, "Constrained and Restrained Refinement in EXAFS Data Analysis with Curved Wave Theory", *Biochem.*, **31** 12117 (1992).

[7] E. Guermen, E. Daniels, J.S. King, "Crystal Structure Refinement of $SrMoO_4$, $SrWO_4$, $CaMoO_4$, and $BaWO_4$ by Neutron Diffraction", *J. Chem. Phys.*, **55** 1093 (1971).

[8] B.M. Gatehouse and R. Same, "The Crystal Structure of a Complex Cerium (III) Molybdate; $Ce_6(MoO_4)_8(Mo_2O_7)$", *J. Solid State Chem.*, **25** 115 (1978).

[9] N. Sawaguchi, T. Yokokawa and K. Kawamura, "Mo K-edge XAFS in Na_2O-K_2O-SiO_2 glasses", *Phys. Chem. Glasses*, **37** 13 (1996).

[10] G. Calas, M. Le Grand, L. Galoisy and D. Ghaleb, "Structural Role of Molybdenum in Nuclear Glasses: an EXAFS Study", *Journal Nuclear Mate*rials, **322** 1 (2003).

[11] P.K. Abraitis, "Dissolution of a Simulated Magnox Waste Glass at Temperatures below 100°C", Ph. D. Thesis, University of Manchester, (1999).

[12] B. Camara and W. Lutze and J. Lux, "An Investigation on the Valency State of Molybdenum in Glasses with and without Fission Products", *Scientific Basis for Nuclear Waste Management*, **2** 93 (1980).

[13] A. Horneber, B. Camara and W. Lutze, "Investigation of the Oxidation State of Molybdenum in Silicate Glasses", *Scientific Basis for Nuclear Waste Management*, **5** 279 (1982).

Ceramic Waste Forms—
Formulation and Testing

ALPHA DECAY DAMAGE IN CERAMIC WASTE FORMS – MICROSTRUCTURAL ASPECT

Jan-Fong Jue, Thomas P. O'Holleran, Steven M. Frank and Stephen G. Johnson
Argonne National Laboratory – West
P. O. Box 2528
Idaho Falls, ID 83403

ABSTRACT

The sodalite-based ceramic waste forms developed by Argonne National Laboratory, an end product of the electro-metallurgical treatment of metallic spent nuclear fuels, contain small amounts of actinides. The major constituents of the ceramic waste forms are binding glass and sodalite. Actinide-bearing phases are generally located in the glass close to the interface between the glass and sodalite. Very few actinide-bearing particles were found within the sodalite areas. Radiation damage in some nuclear waste forms due to the alpha decay of actinides has been reported. The alpha decay damage in the sodalite-based ceramic waste forms is not anticipated to be homogeneous because of the limited penetration range of alpha particles and consequent recoiling nuclei. Simulation of the alpha decay damage using computer software suggests that the extent of alpha decay induced damage strongly depends on the microstructure of the ceramic waste forms. The most important variables are the size and distribution of the actinide containing phases. The predicted damage of the ceramic waste form will be compared with the experimental data obtained from a four-year aging study of a Pu-238 doped ceramic waste form. The effect of the microstructure on alpha decay damage in the ceramic waste form is also discussed.

INTRODUCTION

Argonne National Laboratory has developed an electro-metallurgical technique to treat metallic spent nuclear fuels. This process generates two types of waste forms – a ceramic waste form and a metal waste form. The ceramic waste form developed in this program will contain up to 0.2 weight percent of Pu-239 [1,2]. The alpha decay of actinides in the ceramic waste form is expected to generate radiation damage. Studies on the radiation damage by alpha decay in waste form systems are available in literature [3-9]. A better understanding on the effect of alpha decay on ceramic waste forms developed in this program will be important to predict the long-term durability of these waste forms in a repository [10,11]. It was observed that changing processing parameters could alter the microstructure of the ceramic waste - especially the distribution of actinide bearing phases. In this study, the penetration depth of the alpha particles and the recoiling nuclei was calculated using computer software. By taking into account the range of alpha decay damage, the influence of microstructure on the alpha decay damage in the ceramic waste can be better predicted.

*The submitted manuscript has been created by the University of Chicago as Operator of Argonne National Laboratory under contract No. W-31-109-ENG-38 with the U.S. Department of Energy. The U.S. Government retains for itself, and others acting on its behalf, a paid-up nonexclusive, irrevocable worldwide license in said article to reproduce, prepare derivative works, distribute copies to the public, and perform publicly and display publicly, by or on behalf of the Government.

EXPERIMENTAL PROCEDURE

Two forms of zeolite 4A ($Na_{12}(SiO_2)_{12}(AlO_2)_{12} \cdot xH_2O$) materials were used to occlude the salts resulting from the electrochemical refining process: one is powder (2-4 μm in size) and the other one is granular (45-250 μm in size). After salt occlusion at ~500°C, the salt occluded zeolite was mixed with a binder glass (P57 glass: 66.5wt% SiO_2, 19.1wt% B_2O_3, 6.8wt% Al_2O_3, 7.1wt% Na_2O, 0.5wt% K_2O) then packed in containers for consolidation. Surrogate ceramic waste form was consolidated in three different routes: hot uni-axial pressing (HUP waste form), hot iso-static pressing (HIP waste form) and pressureless consolidation (PC waste form). The consolidation temperatures for HUP, HIP, and PC processes are 750°C, 850°C and 915°C, respectively. In the actual large-scale spent fuel treatment, the starting zeolite will be in granular form and the ceramic waste forms will undergo pressureless consolidation at 915°C. The salt occluded zeolite is converted to sodalite ($Na_8(AlSiO_4)_6Cl_2$) during consolidation.

Surrogate waste forms prepared by HUP, HIP and PC processes were characterized using X-ray Diffraction (XRD), Scanning Electron Microscopy (SEM), Transmission Electron Microscopy (TEM) and Energy Disperive Spectrometry (EDS) in order to determine their phase contents, chemical compositions, and microstructure. Samples prepared by HUP and HIP were also characterized for the size and distribution of the actinide and rare earth bearing phases using transmission and scanning electron microscopy.

Simulation of alpha decay damage was conducted using the SRIM-2003 software. The penetration depth and the amount of displacements generated by both alpha particles and recoiling nuclei were determined in four different matrix materials - sodalite, uranium oxide, P57 binding glass, and the binding glass after consolidation (chemical composition of the binding glass changes after consolidation).

RESULTS

Figure 1 is an optical micrograph of a PC ceramic waste form and Figure 2 is an optical micrograph of a HIP waste form. Both micrographs show that the size of sodalite regions is similar to the starting granular zeolite (70-150μm). The elongated phase is mainly glass. Figure 3 is an SEM micrograph of the PC waste form. The darker phase and brighter phase are primarily the binding glass and sodalite, respectively. EDS data show that the chemical composition of the glass phase has significantly higher aluminum concentrations than the starting P57 binding glass. Since the only other source of aluminum is sodalite, this finding indicates that a reaction between the glass and sodalite has occurred during consolidation. The ratio between Al and Si in the glass phase changed from 1 to 8 in the starting P57 glass to 1 to 2 after pressureless consolidation. Figure 4 is an SEM micrograph of the HIP ceramic waste form. The glass regions have a darker contrast. EDS data show that the ratio between aluminum and silicon in the glass phase ranges between 1 to 8 (the same as the starting P57 binding glass) and 1 to 2 (the same as the glass phase in the PC waste form). Since the PC waste form has a higher processing temperature, these results may suggest that the degree of reaction between the P57 glass and sodalite is a function of consolidation temperature and time. Rare earth bearing phases in the ceramic waste are usually found in the sodalite regions or near the sodalite-glass interface. These particles exist as clusters or small isolated particles. Actinide bearing particles are mainly found in the glass phase, but not too far away from the sodalite-glass interface. Their sizes vary from several nanometers to about a micron. Figure 5 is a TEM micrograph. It shows the interfacial actinide bearing particles in a surrogate ceramic waste form. The dark particles in the glass region are actinide bearing particles (PuO_2).

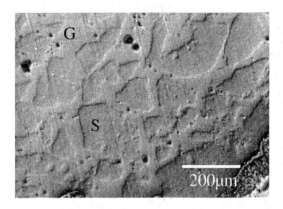

Figure 1. An optical micrograph of a PC waste form. "S" is sodalite region and "G" is the glass region.

Figure 2. An optical micrograph of a HIP waste form. "S" is sodalite region and "G" is the glass region.

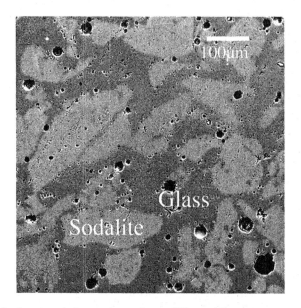

Figure 3. A backscattered electron image from a PC waste form.

Figure 4. A backscattered electron image from a HIP waste form.

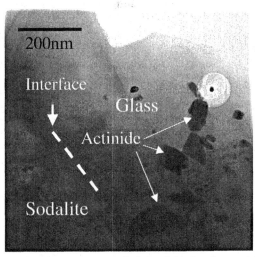

Figure 5. A TEM micrograph shows some interfacial actinide bearing particles in a surrogate ceramic waste form.

An alpha decay in the ceramic waste results in an alpha particle (assumed to have an average energy of 5.5 MeV) and a recoiling particle (assumed to have an average energy of 90keV). The penetration depth of these particles is calculated using the SRIM-2003 software. The matrix materials used in these calculations are: (i) sodalite, (ii) P57 glass, (iii) glass after PC consolidation, and (iv) uranium oxide. The results of the calculations indicate that the penetration depth of the alpha particles varies between 13 and 25 μm and the penetration depth of the recoiling nuclei varies between 16 to 40 nm, depending on the matrix material used. While each alpha particle displaces a couple hundred matrix atoms, each recoiling particle displaces more than 1200 matrix atoms. Table I summarizes the calculated penetration depths of alpha particle and recoiling nucleus. No significant difference was found in penetration depth between the P57 glass and the glass phase after pressureless consolidation.

Table I. Calculated penetration depth and number of displacements for different matrix materials. Displacement energy is assumed to be 25eV.

	Alpha Particle (5.5 MeV)		Recoiling Nucleus (90keV)	
	Penetration Depth (μm)	Number of Displacements	Penetration Depth (nm)	Number of Displacements
Sodalite	24.5	205	40.3	1345
P57 glass	24.6	227	39.6	1318
Interfacial Glass	23.9	234	40.1	1343
UO$_2$	13.1	215	16.6	1241

DISCUSSION

Based on the microstructure observed in the ceramic waste forms, the distribution of actinide bearing phases can be considered to be homogeneous only on the hundred-micron scale. The scale of homogeneity strongly depends on the size of the starting zeolite materials and binding glass. For the damage created by alpha particles, the damage is expected to be relatively homogeneous in most glass regions since they are usually elongated in shape. The middle portion of the sodalite region, however, will not experience much damage from alpha particles because the penetration depth of alpha particle is less than half of the diameter of most sodalite regions. Due to the small penetration depth of recoiling nuclei (less than 50 nm), the damage caused by recoiling nuclei is limited to the actinide bearing phases and their neighboring areas (within 50 nm). Electron microscopy shows that the actinide bearing phases are usually located in the glass phase but seldom more than several microns from the sodalite-glass interface. Based on this observation, the damage from the recoiling nuclei will be concentrated in the actinide bearing phases, the glass regions within several microns from the glass-sodalite interface and the sodalite region within 50 nm from the interface. Since most of the collisions between the matrix atoms and the recoiling nuclei occur at the end of the path of recoiling nuclei, isolated actinide bearing particles with a size less than 10 nm will experience only a small fraction of the displacements caused by recoiling nuclei. Most of the displacements will happen in the neighboring areas around them.

The average displacements-per-atom (dpa) is usually a good indicator for alpha decay damage in nuclear waste systems. However, it may not be a good indicator for the ceramic waste forms developed in this program. The reason is that the distribution of actinide bearing phases in the ceramic waste forms is not homogeneous on the scale of the penetration depth of recoiling nuclei (tens of nanometers). Estimating the alpha decay damage near the sodalite-glass interfacial areas by assuming that the actinide elements are distributed homogeneously throughout the waste form may slightly underestimate the number of matrix atoms displaced by the alpha particles and significantly underestimate the number of matrix atoms displaced by recoiling nuclei.

In an experimental effort to predict the effect of alpha decay on the long term performance of the ceramic waste forms in the repository, a HUP ceramic waste form containing about 1.27 weight percent Pu-238 was fabricated and aged for a four-year accelerated test [12]. The average alpha decay events over the entire waste form correspond to what the actual ceramic waste form will experience in more than five thousand years. The local displacement damage in the actinide bearing phases and the interfacial area between the binding glass and sodalite is expected to be even more severe. The unit cell expansion in the actinide bearing phases as well as amorphization and microcracks around these particles are likely to occur. The conclusions from this four-year aging study were: (i) the ceramic waste form showed no detectable density change after aging, (ii) no helium bubble formation was evidenced, (iii) no amorphization of crystalline phases was observed, (iv) unit-cell expansion in plutonium oxide was detected, and (v) no microcracks were found around the actinide bearing phases. These results indicate that although the radiation damage by alpha decay in the ceramic waste forms is not homogeneous, the segregation of the damage will not create mechanical integrity concerns in the repository.

CONCLUSIONS

Based on the calculations, the change in chemical composition of the glass phase during consolidation will not significantly influence the range and degree of alpha decay damage.

The microstructure of the ceramic waste form has a significant influence on long-term performance because actinides are not homogeneously distributed and the penetration depth of recoil nuclei is limited compared to the dimensions of microstructural features.

The information obtained from the four-year aging study provides an experimental confirmation that the resistance of the ceramic waste forms against alpha decay damage is acceptable.

ACKNOWLEDGMENTS

This work was supported by the Department of Energy, Nuclear Energy Research and Development Program, under contract No. W-31-109-ENG-38. Special thanks Dr. J. I. Cole, and Mr. M. Surchik.

REFERENCES

[1] M. F. Simpson, K. M. Goff, S. G. Johnson, T. J. Bateman, T. L. Moschetti, T. J. Battisti, K. L. Toews, S. M. Frank, T. L. Moschetti, T. P. O'Holleran, and W. Sinkler, "A Description of the Ceramic Waste Form Production Process from the Demonstration Phase of the Electrometallurgical Treatment of EBR-II Spent Fuel," *Nuclear Technology*, **134** 263-277 (2001).

[2] W. Sinkler, D. W. Esh, T. P. O'Holleran, S. M. Frank, T. L. Moschetti, K. M. Goff, S. G. and Johnson, "TEM Investigation of a Ceramic Waste Form for Immobilization of Process Salts Generated during Electrometallurgical Treatment of Spent Nuclear Fuel," *Ceramic Transactions*, **107** 233-240 (2000).

[3] R. C. Ewing, W. J. Weber, and F. W. Jr. Clinard, "Radiation Effects in Nuclear Waste Forms for High-Level Radioactive Waste," *Progress in Nuclear Energy*, **29** 63-127 (1995).

[4] W. J. Weber, "Ingrowth of Lattice Defects in Alpha irradiated UO_2 Single Crystals," *Journal of Nuclear Materials*, **98** 206-215 (1981).

[5] W. J. Nellis, "The Effect of Self-Radiation on Crystal Volume," *Inorganic and Nuclear Chemistry Letters*, **13** 393-398 (1977).

[6] Y. Inagaki, H. Furuya, and K. Idemitsu, "Microstructure of Simulated High-Level Waste Glass Doped with Short-Lived Actinides, [238] Pu and [244]Cm," *Materials Research Society Symposium Proceedings*, **257** 199-206 (1992).

[7] W. J. Weber, "Alpha-Decay-Induced Amorphization in Complex Silicate Structure," *Journal of the American Ceramic Society*, **76** 1729-1738 (1993).

[8] W. J. Weber, "Radiation-Induced Defects and Amorphization in Zircon," *Journal of Materials Research*, **5** 2687-2697 (1990).

[9] S. Utsunomiya, S. Yudintsev, L. M. Wang, and R. C. Ewing, "Ion-Beam and Electron beam Irradiation of Synthetic Britholite," *Journal of Nuclear Materials*, **322** 180-188 (2003).

[10] S. M. Frank, T. L. Barber, T. Disanto, K. M. Goff, S. G. Johnson, J-F. Jue, M. Noy, T. P. O'Holleran, and W. Sinkler, "Alpha-Decay Radiation Damage Study of a Glass-Bonded Sodalite Ceramic Waste Form," *Materials Research Society Symposium Proceedings*, **713** 487-494 (2002).

[11] B. G. Storey, and T. R. Allen, "Radiation Damage of a Glass-Bonded Zeolite Waste Form Using Ion Irradiation," *Materials Research Society Symposium Proceedings*, **481** 413-418 (1998).

[12] J. F. Jue, S. M. Frank, T. P. O'Holleran, T. L. Barber, S. G. Johnson, K. M. Goff, and W. Sinkler, "Aging Behavior of a Sodalite Based Ceramic Waste Form," *Ceramic Transactions*, **155** 361-369 (2004).

CHARGE COMPENSATION IN Ca(La)TiO$_3$ SOLID SOLUTIONS

E.R.Vance, B.D.Begg, J.V.Hanna and
V.Luca
Materials and Engineering Science
ANSTO
Menai, NSW 2234, Australia

J.H.Hadley and F.H.Hsu
Dept of Physics and Astronomy
Georgia State University
Atlanta, GA, USA

ABSTRACT

Different modes of charge compensation for trivalent rare earth ions substituted for Ca in perovskite formed by high-temperature sintering are discussed. Cation vacancies were deduced from positron annihilation lifetime spectroscopy (PALS) measurements to exist in Ca$_{(1-x)}$La$_x$TiO$_3$ solid solutions (x = 0.001- 0.2) fabricated in air at ~1500°C whether or not they were further heated in reducing atmospheres. Electron paramagnetic resonance measurements have shown the presence of paramagnetic species in Ca$_{(1-x)}$La$_x$TiO$_3$ samples fabricated in air, but their identities are not yet clear (other than Mn^{2+} impurities), although their intensities are not linear in La content, so the species of interest are unlikely to be Ti^{3+}. ^{139}La static nuclear magnetic resonance (NMR) measurements show the La ions in Ca$_{(1-x)}$La$_x$TiO$_3$ samples fabricated in air to yield very broad featureless line shapes, indicative of low site symmetry and/or interactions with paramagnetic species. However, corresponding broad resonances were also observed in vacancy-compensated samples; this line broadening observed in the compensated and uncompensated systems is attributed to low site point symmetry and short-range disorder in bond angles and bond lengths around each La^{3+} position. This proposition is supported by the observation of narrow ^{139}La NMR resonances in the vacancy-compensated system Sr$_{(1-3x/2)}$La$_x$TiO$_3$ (for x • 0.1), where the point symmetry of the Sr site was cubic rather than orthorhombic, as encountered in the analogous Ca$_{(1-x)}$La$_x$TiO$_3$ system. Samples were fabricated in air with Ca$_{(1-x)}$La$_x$Ti$_{(1-x/4)}$O$_3$ stoichiometry to promote Ti vacancies, and the solid solubility limit corresponded to 0.10 f.u. of such vacancies.

INTRODUCTION

Charge compensation models in RE-doped CaTiO$_3$ (and more particularly BaTiO$_3$) have been widely discussed in the literature. There appears to be agreement that single-phase air-fired stoichiometries of Ca$_{(1-3x/2)}$RE^{3+}$_{(x)}$TiO$_3$ contain x/2 formula units (f.u.) of Ca vacancies [1,2]. Compensation in air-fired single-phase Ca$_{(1-x)}$RE^{3+}$_{(x)}$TiO$_3$ compositions is more contentious [3-5,6]. Vance et al. [3] argued that x f.u. of Ti^{3+} were the compensation species, but have recently concluded from positron annihilation lifetime spectroscopy (PALS) that an equal mixture of Ca and Ti vacancies are applicable to this case [7]. This assignment was based on the ~ 20% increase in the annihilation lifetime of the major component of the PALS spectrum, induced by La doping. By contrast, Vashook et al. [8] argued that Ca vacancies only were present in such materials. In the present work, we have made: (a) initial studies by electron paramagnetic resonance to bear on the question of whether Ti^{3+} could be a significant charge compensating species in the air-fired Ca$_{(1-x)}$La$_x$TiO$_3$ samples (not withstanding the above); (b) solid-state nuclear magnetic resonance studies of ^{139}La in Ca$_{(1-x)}$La$_x$TiO$_3$, Ca$_{(1-3x/2)}$La$_x$TiO$_3$ and Sr$_{(1-3x/2)}$La$_x$TiO$_3$ samples. We have also investigated whether La-doped CaTiO$_3$ could be fabricated with Ti vacancies as the charge compensators via a stoichiometry of Ca$_{(1-x)}$La$_x$Ti$_{(1-x/4)}$O$_3$.

EXPERIMENTAL

Samples of $Ca_{1-x}La_xTiO_3$ stoichiometry, containing 0.00, 0.001, 0.005, 0.01, 0.02, 0.05, 0.1 and 0.15 formula units of La, were made up by the standard alkoxide/nitrate route [9], with a final firing treatment of 4 days at 1550°C in air. Other samples nominally containing Ti vacancies, were similarly fired and these had compositions of $Ca_{(1-x)}La_xTi_{(1-x/4)}O_3$, where x = 0.05 and 0.1. Further Ca-vacancy-bearing samples of $Ca_{(1-3x/2)}La_xTiO_3$ composition were similarly prepared, as well as a $Sr_{0.94}La_{0.04}TiO_3$ sample.

EPR studies at a frequency of ~ 9.3 GHz were made using a Bruker EMX spectrometer with a modulation frequency of 100 kHz and a modulation field of 10G. A power of ~ 0.5 mW was generally used. The temperature was normally 77K, but some measurements at temperatures in the 10-77K range were performed, using a liquid He cryoflow device. Concentrations of paramagnetic centres in the perovskites were estimated by comparison with signals due to Cu^{2+} in aqueous $CuSO_4$ solutions, cooled to 77K, and of similar volumes to those of the perovskites.

Scanning electron microscopy was carried out with a JEOL JXA-480 instrument run at 15 keV, and fitted with a Tracor Northern energy-dispersive TN5540 microanalysis facility [3].

X-ray diffraction data were obtained with a Siemens D500 diffractometer.

For positron annihilation lifetime spectroscopy a conventional fast-fast lifetime spectrometer employing plastic scintillators was used, with a ^{22}Na source sandwiched between two pieces of sample material. A time resolution of 0.320 ns was measured for a ^{60}Co "prompt" at FWHM under actual experimental conditions. The sample temperature was held at 21°C using an Oxford Instruments closed-cycle helium refrigerator. The temperature is measured and controlled using a Lakeshore controller with an IEEE interface to the computer. The computer controls counting times, stopping and starting, and temperature changes. A commercially available computer program, POSITRONFIT-EXTENDED is used to analyse the data. This program determines background and also deconvolutes to remove the "prompt curve" (instrumental broadening) before fitting the data (counts versus time) to obtain the lifetime parameters (lifetimes and their corresponding intensities) of up to three spectral components (exponential decays).

Solid-state static ^{139}La nuclear magnetic resonance (NMR) measurements were performed at ambient temperatures on a Bruker MSL-400 spectrometer operating at the ^{139}La frequency of 56.52 MHz. The solid echo θ-τ-θ-τ-acquire sequence was employed to obtain undistorted broadline spectra; this experiment was implemented on Bruker static probe which consisted of a 7.5 mm horizontal solenoid arrangement. ^{139}La pulse time measurement and chemical shift referencing was performed using a 0.01 M $LaCl_3$ solution. Non-selective pulse times of 4.5 μs were obtained for this I=7/2 quadrupolar system, which corresponded to selective pulse times of 1.0 μs for the central transition. This solution resonance was arbitrarily set to 0.0 ppm.

RESULTS AND DISCUSSION

$Ca_{(1-x)}La_xTiO_3$ and $Ca_{(1-3x/2)}La_xTiO_3$ Stoichiometries

From the electron paramagnetic resonance results on $Ca_{(1-x)}La_xTiO_3$ samples (Figure 1), it seems clear that there are paramagnetic centres associated with La-doping in single-phase $Ca_{(1-x)}La_xTiO_3$ samples heated in air. The spectra of the $La_{0.01}Ca_{0.99}TiO_3$ and $La_{0.10}Ca_{0.90}TiO_3$ samples are shown in Figure 1a and 1b and appear to be quite similar. The spectra consist of an intense, broad spectral envelope centered at about 3500 G. Superimposed on this broad intense resonance is a pronounced fine structure. When the spectrum of the $La_{0.01}Ca_{0.99}TiO_3$ sample is converted to the second derivative (Figure 2) it becomes apparent that the fine structure results from the superposition of the six-line hyperfine spectrum of the Mn^{2+} ion which is ostensibly present as an impurity in some of the reactants used for the preparations.

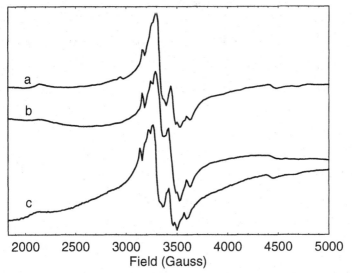

Figure 1.

77 K EPR spectra of $La_xCa_{1-x}TiO_3$; (a) $x = 0.1$, (b) $x = 0.01$ and (c) $La_{0.01}Ca_{0.99}{}^{49}TiO_3$.

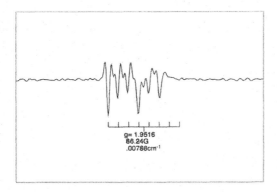

g= 1.9516
86.24G
.00786cm⁻¹

Figure 2. Second derivative spectrum of $La_{0.01}Ca_{0.99}TiO_3$ sample showing the six-line hyperfine pattern of a Mn^{2+} (I=5/2) impurity.

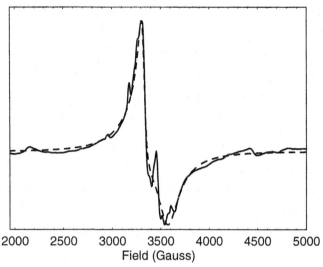

Figure 3.

Simulation of the spectrum of $La_{0.10}Ca_{0.90}TiO_3$ (ν = 9.4890 GHz) ignoring the fine structure and only fitting the main envelope. The spectrum was fit (dashed line) using an orthorhombic g-tensor with components g_{xx} = 2.0324, g_{yy} = 2.035, and g_{zz} = 1.894 giving g_{ave} = 1.987.

If the fine structure is ignored, then the intense central line can be simulated (Figure 3) to give g-tensor parameters g_{xx} = 2.0324, g_{yy} = 2.035, and g_{zz} = 1.894 (g_{ave} = 1.987). Such an average g-value appears not to be consistent with assignment to the Ti^{3+} ion in a perovskite matrix. Isolated Ti^{3+} in $BaTiO_3$ has been reported to have g_{ave} = 1.963 – 1.974 (9). In $PbTiO_3$ two light-induced Ti^{3+} centers have been observed with g_{ave} = 1.911 for one center and g_{ave} = 1.862 for the other. The former center shows clearly resolved ^{49}Ti and ^{47}Ti hyperfine structure while the latter does not.

In order to further test the hypothesis that the broad resonance is indeed not attributable to isolated Ti^{3+}, we have prepared samples of $La_{0.01}Ca_{0.99}{}^{49}TiO_3$ using 100% ^{49}Ti-enriched oxide. The spectrum of this sample is shown in Figure 1c and it is apparent that it is very similar to that of the sample prepared using natural abundance $^{47/49}Ti$. Moreover, the ^{49}Ti hyperfine structure that would be expected of a Ti^{3+} center is not observed, supporting the assignment to other defects.

The weight loss upon reduction in hydrogenous atmospheres at temperatures of ~1400°C of initially air-fired $Ca_{(1-x)}La_xTiO_3$ samples corresponds to expectations based upon x f.u. of Ti^{4+} being replaced to Ti^{3+}[6]. On this model it would be expected that all the cation vacancies would disappear upon reduction. However no change in the PALS results (which indicated cation vacancies were the principal charge compensators in the air-fired samples [6]) were obtained when the samples were reduced in H_2/N_2 at 1200 or 1400°C. This result however may be explainable in terms of the ionic model being unsuitable for the reduced samples, especially insofar as the reduced samples are semiconducting [8].

Table 1. Number of electron spins per Ti ion deduced to
contribute to area under the broad EPR signal near g = 2.

La (f.u.)	Spins/ La	Spins/Ti
0	n.a.	?
0.001	1.6	1.6×10^{-3}
0.01	0.11	1.2×10^{-3}
0.05	0.04	2.1×10^{-3}

[139]La static NMR studies of these samples yielded very broad, featureless line shapes even for values of x as low as 0.001; this is again consistent with the proposal that the La ions occupied sites of low point symmetry and/or were adjacent to paramagnetic centres (see Figure 4). The [139]La static NMR technique was also employed to investigate vacancy-compensated stoichiometries within this system, however, narrower [139]La resonances were not observed (spectra not shown). The full-width-at-half-maximum (fwhm) of these [139]La central-transition resonances for the uncompensated and compensated systems was constant at ~185 kHz (~3200 ppm), irrespective of the value of x, thus suggesting that the line width of the broad [139]La resonance is dominated by low point symmetry and short-range disorder about the La^{3+} site. This short-range disorder represents the statistical variation in bond angles and bond distances about each La^{3+} position, subsequently invoking a distribution of quadrupolar coupling constants (C_q) and asymmetry parameters (η) defining the quadrupolar interaction experienced by the $I=7/2$ quadrupolar [139]La nucleus. These inferences are supported by the behaviour of the [139]La static spectra exhibited by the vacancy-compensated $Sr_{(1-3x/2)}La_xTiO_3$ system. From Figure 5 the [139]La resonance from the cubic $Sr_{0.94}La_{0.04}TiO_3$ sample was much narrower in comparison, with the fwhm line width of this central-transition resonance measured at ~1.2 kHz (~21 ppm). This is expected for systems exhibiting cubic site symmetry with minimal or no disorder about the La^{3+} position. A similar (but slightly broader) result was observed for the cubic $Sr_{0.85}La_{0.1}TiO_3$ sample. As the value of x increases, this cubic character is reduced, with tetragonal and (eventually) orthorhombic phase transitions now emerging [11]. From Figure 5, the [139]La line width increases monotonically as x increases from 0.04 – 0.67 in accordance with these structural phase transitions. In addition to reduced La^{3+} point symmetry invoked by these emerging structural forms, an increased statistical distribution of bond angles and bond distances will occur from the short-range disorder introduced by the multiplicity of nearest-neighbour and second-nearest-neighbour combinations. This disorder is responsible for featureless [139]La line shapes with a degree of elongated asymmetric tailing on the high field (low ppm) side. Previous multinuclear solid state NMR studies on glasses and ceramics have clearly demonstrated this phenomenon [12-18].

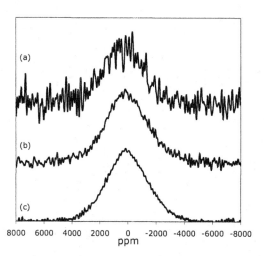

Figure 4. ^{139}La static NMR spectra of $Ca_{(1-x)}La_xTiO_3$: (a) x = 0.02; (b) x = 0.15; (c) x = 0.3

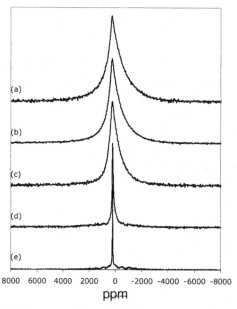

Fig. 5. ^{139}La static NMR spectra of $Sr_{(1-3x/2)}La_xTiO_3$; (a) x = 0.67; (b) x = 0.6; (c) x = 0.4; (d) x = 0.1; (e) x = 0.04.

$Ca_{(1-x)}La_xTi_{(1-x/4)}O_3$ Stoichiometry

The $Ca_{0.8}La_{0.2}Ti_{0.95}O_3$ composition (x = 0.2) was found to be single-phase, while $Ca_{0.6}La_{0.4}Ti_{0.9}O_3$ was found to contain a small amount of La-rich second-phase material. X-ray diffraction showed weak "extra" lines at d spacings of 0.308 and 0.303 nm, and these suggested that the second phase was $La_2CaTi_6O_{15}$., the La-Ca-Ti-O compound most dilute in La [10]. PALS studies remain to be performed to see whether the Ti vacancies gave a greater lifetime enhancement than Ca vacancies. This might be expected from the greater charge on the Ti vacancy (assuming a pure ionic model of perovskite), even though the Ti vacancy would on the basis of ionic size appear to be smaller than the Ca vacancy.

CONCLUSIONS

From PALS and electron microscopy measurements, we have deduced that ~ 0.1 formula units of Ti vacancies can be accommodated in $CaTiO_3$ doped with La ions substituted for Ca, depending on the overall stoichiometries. Electron paramagnetic resonance measurements have shown the presence of as yet unknown paramagnetic species in $Ca_{(1-x)}La_xTiO_3$ samples fabricated in air. ^{139}La nuclear magnetic resonance (NMR) measurements show the La ions in both $Ca_{(1-x)}La_xTiO_3$ and $Ca_{(1-3x/2)}La_xTiO_3$ samples fabricated in air to yield very broad signals. Much sharper ^{139}La NMR signals were observed in dilute vacancy-compensated $Sr_{(1-3x/2)}La_xTiO_3$ samples in which the point symmetry of the Sr site was cubic, rather than orthorhombic as in the analogue Ca(La) titanate samples.

ACKNOWLEDGEMENT

We thank M. Bhati, D. Caudle,S. Brodala and E. Roach for sample fabrication, and H. Li for SEM measurements.

REFERENCES

[1] E.R. Vance, R.A. Day, Z. Zhang, B.D. Begg, C.J. Ball and M.G. Blackford, "Charge Compensation in Gd-doped $CaTiO_3$", *Journal of Solid State Chemistry*, **124**, 77-82 (1996).

[2] U. Balachandran and N. G. Eror, "Self-Compensation in Lanthanum-Doped Calcium Titanate", *Journal of Materials Science*, **17**, 1795-1800 (1982).

[3] E.M. Larson, P.G. Eller, J.D. Purson, C.F. Pace, M.P. Eastman, R.B. Greegor, and F.W. Lytle, "Synthesis and Structural Characterization of $CaTiO_3$ Doped with 0.05-7.5 Mole% Gadolinium (III)", *Journal of Solid State Chemistry*, **73**, 480-7 (1988).

[4] M. S. Dadachov, "Does Lanthanide Substitution Reduce Ti^{4+} to Ti^{3+} in $CaTiO_3$ Fired in Air at 1550°C?", *Journal of Solid State Chemistry*, **137**, 355-6 (1998).

[5] E.R. Vance, R.A. Day, Z. Zhang, B.D. Begg, C.J. Ball and M.G. Blackford, "Reply to M. Dadachov: Does Lanthanide Substitution Reduce Ti^{4+} to Ti^{3+} in $CaTiO_3$ Fired in Air at 1550°C?", *Journal of Solid State Chemistry*, **137**, 357-8 (1998).

[6] J.H. Hadley, Jr., F.H. Hsu, E.R. Vance, and B.D. Begg, "Positron Annihilation Lifetime Spectroscopy of Air-Fired $Ca_{(1-x)}(La)_xTiO_3$ Perovskites", submitted for publication, *Journal of the American Ceramic Society* (2004).

[7] V. Vashook, L. Vasylechko, M. Knapp. H. Ullmann and U. Guth, "Lanthanum Doped Calcium Titanates: Synthesis, Crystal Structure, Thermal Expansion and Transport Properties", *Journal of Alloys and Compounds*, **354**, 13-23 (2003).

[8] B.D. Begg, E.R. Vance, B.A. Hunter and J.V. Hanna, "Zirconolite Transformation Under Reducing Conditions", *Journal of Materials Research*, **13**, 3181-90 (1997).

[9] Joint Committee on Powder Diffraction Standards, Card #27-1057.

[10] T. Kolodiazhnyi and A. Petric, "Analysis of Point Defects in Polycrystalline $BaTiO_3$ by Electron Paramagnetic Resonance", *Journal of Physics and Chemistry of Solids*, **64**, 953-60 (2003)

[11] C.J. Howard, G.R. Lumpkin, R.I. Smith, and Z. Zhang, "Crystal Structures and Phase Transition in the System $SrTiO_3$-$La_{2/3}TiO_3$", submitted for publication, *Journal of Solid State Chemistry*.

[12] S.C. Kohn, R. Dupree, M.G. Mortuza, and C.M.B. Henderson, "NMR Evidence for Five- and Six-Coordinated Aluminum Fluoride Complexes in F-Bearing Aluminosilicate Glasses", *American Mineralogist*, **76**, 309-12 (1991).

[13] S.C. Kohn, R. Dupree and M.E. Smith, "A Multinuclear Magnetic Resonance Study of the Structure of Hydrous Albite Glasses", *Geochimica Cosmochimica Acta*, **53**, 2925-35 (1989).

[14] B.L. Phillips, R.J. Kirkpatrick, A.P. Taglialavore and B. Montez, "Solid State NMR Evidence of 4-, 5-, and 6-Fold Aluminum Sites in Roller-Quenched SiO_2-Al_2O_3 Glasses", *Journal of the American Ceramic Society*, **70**, C10-C12 (1987).

[15] B.L. Phillips, R.J. Kirkpatrick and G.L. Hovis, *Physics and Chemistry of Minerals*, **16**, 262-275 (1988).

[16] G. Kunath, P. Losso, S. Steuernagel, H. Schneider and C. Jäger, "[27]Al Satellite Transition Spectroscopy (SATRAS) of Polycrystalline Aluminium Borate $9Al_2O_3 \cdot 2B_2O_3$", *Solid State Nuclear Magnetic Resonance*, **1**, 261-266 (1992).

[17] C. Jäger, G. Kunath, P. Losso and G. Scheler, "Determination of Distributions of the Quadrupole Interaction in Amorphous Solids by [27]Al Satellite Transition Spectroscopy", *Solid State Nuclear Magnetic Resonance*, **2**, 73-82 (1993).

[18] G. Kunath-Fandrei, T.J. Bastow, J.S. Hall, C. Jäger and M.E. Smith, "Quantification of Aluminium Coordinations in Amorphous Aluminas by Combined Central and Satellite Transition Magic Angle Spinning NMR Spectroscopy", *Journal of Physical Chemistry*, **99**, 15138-41 (1995).

HOLLANDITE CERAMICS: EFFECT OF COMPOSITION ON MELTING TEMPERATURE

M.L. Carter[1,2], E.R. Vance[1], D.J. Cassidy[1], H. Li[1] and D.R.G. Mitchell[1]
[1]Australian Nuclear Science and Technology Organisation
New Illawarra Rd
Lucas Heights, NSW 2234, Australia
[2]School of Chemistry
The University of Sydney
Sydney, NSW 2006, Australia

ABSTRACT

Hollandite-bearing (30-60 wt%) ceramic melts incorporating varying amounts of Cs_2O (1.96-7.5 wt%) have been prepared in air by melting. Minor phases included zirconolite, perovskite and rutile. Detailed analysis of the phase assemblage of the samples by electron microscopy is presented on materials in which Fe, Co, Cr, Ni or Mn is targeted towards the B-site of the titanate hollandites, and the Cs in the hollandite A-site. Mn, Ni and Co entered the hollandite as divalent species while Fe and Cr in the hollandite were trivalent. DTA measurements showed that the melting temperatures of the differently substituted hollandite-rich ceramic melts varied between 1315 and 1450°C. The effect on melt temperature and phase assemblage of substituting K for Ba in the hollandite structure of the melts was also examined.

INTRODUCTION

Immobilization of high-level radioactive wastes in stable matrices for long term geological disposal is important in closing the nuclear fuel cycle. The immobilization of Cs in borosilicate glass is well known and in recent years development of new vitreous matrices has taken place to improve leach resistance. This has resulted in a 10-fold improvement in leach rates[1]. To further improve leach resistance work has been carried out on crystalline matrices.

The hollandite group of minerals has the general formula $A_xM_y B_{8-y} O_{16}$. The M and B cations are surrounded by octahedral configurations of oxygens. Each of these $(M,B)O_6$ octahedra share two edges to form paired-chains running parallel to the c-axis. These chains are corner-linked to neighbouring paired-chains to form a 3-dimensional framework with tunnels running parallel to the c-axis. The large A cations are located in these tunnels.

The titanate synroc phase hollandite ($Ba_xCs_yM^{3+}_{2x+y}Ti_{8-2x-y}O_{16}$ where trivalent M = Al in oxidizing conditions and Al and Ti in reducing conditions) is well known for its ability to incorporate Cs with excellent leach resistance when produced by hot pressing [2,3]. We have previously developed melted hollandite-rich materials processed in air, to immobilize Cs [4,5]. These contained 7.5 wt% Cs and the hollandites, substituted with Co, Mn, Fe, Zn, Cr and Ni, all had excellent leach resistance.

In the current work we investigated melt temperatures and phase development in a series of hollandite-bearing materials in which Fe, Co, Cr, Ni or Mn was targeted towards the B-site and the Cs_2O content was 1.96-7.5 wt%. The ratios of the Ca, Ti, Zr, Ca and Ba were kept constant. The effect on melt temperature of substituting Ba for K in the hollandite was also investigated. When substituting the K for Ba the ratios of all other elements were kept constant. The valence state of the 3d transition metals was deduced from stoichiometric measurements on the hollandites and electron energy loss spectroscopy (EELS) was also used for this purpose for Mn.

As ^{137}Cs and [135]Cs are (beta, gamma) emitters structural long-term radiation damage effects would be insignificant. Sr was not added to the formulations (Sr is very likely to be present in Cs rich waste streams) as it would enter the perovskite in air-melted titanate wasteforms [4-6].

EXPERIMENTAL

Samples (see Table I) for melting were produced by the alkoxide-route[6]. The correct molar quantities of titanium (IV) isopropoxide and zirconium tertbutoxide in ethanol were mixed with the other metal nitrates dissolved in water, while continuously stirring. The mixture was then heated to ~110°C to drive off the alcohol and water. The dry product was then calcined in air for two hours at 750°C. The samples were placed in Pt crucibles and melted in air at 1450-1550°C in an electric furnace.

A JEOL JSM6400 scanning electron microscope (SEM) equipped with a Noran Voyager energy-dispersive spectroscopy system (EDS) was operated at 15 keV for microstructural work. X-ray diffraction was performed with a Siemens D500 diffractometer and CoKα radiation

The melting point of the hollandite was determined by using a Setaram TAG24 (France) differential thermal analysis (DTA) instrument to heat approximately 70mg of sample at 10°C min^{-1} to 1500°C in a platinum crucible under an air atmosphere.

EELS specimens were prepared by crushing powders in methanol and dispersing them on holey carbon support films. Oxide reference materials were obtained from Aldrich and included MnO, Mn_2O_3 and MnO_2. EELS spectra were obtained using a JEOL 2010F/GIF 2000 electron microscope system operating at 200 keV. Convergence and collection angles were 10.85 and 27.6 mrad respectively. All samples were cooled to liquid nitrogen temperature during data acquisition, to minimize electron beam damage effects. Spectra were acquired at spectrometer dispersions of 0.5 eV and 0.2 eV. EELS spectra were processed by taking the second derivative of the spectrum, extracting only the positive component thereof and determining the integral for the relevant L_2 and L_3 peaks. The L_3/L_2 ratio of transition metal species has been shown to be sensitive to the valence state [7]. An alternative parameter is the so called Branching Ratio, defined as $L_3/(L_2+L_3)$ [8]. The extremely high spatial resolution of EELS makes this an attractive method for studying valence state, and a number of transition metal systems including Mn [9,10] have been studied with the method.

The following method was used to calculate the 3d transition metal distribution in the hollandite bearing ceramics. It was assumed in every case that all the Cs and Ba entered the hollandite. The Ba+Cs in the hollandite was assumed to constitute 1.2 formula units (f.u.) [4] and when the 3d metal was trivalent the previously mentioned formula was used. In the case of the divalent 3d metal ion the following formula was used: $Ba_xCs_yM^{2+}_{x+y/2}Ti_{8-x-y/2}O_{16}$ [5]. Where there was no Ba in the formulations and K was present, K + Cs in the hollandite constituted 1.5 f.u . Note that the occupancy of K and Cs is higher in K and Cs substituted hollandites than Ba substituted hollandites [11-13]. In these cases the following hollandite formula, $K_xCs_yM^{3+}_{x+y}Ti_{8-x-y}O_{16}$, was used. Thus for each case the amount of MO_x needed in the hollandite could be calculated. If there was 'left-over' MO_x, it could then be found in other phases, either the synroc-type phases or new phases. From the stoichiometric measurements, it could be deduced that the Ni, Co and Mn in the B site of the hollandite were divalent, while the Fe and Cr in the B site were trivalent.

Table I: Compositions of samples for melt experiments

Sample	Composition	wt% oxide	Sample	Composition	wt% oxide
Group I	M_2O_3/MO_2	6.74	Group II	Fe_2O_3	10.32
M = Cr, Mn	BaO	5.08	Hollandite	BaO	5.83
Co, and Ni	CaO	9.97	Fe	CaO	7.34
	ZrO_2	5.98		ZrO_2	4.61
	TiO_2	64.72		TiO_2	64.40
	Cs_2O	7.50		Cs_2O	7.50
Group III	Fe_2O_3	10.30	Group IV	Cr_2O_3	6.74
Hollandite	K_2O	1.80	Hollandite Cr/K	K_2O	1.55
Fe/K	CaO	7.30		CaO	9.97
	ZrO_2	4.60		ZrO_2	5.98
	TiO_2	68.4		TiO_2	68.3
	Cs_2O	7.50		Cs_2O	7.50
Group V	M_2O_3/MO_2	6.94	Group VI	M_2O_3/MO_2	7.15
M = Mn, Co ,	BaO	5.24	M = Mn, Co Cr	BaO	5.38
Cr and Fe	CaO	10.27	and Fe	CaO	10.57
	TiO_2	66.63		ZrO_2	6.34
	ZrO_2	6.15		TiO_2	68.60
	Cs_2O	4.76		Cs_2O	1.96

RESULTS AND DISCUSSION

Samples in Group I, II, III and IV

SEM investigation of Group I samples incorporating 7.5 wt% Cs_2O with M = Cr, Mn, Co, and Ni showed them to contain major Cs-bearing hollandite (55-60 wt%), zirconolite (15-20   wt%), perovskite (15-20 wt%) and   rutile(~5 wt%).   Figure 1 shows the microstructure for the M = Co sample. The microstructures for M = Mn and Fe are shown elsewhere [4] and those for M = Cr and Ni are given in [5]. The oxidation state of the Mn was measured using EELS as Mn is known to exist in hollandite in several oxidation states [14-16]. The L_3/L_2 and branching ratios were determined for pure Mn oxide phases and the Mn containing Group I sample and these are shown in Table II. Errors reported are one standard deviation of the data set. The branching ratio is very sensitive to Mn valence state and the general relationship and L_3/L_2 values for the oxides are very similar to those determined by Wang et al. [10]. The branching ratio shows the Mn valence in the Group I sample with M = Mn sample was very close to that of the MnO (+2), within experimental error.

SEM investigation of the Group II sample (Fe) showed a major Cs-bearing hollandite (~61 wt%), zirconolite (~13 wt%), perovskite(~13 wt%) and rutile(~13 wt%).

If the 3d elements in Group I were present as divalent metal oxides around 4.5 wt% of the metal oxide would be required to incorporate all of the Ba and Cs into the hollandite. This was the case for Ni, Mn and Co. The excess (Ni, Mn and Co) ions entered the zirconolite (~0.1 formula units f.u.) and the remainder formed a small amount (~3 wt%) of $(Ca,Ba)_2M_4Ti_{15}O_{38}$, determined by X-ray diffraction and EDS in the case of Mn and Co and the excess Ni formed $NiTiO_3$.

Table II. Summary of EELS data for pure oxides Sample from Group I with M = Mn

Species	L_3/L_2	Std Dev	$L_3/(L_3+L_2)$	Std Dev
MnO (II)	3.9	0.2	0.80	0.02
Mn_2O_3 (III)	2.54	0.02	0.718	0.002
MnO_2 (IV)	2.2	0.1	0.69	0.01
Group I sample with M = Mn	3.8	0.4	0.79	0.01

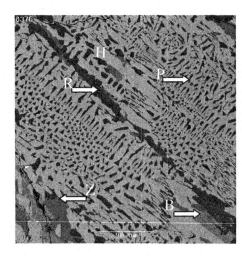

Figure 1: Backscattered electron image of hollandite from Group I with M = Co. H = hollandite, Z = zirconolite, P= perovskite, R = rutile and B = $(Ca, Ba)_2Co_4Ti_{15}O_{38}$. Bar = 90 μm

The Group I and II samples containing the 3+ metal ions (Fe and Cr) required around 10.3 wt% of the metal oxide to incorporate all the Ba and Cs in the hollandite. In the formulations used in this study this is true of the Group II, Fe sample (see Table 1), but there would not be enough Cr (only 65%) in the Group I mix to fully incorporate the Cs and Ba in the hollandite structure. We reported earlier [5] that the sample with M = Cr had low leach rates (0.15g/l) and no undesirable Cs phases were present. As we found no secondary Ba and Cs phase we would have to conclude that there was Cs loss by volatilization during heating of this sample. The EDS analysis of the hollandite in the Group 1 Cr containing sample indicated that the Ba/ Cs ratio had increased from ~0.6 to ~1 verifying Cs loss, and explaining why no secondary Cs-bearing phase was found.

The SEM analysis of the Group III sample (Fe/K) contained hollandite (~50 wt%) with perovskite (~10 wt%), zirconolite (~10 wt%) and rutile (~15wt%) - see Figure 2a. The sample would only require around ~ 7.2 wt% Fe oxide to fully incorporate all the K and Cs into the hollandite. Thus there is 3.1 wt% excess Fe oxide available to enter other phases. In this case ~0.1 f.u. entered the zirconolite and a calcium iron titanate phase (~ 14 wt%) in which the elemental ratios of Ca:Fe:Ti were 1:1:3.5 (from EDS) was found.

The Group IV sample (Cr/K) contained hollandite (~45 wt%) with perovskite (15-20 wt%), zirconolite (15-20wt%) and rutile (~20 wt%) - see Figure 2 (b-e). The Group IV sample (Cr/K) samples would require around 6.6 wt% of Cr oxide to incorporate all the K and Cs in the hollandite using the above formula, and the slight excess of Cr oxide (~0.1 wt%) entered the zirconolite (~0.1 f.u.). There is no contrast between the rutile and perovskite in this image and the X-ray maps of Ti, Ca and Zr clearly depict the different phases. The Ti x-ray map clearly shows the rutile and hollandite but the zirconolite and perovskite have the same grey scale (the lighter the grey, the higher the ion content). The Ca x-ray map clearly shows the difference between the perovskite and zirconolite (perovskite containing more Ca) and the Zr map only shows the zirconolite.

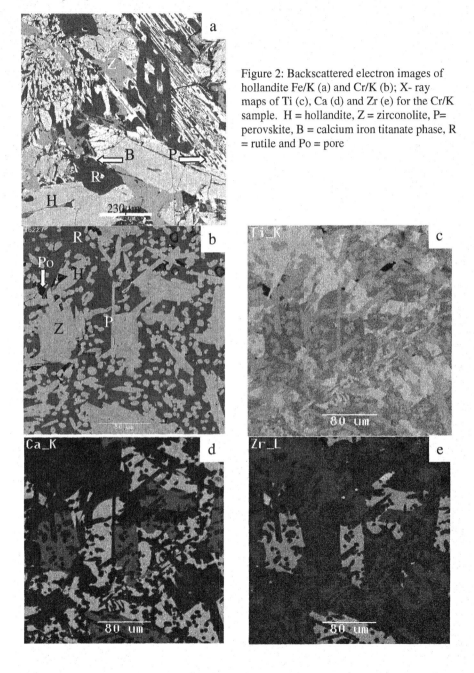

Figure 2: Backscattered electron images of hollandite Fe/K (a) and Cr/K (b); X- ray maps of Ti (c), Ca (d) and Zr (e) for the Cr/K sample. H = hollandite, Z = zirconolite, P= perovskite, B = calcium iron titanate phase, R = rutile and Po = pore

The DTA plots in Figures 3-6 show the endothermic reactions of the samples melting and the exothermic reactions of the crystallization of the samples containing 7.5 wt% Cs_2O on cooling. Table III lists the melt temperatures (position of endothermic reaction) of all sample measured. The melting temperature is the highest for the Group I sample containing the Cr at approximately 1450°C (Figure 5) and this high melting temperature could lead to losses of Cs which would help explain the results deduced in the SEM analysis of this sample (see above). The other samples had melting temperatures between 1315°C and 1375°C, with the Mn-bearing sample having the lowest. The effect of adding K instead of Ba in the samples with Fe and Cr had little effect on the melting temperature but the crystallization temperature in both samples moved to a higher value (see Figure 5-6).

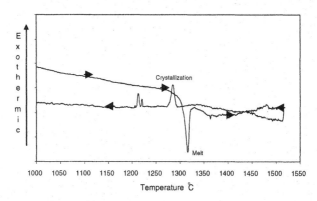

Figure 3: DTA of Group I sample with M = Mn

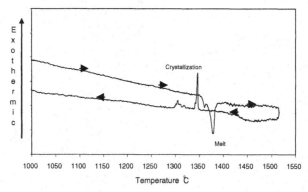

Figure 4: DTA of Group I sample with M = Ni

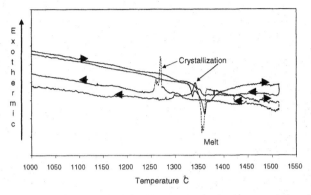

Figure 5: DTA of Group II sample with M = Fe (line broken only on peaks) and Group IV sample (solid line)

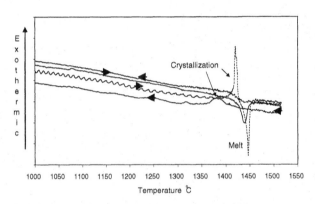

Figure 6: DTA of Group I sample with M = Cr (solid line) and Group III sample (line broken only on peaks)

Table III. Summary of DTA results.

Group	I	I	I	I	II	III	IV	VI
Sample	Mn	Ni	Co	Cr	Fe	Fe/K	Cr/K	Co
Melt Temp. °C	1315	1380	1350	1445	1350	1355	1450	1350

Samples from Group V

The Mn and Co formulations resulted in less hollandite (~45 wt%) than the samples from Group I, with zirconolite (15-20 wt%) perovskite (~15 wt%) and minor rutile (~3 wt%) also being formed. Figure 7 shows the backscattered electron image for the Group V sample with M = Co. Only around 3.6 wt% of the divalent 3d metal oxide is needed to incorporate all the Ba and Cs into the hollandite. The excess 3d divalent metal ion (Mn or Co) partitioned into the zirconolite (~0.1 f.u.) and the remainder formed $(Ca, Ba)_2M_4Ti_{15}O_{38}$ (~18 wt%) which accounts for the much reduced rutile in the Mn and Co samples compared to those in Group I.

The Cr and Fe containing samples were found to contain hollandite (~40 wt%) with major zirconolite (~13 wt%), perovskite (~20 wt%) and rutile (~26 wt%). As more of the trivalent 3d metal oxide is needed (~8 wt%) than is available to incorporate all the Ba and Cs into the hollandite, it would be expected that undesirable phases would be present and indeed Cs titanate was found in the Fe-bearing sample. In the case of the Cr-bearing sample the high melting temperatures may have contributed to Cs loss. This would reduce the amount of Cr needed and so it was not surprising that no Cs-bearing phase was found other than hollandite.

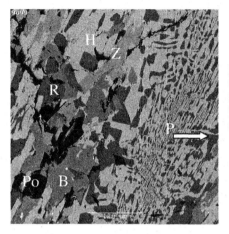

Figure 7: Backscattered electron image of a Group V sample with M = Co. H = hollandite, Z = zirconolite, P = perovskite, B = $(Ca, Ba)_2Co_4Ti_{15}O_{38}$, R = rutile, and Po = pore.

Samples from Group VI

Mn and Co resulted in hollandite (~30 wt%) with zirconolite(~17 wt%), perovskite (~15 wt%), and rutile (~15 wt%). Only around 3.6 wt% of the divalent 3d metal oxide is needed to incorporate all the Ba and Cs into the hollandite. The excess metal ion (Mn or Co) partitioned into the zirconolite (~0.1 f.u.) and the remainder formed around 22 wt% (Ca, Ba)$_2$M$_4$Ti$_{15}$O$_{38}$. The backscattered electron image of the Group VI sample with M = Co is shown in Figure 8. This sample contained rutile but this is not shown in backscattered electron image shown. The rutile in this sample formed large >100 μm crystals. If shown in a micrograph of this size, it would fill the field of view and all other phases would not be seen.

The Cr and Fe containing samples were found to contain hollandite (~33 wt%) with zirconolite (~17 wt%) perovskite (~20 wt%) and an increased amount of rutile (~30 wt%) compared to the Group I samples. 6.6 wt% of trivalent metal oxide was incorporated in the hollandite and only a very small amount of excess trivalent 3d metal oxide (0.5 wt%) was available to enter other phases, ~0.1 f.u were found in the zirconolite.

The DTA of the Group VI sample with M = Co is shown in figure 9 with the DTA of the group I sample with M = Co. This shows that the melt temperature did not significantly change with the decrease in the Cs content.

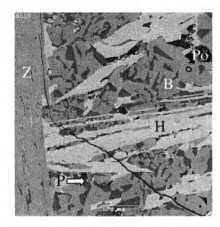

Figure 8: Backscattered electron image of a Group V sample with M = Co. H = hollandite, Z = zirconolite, P = perovskite and B = $(Ca, Ba)_2Co_4Ti_{15}O_{38}$ and Po = pore

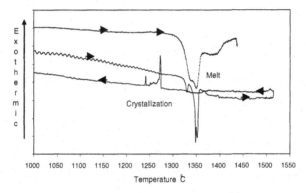

Figure 9: DTA of Group I sample with M = Co (solid line) and Group VI sample with M = Co (line broken only on peaks)

CONCLUSIONS

Hollandite-bearing (30-61 wt%) ceramic melts incorporating varying amounts Cs_2O (1.96-7.5 wt%) have been prepared in air by melting. Minor phases included zirconolite, perovskite and rutile. When the transition 3d metal ion that was targeted towards the B site in the hollandites was in excess a small amount entered the zirconolite (0.1 f.u.) and the remainder formed a variety of phases. For the divalent ions M = Mn and Co the extra phase formed was $(Ca, Ba)_2M_4Ti_{15}O_{38}$ and when the ion was Ni, Ni titanate was formed. For the samples containing excess Fe a calcium iron titanate phase was formed. DTA measurements showed that the melting temperatures of the differently substituted ceramic melts varied between 1315 and 1450°C. The substitution of K for Ba in the melted ceramics had little effect on the melt temperature and neither did reducing the Cs content.

REFERENCES

[1] F. Bart, S Sounihac, J.L. Dussossoy and A. Bonnetier, "Development of Vitreous Matrices for Conditioning Cesium"; pp. 353-360 in *Environmental Issues and Waste Management Technologies VI*. Edited by D. R. Spearing, G.L. Smith and R.L. Putnam. American Ceramic Society, Westerville, Ohio 2000.

[2]S.E. Kesson, "The Immobilisation of Cesium In Synroc Hollandite", *Radioactive Waste Management and the Nuclear Fuel Cycle*, **4**, 53-71 (1983)

[3]M.L. Carter, E.R. Vance, D. R.G. Mitchell, J.V. Hanna, Z. Zhang and E. Loi, "Fabrication, Characterisation, and Leach Testing of Hollandite $(Ba,Cs)(Al,Ti)_2Ti_6O_{16}$", *Journal of Materials Research*, **17** [10] , 2578-89 (2002)

[4]M.L. Carter, E.R. Vance and H. Li, "Hollandite-Rich Ceramic Melts"; pp. 21-30 in *Environmental Issues and Waste Management Technologies IX*. Edited by J.D. Vienna and D.R. Spearing. American Ceramic Society, Westerville, OH 2004.

[5]M.L. Carter, E.R. Vance and H. Li, "Hollandite-rich Ceramic Melts for the Immobilisation of Cs"; pp. 249-54 in *Scientific Basis for Nuclear Waste Management XXVII*. Edited by V.M. Oversby and L.O. Werme. Materials Research Society, Warrendale, PA 2004.

[6]A.E. Ringwood, S.E. Kesson, K.D. Reeve, D.M. Levins and E.J. Ramm, "Synroc"; pp. 233-334 in *Radioactive Waste Forms for the Future*. Edited by W. Lutze and R.C. Ewing, North-Holland. New York 1988.

[7]R.D. Leapman, L.A. Grunes and P.L. Fejes, "Study of the L_{23}Edges in the 3d Transition Metals and Their Oxides by Electron-Energy-Loss Spectroscopy with Comparisons Theory", *Physical Review* B, **26**, 614-635 (1982).

[8]L.A.J. Garvie and P.R. Buseck, "Determination of Ce^{4+}/Ce^{3+} in Electron-Beam-Damaged CeO_2 by Electron Energy-Loss Spectroscopy", *Journal of Physics and Chemistry of Solids*, **60**, 1943-1947 (1999).

[9]L.A.J. Garvie, A.J. Craven and R. Brydson, "Use of Electron-Energy Loss Near-Edge Fine Structure in the Study of Minerals", *American Mineralogist*, **79**, 411-425 (1994)

[10]Z. L. Wang, J. Bentley and N. D. Evans, "Mapping the Valence States of Transition-Metal Elements using Energy-Filtered Transmission Electron Microscopy", *Journal of Physical Chemistry B*, **103**, 751-753 (1999).

[11]H.U. Beyeler and C. Schüler, "Cyrstal-Growth and Structural Properties of Some Hollandites", *Solid State Ionics,* **1**, 77-86 (1980).

[12]W. Sinclair and G.M. McLaughlin, "Structure Refinement of Priderite", *Acta Crystallographica* B**38**, 245-246 (1982).

[13]S.A. Petrov, L.F. Grigor'eva, I.Yu. Sazeev and S.K. Filatov, "Some Crystallochemical Features of Hollandite Phases in the Systems K_2O-MO(M_2O_3)-TiO_2 (M= Mg, Zn, Ga) and M'$_2$O-MgO- TiO_2 (M'=li, K, Rb, Cs)", *Inorganic Materials*, **30**, 892-895 (1994).

[14]A. Byström and A.M. Byström, "The Crystal Structure of Hollandite, and the Related Manganese Oxide Minerals, and α-MnO_2", *Acta Crystallographica* **3**, 146-154 (1950).

[15]J.E. Post, R.B. Von Dreele and P.R. Buseck, "Symmetry and Cation Displacement in Hollandite; Structure Refinements of Hollandite, Crytomelane and Priderite", *Acta Crystallographica* B**38**, 1056-1065 (1982).

[16]A.L. Prieto, T. Siegrist and L.F. Schneemeyer, "New Barium Manganese Titanates Prepared under Reducing Conditions", *Solid State Sciences* **4**, 323-327 (2002).

CHEMICAL DURABILITY OF IRON-SUBSTITUTED HOLLANDITE CERAMICS FOR CESIUM IMMOBILIZATION

F. Bart*, G. Leturcq*, and H. Rabiller**
CEA Valrho Marcoule, Nuclear Energy Division
DEN/DTCD, *SECM/LM2C, **SCDV/LEBV
BP 17171, 30207 Bagnols-sur-Cèze France

ABSTRACT

In France, there is strong interest in developing new ceramic wasteforms for containment of separated long-lived radionuclides. Barium hollandite ($BaAl_2Ti_6O_{16}$) ceramics are suitable for cesium immobilization by virtue of their loading capacity and high chemical durability. A 5 wt% Cs_2O-doped Fe-substituted hollandite was synthesized at laboratory scale (100 g), using an alkoxide process. Hollandite crystallization was achieved after calcination for 5 hours at 1000°C in air. After milling and cold pressing, ceramic pellets were sintered in air for 15 hours at 1250°C. The resulting materials were dense single-phase pellets. Leaching experiments were carried out to assess the chemical durability of this specific hollandite. Initial leaching rates were measured at 100°C using a Soxhlet apparatus, and long-term alteration rates at 90°C were determined by static leaching of powder samples. The chemical durability of iron hollandites, loaded with 5 wt% cesium oxide was very good, and comparable to that of Synroc hollandite.

INTRODUCTION

The French nuclear waste management act of 30 December 1991 outlined three areas of research to develop enhanced separation and transmutation processes for long-lived radioelements, to investigate waste disposal in a geological formation, and to develop concepts for conditioning and reversible long-term interim storage of nuclear waste. Cesium is one of the main uranium fission products; it is present in concentrated nitric acid solutions after spent fuel reprocessing. Cesium is currently conditioned industrially in borosilicate glass materials, which have been adopted on an industrial basis for immobilizing fission product solutions in France and around the world.

Enhanced cesium separation processes have been investigated by the CEA to isolate this element from the other fission products. Cesium transmutation would be difficult to implement, however, because its complex isotopic distribution would require prior separation of the long-lived isotope. In addition to the stable isotope 133, the isotopic composition of cesium includes isotope 135 with a half-life of 2.3 million years, and isotopes 137 and 134 with shorter half-lives of 30 years and 2 years, respectively. Separate conditioning of cesium has also been examined with the objective of developing dedicated containment materials for this element. The combined presence of isotopes with very different half-lives requires a conditioning matrix with both very high chemical durability, considering the mobility of cesium in aqueous media, and good thermal stability to withstand the heating due to radioactive decay of the short-lived isotopes. The selected matrix must also be compatible with the presence of radiogenic barium. Cesium being a volatile element at relatively low temperatures, the matrix fabrication process must very effectively incorporate it in the material.

While developing aluminosilicate glass compositions specifically adapted for immobilizing cesium[1], the CEA also investigated ceramics such as Synroc[2], a multiphase material developed by the Australian Nuclear Scientific and Technological Organization (ANSTO) in the 1980s for containment of fission product solutions. One of the four major constituent phases of Synroc is a hollandite phase $(Ba,Cs)Al_2Ti_6O_{16}$ dedicated to loading the cesium and barium found in fission product solutions.

The study described here concerned the development of a ferriferous hollandite ceramic conditioning material by natural sintering in air. After a description of the synthesis method used in the laboratory, the static and dynamic leach test results are discussed.

SYNTHESIS, COMPOSITION AND MICROSTRUCTURE OF THE CERAMIC

The hollandite compositions largely described in the literature[3] for waste conditioning are basically $(Ba,Cs)Al_2Ti_6O_{16}$. Structurally, hollandite consists of a chain of edge-sharing pairs of Ti and Al octahedra forming tunnels parallel to the c axis. The tunnels contain octacoordinated sites in which Cs and Ba cations can be loaded. In Synroc, hollandite can contain not only trivalent aluminum, but also trivalent titanium whose structural role is to enlarge the tunnels to accommodate cesium, which has a larger ionic radius than barium[2].

We studied a series of hollandite compositions containing no trivalent titanium but instead trivalent iron: $Ba_xCs_y(Al,Fe)^{3+}{}_{2x+y}Ti^{4+}{}_{8-2x-y}O_{16}$, where x ranged from 0.5 to 1. The single-phase ceramic composition selected on completion of the formulation study[4,5] was $Ba_1Cs_{0.28}(Fe_{0.82}Al_{1.46})Ti_{5.72}O_{16}$, corresponding to the following weight percentage composition: 19.42% BaO; 5% Cs_2O; 9.43% Al_2O_3; 8.29% Fe_2O_3; 57.87% TiO_2. The actual cesium oxide loading obtained was the target value of 5 wt% cesium oxide, as confirmed by X-ray fluorescence analysis.

The experimental protocol specified for fabricated the ceramics at laboratory scale in hundred-gram quantities consisted in natural sintering of the powder at 1250°C in air after calcining at 1000°C. It included the following steps:
- Wet-route synthesis of the precursor (100–200 g) by blending an aqueous solution of Cs and Fe nitrates and Ba acetate with a solution containing Ti and Al alkoxides diluted in ethanol. When the two solutions were mixed, the hollandite precursor precipitated.
- Evaporation of the solvent in a rotary evaporator, followed by recovery and drying of the powder overnight at 120°C.
- Calcining of the powder for 5 hours at 1000°C to form the hollandite phase.
- Wet milling for 1 hour at 300 rpm
- Pelletizing by uniaxial press compaction at 100 MPa at room temperature
- Natural sintering of the pellets in air for 15 hours at 1250°C.

The protocol produced ceramics consisting exclusively of hollandite in the form of samples 3 mm thick and 22 mm in diameter weighing about 15 grams, densified to 95%TD, as shown by helium pycnometry. The following figures show the structural characteristics under X-ray diffraction (Figure 1) and the microstructure of the material on a scanning electron microscope image obtained by secondary electron detection (Figure 2).

Chemical Durability of Iron-Substituted Hollandite Ceramics

Test objective and description: The chemical durability of Synroc ceramic hollandite has been extensively investigated by ANSTO[6,7]. We characterized the effect of iron substitution for a fraction of the titanium on the ceramic alteration rate in pure water. Two series of tests were carried out in pure water to measure the initial alteration rate of the ceramic in a medium with a high renewal rate at 100°C, and to measure the longer-term alteration rates under static conditions over one year at 90°C.

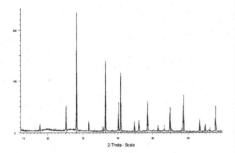

Figure 1. X-ray diffraction diagram of a sintered hollandite ceramic with the reference composition $Ba_1Cs_{0.28}(Fe_{0.82}Al_{1.46})Ti_{5.72}O_{16}$; the vertical lines correspond to the reference JCPDS 33-133

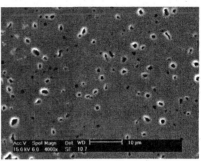

Figure 2. Morphology of a sintered hollandite ceramic with reference composition $Ba_1Cs_{0.28}(Fe_{0.82}Al_{1.46})Ti_{5.72}O_{16}$ (electron micrograph by secondary electron detection)

Open system tests: Soxhlet tests lasting 56 days were carried out at 100°C on dense test samples 22 mm in diameter. Leachate samples were taken from the boiler at regular intervals for ICP-MS analysis to determine the normalized mass losses* of the material over time.

Closed system tests: A test was carried out under static conditions at 90°C with hollandite powder at a material-surface-area-to-water-volume (S/V) ratio of 80 cm^{-1}; the surface area was measured by the BET adsorption method on the 63–125 μm grain size fraction. The test was conducted in initially pure water in PTFE vessels placed in a controlled temperature chamber at 90°C, and lasted one year. As for the Soxhlet tests, leachate samples were taken at regular intervals and analyzed by ICP-MS.

Soxhlet Test Results: Alteration Rate Variations with a High Renewal Rate

Figure 3 shows the normalized mass losses for Cs (squares) and Ba (diamonds) obtained with polished samples during Soxhlet tests. Hollandite dissolution was incongruent: the cesium release was slightly greater than for barium and aluminum [4,6,7]. Titanium was released from matrix at a very low rate. We therefore selected the normalized cesium mass loss to calculate the material alteration rate.

The total cesium release was less than 0.25 $g \cdot m^{-2}$, which corresponds to a maximum altered thickness of about 40 nm. Furthermore, the leach rate dropped rapidly during the first few days of the test, despite the high renewal rate: it was about 2×10^{-2} $g \cdot m^{-2}d^{-1}$ between 0 and 1 day, but after the initial transient phase, the alteration rate decreased by one or two orders of magnitude, even with very high solution renewal rates. Under steady-state conditions, (between 25 and 56 days under the test conditions) the measured values were about 6×10^{-4} $g \cdot m^{-2}d^{-1}$. The mean measured rate was 2×10^{-3} $g \cdot m^{-2}d^{-1}$ when the total mass loss was taken into account.

---

* The normalized mass loss of an element i corresponds to the quantity ($g \cdot m^{-2}$) of the matrix that would have to be altered to obtain the measured concentration of element i in the leachate. As the device used is made of stainless steel, no data could be obtained on the iron concentration in solution.

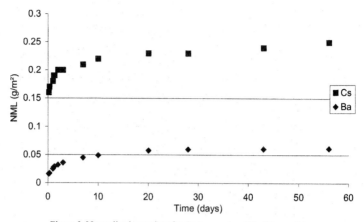

Figure 3. Normalized mass loss for Cs (squares) and Ba (diamonds)
obtained on a polished sample during Soxhlet testing
(hollandite ceramic with reference composition $Ba_1Cs_{0.28}(Fe_{0.82}Al_{1.46})Ti_{5.72}O_{16}$)

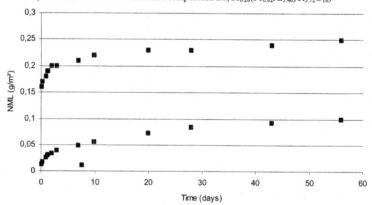

Figure 4. Normalized Cs mass loss versus time for a polished sample (upper curve)
and an as-sintered sample (lower curve) during Soxhlet testing
(hollandite ceramic with reference composition $Ba_1Cs_{0.28}(Fe_{0.82}Al_{1.46})Ti_{5.72}O_{16}$)

Figure 4 compares the normalized cesium mass losses obtained for the as-sintered sample (lower curve) and the polished sample (upper curve) during Soxhlet tests in pure water at 100°C. The initial release appears to be substantially lower for the as-sintered sample than for the polished sample. This effect has already been observed[4] but is only an artifact related to a slight difference in composition between the pellet outer surface and the remaining volume. Time-of-Flight Secondary Ion Mass Spectroscopy (ToF-SIMS)* was used to characterize the two types of sample surfaces (as-sintered and polished) under identical conditions according to the same protocol used for Soxhlet testing. The ToF-SIMS technique consists in measuring the relative concentration of matrix constituents (Cs, Ba, Fe, Al and Ti) versus the sample sputtering time, and thus to establish qualitative sample composition profiles from the surface to the bulk ceramic, over a total thickness of about 1 μm.

---

* Experiments performed by Biophy Research S.A. using an Iontof "ToF-SIMS IV" device.

The profiles were obtained by alternating the sputtering sequences (sputtered area 300×300 μm^2, 3 kV Ar^+ primary ions) and analysis sequences (analyzed surface area 100×100 μm^2, 25 kV Ga^+ primary ions, current 0.25 pA, surface neutralization by pulsed low-energy electron flood gun, $E < 20$ eV). Under the analysis conditions, the maximum depth profiling rate determined by calibration on SiO_2 was estimated at 0.28 nm/second, i.e. 1 micrometer per hour.

Figures 5 and 6 illustrate the results obtained by ToF-SIMS by plotting the ratios of the Cs, Ba, Fe, Al concentrations to the Ti concentration, which remained constant throughout the analysis. Figure 5 corresponds to the as-sintered, unpolished surface, and Figure 6 to a polished specimen. The extreme outer surface of the as-sintered specimen is thus clearly depleted* in cesium; a cesium concentration gradient can be observed in the sample to a depth estimated at about 1 to 2 micrometers (Figure 5). No similar phenomenon was observed for the polished specimen (Figure 6).

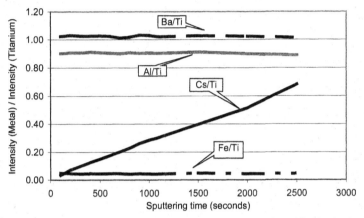

Figure 5. Metal / titanium ratio measured by ToF-SIMS versus sputtering time for a hollandite sample with as-sintered surface finish (2500 seconds correspond to a profiling depth of about 0.7 μm) (hollandite ceramic with reference composition $Ba_1Cs_{0.28}(Fe_{0.82}Al_{1.46})Ti_{5.72}O_{16}$).

It is extremely complex to determine the cesium concentration quantitatively by ToF-SIMS in the extreme outer surface layer of the sample: it depends on many physical and chemical factors, and in particular on the sputtering yield, which is difficult to compute for matrices with complex compositions such as hollandite. Nevertheless, the thickness affected by cesium depletion (about one micrometer) is much greater than the sample thickness altered by Soxhlet testing (no more than one tenth of a micrometer). This phenomenon accounts for the apparently lower alteration rate of the as-sintered specimen when measured by the cesium mass loss. In fact, the cesium concentration in the extreme outer surface of the test sample (to a depth of about 1 to 2 μm) was below 5 wt%; normalizing the mass loss to a theoretical value of 5% thus artificially underestimates the matrix alteration.

---

* The depletion is very slight, and is undetectable by overall XRF chemical analysis of the sample.

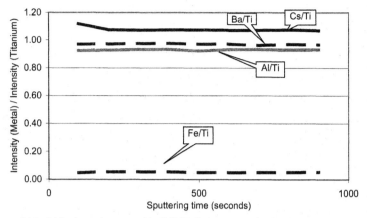

Figure 6. Metal / titanium ratio measured by ToF-SIMS versus sputtering time for a hollandite sample with polished surface finish (900 seconds correspond to a profiling depth of about 0.3 μm) (hollandite ceramic with reference composition $Ba_1Cs_{0.28}(Fe_{0.82}Al_{1.46})Ti_{5.72}O_{16}$).

Results of Static Tests at 90°C

Iron-substituted hollandite: The normalized mass losses curves (Figure 7) show that saturation conditions were reached very rapidly, after only a few days. Cesium is more mobile than the other matrix constituents.

The mean alteration rate measured at 90°C between 0 and 7 days was about 2×10^{-3} $g \cdot m^{-2} d^{-1}$, diminishing to 8×10^{-5} $g \cdot m^{-2} d^{-1}$ after 30 days. The residual rate measured thereafter was less than 10^{-5} $g \cdot m^{-2} d^{-1}$, a value representative of the uncertainties on the leachate analysis. This behavior is comparable to the results obtained by ANSTO for hollandite without iron [6,7].

Comparison with titanium hollandite: The same test was carried out, under static conditions for a hollandite sample given by ANSTO, presenting the following formula $Ba_1Cs_{0.09}Al_{1.5}Ti_{6.5}O_{16}$ + 0,5 TiO_2 ; the corresponding weight percentage composition is: BaO 19.9 % ; Cs_2O 1.75% ; Al_2O_3 9.5% ; TiO_2 69.6%. The experimental protocol specified for the fabrication of this sample (hot pressing), and the corresponding microstructure are described in reference 7.

The normalized mass losses curves (Figure 8) show that saturation conditions were reached very rapidly, after only a few days. Cesium is more mobile than the other matrix constituents, for both hollandite composition. The barium release is slightly greater for pure titanium hollandite than for iron-substituted hollandite, as shown by this work, and by Carter [6,7].

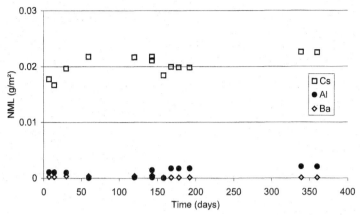

Figure 7. Normalized Cs, Ba and Al mass losses versus time (1 year) during static testing at 90°C with S/V=80 cm^{-1} (hollandite ceramic with reference composition $Ba_1Cs_{0.28}(Fe_{0.82}Al_{1.46})Ti_{5.72}O_{16}$).

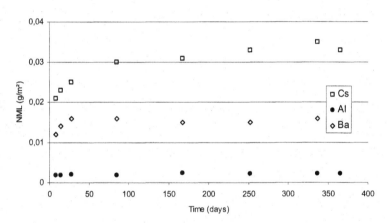

Figure 8. Normalized Cs, Ba and Al mass losses versus time (1 year) during static testing at 90°C with S/V=80 cm^{-1} for a hollandite sample presenting the formula $Ba_1Cs_{0.09}Al_{1.5}Ti_{6.5}O_{16} + 0.5\ TiO_2$ (sample given by ANSTO)

Conclusion of the static tests: These results indicate that the alteration mechanism of hollandite (iron substituted or not) begins first by a release of cesium, and in a less quantity of barium, concerning a few tens of nanometers. After this first step, the alteration is very slow, with residual rate over longer durations was less than $10^{-5}\ g \cdot m^{-2} d^{-1}$.

CONCLUSION

Substituting iron for a fraction of the titanium in hollandite derived from Synroc allowed homogeneous single-phase ceramics to be fabricated by natural sintering in air. Substituting Fe^{3+} for Ti^{3+} in the mineral structure did not modify its chemical durability. In pure water at 100°C, alteration began with a transient phase at a rate of about $2 \times 10^{-2}\ g \cdot m^{-2} d^{-1}$ between 0 and 1 day. The alteration rate then diminished by one or two orders of magnitude, even with very high solution renewal rates. Under steady-state conditions the measured rate was about 6×10^{-4} $g \cdot m^{-2} d^{-1}$.

Tests were also conducted under static conditions for 1 year at 90°C. Under saturation conditions the mean alteration rate measured between 0 and 7 days was about 2×10^{-3} $g \cdot m^{-2} d^{-1}$. Between 7 and 30 days the rate dropped to about 8×10^{-5} $g \cdot m^{-2} d^{-1}$. The residual rate over longer durations was less than 10^{-5} $g \cdot m^{-2} d^{-1}$.

These results confirm the cesium-specific immobilization potential of iron-substituted hollandite ceramics. Characterization of the material performance is now continuing, with emphasis on the effects of the temperature, pH and S/V ratio on the alteration rate, and on the self-irradiation behavior of the material[8].

Other work is also in progress to determine the parameters related to the technological feasibility of these matrices, and in particular to establish a composition range around the reference composition for suitable management of chemical impurities liable to be present in the waste stream.

ACKNOWLEDGMENTS
The authors thank M. Carter and L. Vance from ANSTO for their precious help and comments in this work.

REFERENCES
[1] F. Bart, S. Sounilhac, J.L. Dussossoy, A. Bonnetier and C. Fillet, "Development of new vitreous matrices for conditioning cesium"; pp. 353-360 in American Ceramic Society Conference Proceedings 1999, *Ceramic Transactions*, vol 119, 2001.

[2] A.E. Ringwood and S.E. Kesson, "Synroc"; pp. 233-334 in *Radioactive Waste Forms for the Future*, Edited by W. Lutze et R. Ewing, 1988.

[3] S.E. Kesson and T.J. White, "[BaxCsy][(Ti,Al)3+2x+yTi4+8-2x-y]O16 Synroc-types hollandites Part I: Phase chemistry", *Proceeding of the Royal Society of London*, pp. 73-101 A 405 (1986), and "[BaxCsy][(Ti,Al)3+2x+yTi4+8-2x-y]O16 Synroc-types hollandites Part II: Structural Chemistry", *Proceeding of the Royal Society of London*, pp. 295-319 A 408 (1986).

[4] F. Bart, G. Leturcq and H. Rabiller, "Iron-substituted Barium hollandite ceramics for cesium immobilization" presented at the American Ceramic Society Conference 2003, Proceedings *Ceramic Transactions*, (to be published).

[5] G. Leturcq, F. Bart and A. Compte, *Céramique de structure hollandite incorporant du césium utilisable pour un éventuel conditionnement de césium radioactif et ses procédés de synthèse*, French patent No. E.N. 01/15972 (December 2001).

[6] M.L. Carter and E.R. Vance, "Leaching studies of Synroc-types hollandites" in *PacRim2* CD, edited by P. Walls symposium 16 (1998)

[7] M.L. Carter, E.R. Vance, D.R.G. Mitchell, J.V. Hanna, Z. Zhang and E. Loi, "Fabrication, characterization, and leach testing of hollandite, $(Ba,Cs)(Al,Ti)_2Ti_6O_{16}$" *J. Mater. Res.*, **17**, 2578 (2002).

[8] V. Aubin, D. Caurant, D. Gourier, N. Baffier, T. Advocat, F. Bart, G. Leturcq, and J. M. Costantini, "Synthesis, Characterization and Study of the Radiation Effects on Hollandite Ceramics developed for Cesium Immobilization". *Mat. Res. Symp. Proc*, Kalmar, Suède, 10-13 June 2003.

TITANATE CERAMICS FOR IMMOBILIZATION OF U-RICH WASTES

E. R. Vance, M. L. Carter, S. Moricca and T. L. Eddowes
Materials and Engineering Science
ANSTO
Menai, NSW 2234, Australia

ABSTRACT

U-rich liquid wastes arise from UO_2 targets which have been neutron-irradiated to generate medical radioisotopes such as ^{99m}Tc. A modified synroc-F hot-pressed titanate ceramic will accommodate at least 40 wt% of such waste, expressed on an oxide basis. Leach rates are comparable with those from Synroc-C. Another option using a synroc-B precursor, with a waste loading of ~ 30 wt%, was studied.

INTRODUCTION

Intermediate-level uranium-rich nuclear wastes arise when reactor irradiation of enriched UO_2 is carried out, with the targets being dissolved and selected radioisotopes, notably ^{99m}Tc, separated out for medical use. Options for the waste could include cementation when the waste is many years old, but in the present work we focus on titanate ceramics of the synroc type. On an oxide basis, (neglecting water, nitric acid and other entities that can be removed by calcination) the waste would typically contain >90 wt% of UO_x, process chemicals, and ~ 0.2 wt% fission products. The nearest type of waste for which the synroc family of titanate ceramics has been previously targeted is dissolved spent fuel, and the corresponding synroc variant is synroc-F (1). Synroc-F consists of 85 wt% of pyrochlore-structured $CaUTi_2O_7$, plus 5 wt% each of perovskite, hollandite and rutile. In the present work we have aimed at 80 wt% $Ca(U,Zr)Ti_2O_7$, plus 10 wt% each of hollandite and rutile.

The difference between the two applications is that spent fuel contains roughly 3 wt% of fission products, necessitating the presence of perovskite to incorporate the Sr, whereas the radioisotope application has much fewer fission products. Hence Sr can be incorporated in the hollandite or, if not, can react with the excess rutile to form a very small amount of perovskite if necessary. We have also looked at using synroc-B as a precursor to mix with say 30 wt% of U-rich radioisotope waste.

EXPERIMENTAL

Ceramic samples were made by three somewhat different routes. In (a) the precursor was made by our standard alkoxide route in which Ti and Al alkoxides are mixed with aqueous Ca, Zr, and Ba nitrate solutions and stir-dried. The simulated waste was made with nitrates of U, Cs, Sr, Fe and Mg, dried and calcined (1 hr at 750°C in 3.5% H_2/N_2). The precursor and simulated waste were wet-milled for 24 hours and then calcined in 3.5% H_2/N_2 for 1 hr. In (b) the standard alkoxide/nitrate route was used in which Ti and Al alkoxides are mixed with aqueous Ca, Zr, Ba, Sr, and Cs nitrate solutions and stir-dried and calcined (1 hr at 750°C in 3.5% H_2/N_2). In (c) an aqueous solution of the simulated waste was mixed with microspheres of synroc B precursor (Table I). The method for microsphere production is described elsewhere (2). The mixture was then dried and calcined at 750°C in air. 5 wt% Ti metal was mixed with the samples.

Consolidation was achieved either by hot pressing at ~ 1200°C, uniaxially at a pressure of 20MPa or isostatically at 100 MPa. The samples produced were studied by scanning electron microscopy (SEM) and X-ray diffraction to check on the materials design. Suitable polished samples were also studied by MCC-1 leach tests. Sample compositions can be derived from Table I in which the waste loading was varied from 25-44 wt% by changing the amounts of

Zr and U in the $Ca(U,Zr)Ti_2O_7$ pyrochlore majority phase, noting that both U^{4+} and U^{5+} can be incorporated in the Zr site and U^{4+} in the Ca site (3), giving considerable chemical flexibility.

Scanning electron microscopy (SEM) was carried out with a JEOL 6400 instrument operated at 15 keV, and fitted with a Tracor Northern Voyager IV X-ray microanalysis System (EDS). Powder X-ray diffraction analyses were conducted with a Siemens D500 diffractometer using Co Kα radiation.

Table I: Synroc B precursor composition

Oxide	Wt %	Oxide	Wt %
TiO_2	57.0	BaO	4.4
ZrO_2	5.4	CaO	8.9
Al_2O_3	4.3		

Table II: Ceramic compositions

	Waste loadings wt%		
Oxide	25	35	44
TiO_2	47.3	44.9	42.4
ZrO_2	12.8	6.2	-
Al_2O_3	1.6	1.6	1.6
BaO	2.3	2.3	2.3
CaO	10.9	10.1	9.2
$UO_{2=}$	25	35	44

RESULTS AND DISCUSSION

The samples targeted to contain 25-44 wt% of waste (see Table II) displayed approximately the target mineralogies and the microstructures of some samples are illustrated in Figure 1.

25% Waste Loading

The initial target loading of ^{99}Mo waste in Synroc was 25 wt%, using the zirconolite-rich Synroc approach. The amount of U included in the $ABTi_2O_7$ phase (calculated stoichiometry $CaU_{0.47}Zr_{0.53}Ti_2O_7$) lies close to that of the 4M zirconolite polytype (3). These samples were made using method (a) and two slightly different simulated wastes were used (see Table III). The calcined samples were placed in stainless steel bellows and hot-pressed for 2 hr at 21 MPa and 1250°C. The lid was removed and samples removed by core-drilling, to produce a 10 mm diameter, 2 mm thick disc for leaching, together with an XRD/SEM sample.

SEM (see Figure 1) showed that the desired phases had formed, but that, especially for Mix 1, ~5 μm-sized regions of brannerite (UTi_2O_6) surrounding micron-sized UO_2 grains were also present. The brannerite contained a very small amount (~ 0.01 formula unit) of Sr. These features obviously derived from incomplete reaction of the majority UO_2 component of the simulated waste with the precursor. This could have been remedied in principle by increasing the hot-pressing temperature. XRD confirmed the phases observed in SEM and that the major phase was zirconolite.

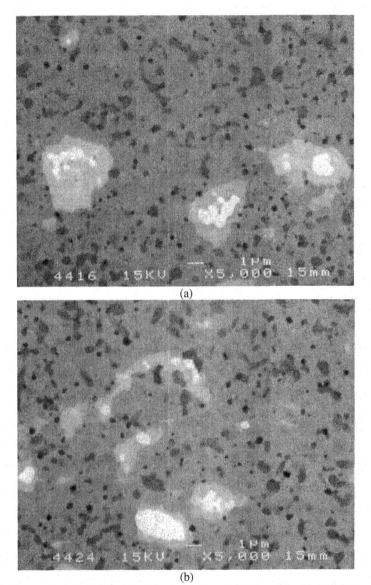

(a)

(b)

Figure 1: Backscattered electron micrographs of Synroc samples (25wt% loading) made from Mix 1 (a) and Mix 2 (25 wt%) (b). Both micrographs show the zirconolite matrix, hollandite and rutile (both dark grey), and light grey regions of brannerite (UTi_2O_6) surrounding micron-sized (bright) UO_2 grains. Small black regions = pores.

Table III: Composition of Mixes 1 and 2

Element	Simulated Waste/Mix 1 g/L	Simulated Waste/Mix 2 g/L
U	120	120
Cs	0.003	0.003
Sr	0.003	0.003
Fe		0.5
Mg		0.5

Leaching data (see Table IV) on these samples yielded results consistent with those obtained on reference grade Synroc-C (4) in the case of Mix 2, although the Cs/Sr results for Mix 1 were an order of magnitude higher than those for Synroc-C. The Ba results for Mix 1 were a factor of 2 higher. Nevertheless, the fact that all leach results were within a factor of 10 (except for 7-28 day values for Mix 1) of those for Synroc-C was very encouraging. The addition of Ti metal to the mix before HUPing (4) would likely have improved the leach results.

Table IV: Leach data (g/m^2/day) at 90°C on titanate ceramics containing 25wt% simulated ^{99}Mo waste calcine.

Mix	1				2			
Days	0-7	7-28	28-56	56-84	0-7	7-28	28-56	56-84
Element								
Ba	0.11	<0.01	<0.01	<0.01	0.04	<0.01	<0.01	<0.01
Cs	1.1	0.67	<0.3	<0.3	0.03	0.07	<0.01	<0.01
Sr	1.4	<0.01	<0.03	<0.03	0.04	0.01	<0.01	<0.01
Ca	<0.01	<0.01	<0.05	<0.01	<0.01	0.1	<0.01	<0.01

All Al, Ti, Zr, U values < 0.01 g/m^2/d.

44% Waste Loading

The waste loading was then increased to the maximum possible (44wt%; see Table II), to make 80 wt% of pyrochlore-structured CaUTi$_2$O$_7$ + 10 wt% each of hollandite and rutile. The samples were made using Method (a) but milling was done after calcination this time. To facilitate Cs and Sr detection in sample leachates, an additional 0.1 wt% each of Cs$_2$O and SrO was added to the precursor. Higher loadings would have made detection easier but could have compromised the overall sample chemistry, especially when the Sr needs to be dissolved in the Ca site of pyrochlore, in which its solubility is not high. Of course it was possible that Sr "undissolved" in the pyrochlore phase would combine with rutile to make small amounts of (durable) perovskite, which could easily escape detection. Two varieties of simulated waste, designated Mix 1 and Mix 2 were used - see Table III.

XRD results were in accord with the design formulation, but a very small amount of brannerite was seen by SEM- see Figure 2b. The SEM investigation showed the brannerite, surrounding micron-sized UO$_2$ "cores" was as observed in the preparations with 25% waste loading (see Figure 1a-b). No evidence of rutile in the XRD pattern was present, but this would be in keeping with the reducing conditions employed in consolidation- Magneli phases are essentially undetectable by XRD in synroc preparations.

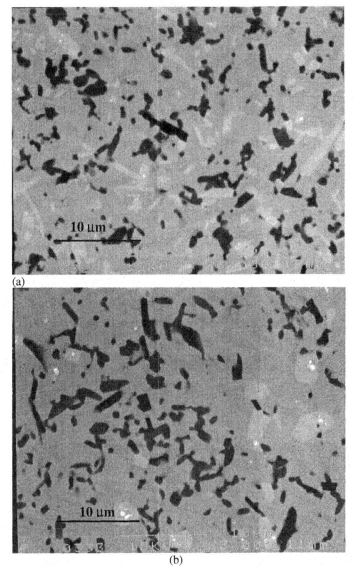

(a)

(b)

Figure 2: Backscattered electron micrographs of samples (44 wt% loading) made from Mix 1 (a) and Mix 2 (b). Both micrographs show the desired phases, pyrochlore (matrix phase), hollandite and rutile (dark), and regions of brannerite (UTi_2O_6; light grey) surrounding bright micron-sized UO_2 grains in (b).

Leaching data (see Table V) on these samples yielded results consistent with those obtained on reference grade Synroc-C (4). As mentioned above, the addition of Ti metal to the mix before HUPing should further improve the leach results.

Table V: Leach data (g/m²/day) at 90°C on titanate ceramics containing 44wt% simulated ⁹⁹Mo waste calcine

Mix	1				2			
Days	0-7	7-28	28-56	56-84	0-7	7-28	28-56	56-84
Element								
Al	n.m	<0.01	<0.01	<0.003	n.m	<0.01	<0.01	<0.01
Ba	0.015	0.003	<0.001	<0.001	0.02	<0.002	<0.001	<0.001
Ca	n.m	<0.003	0.001	<0.002	n.m	<0.002	0.001	<0.001
Cs	0.10	0.014	<0.02	<0.02	0.13	0.02	<0.02	<0.02
Sr	0.08	< 0.01	<0.01	<0.01	0.13	0.01	<0.01	0.006

Ti, U and Zr all <0.0001; n.m. = not measured.

35% Waste Loading

A "half-way" option between the two previous loadings was also studied. A sample of Synroc was made with 35 wt% of waste (see Table II). 0.1 wt% each of Cs_2O and SrO were also added. This sample was made by method (b), rather than starting with a calcined simulated waste. SEM of the HUPed material (2 h at 1250°C/20MPa) showed that no UO_2 particles were present. Some small amount of brannerite was present, in conjunction with the major pyrochlore phase and minor zirconolite, hollandite and rutile (see Figure 3).

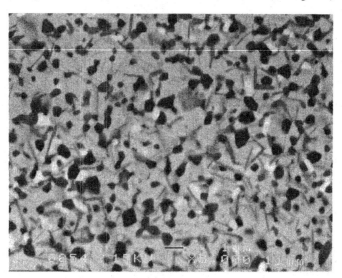

Figure 3: Secondary electron micrograph of Synroc with 35wt% UO_2. (X5000). Pyrochlore matrix (grey) contains hollandite and rutile (both black). Dark grey material is 4M zirconolite. Brighter phase is brannerite.

Leaching data (see Table VI) on these samples yielded results consistent with those obtained on reference grade Synroc-C (4).

Table VI: Leach data (g/m^2/day) at 90°C on Synrocs containing 35 wt% simulated waste.

Days	0-7	7-28	28-84
Element			
Al	*0.05	*0.0009	*0.05
Ba	*0.005	*0.0004	0.0005
Ca	*0.009	*0.0012	0.001
Cs	<0.08	0.097	0.006
Sr	*0.08	<0.0006	*0.0004

* The result was not significantly different from the blank
Ti, U and Zr all <0.0001 g/m^2/day.

Synroc B with 20 –40 wt% Waste Lloading
    The simulated waste - see table VI - was mixed at 20, 30 and 40 wt% oxide loadings with the synroc-B microspheres using method (c).   The samples were then HIPed at 1260°C/100MPa/2hrs.

Table VII : Composition of simulated waste

Element	g/L	Element	g/L
U	120	Fe	0.5
Cs	0.3	Mg	0.2
Sr	0.3		

    The samples were examined by XRD and SEM and all samples contained pyrochlore, hollandite, rutile and perovskite.   The samples loaded with 30 and 40 wt% waste also contained brannerite. Calculations qualitatively bear out this phase assemblages. Table VIII lists the calculated phase assemblages of the samples.

Table VIII: Calculated phase abundances in the Synroc-B loaded samples

Phases	wt % Phase in 20 wt% Loading	wt % Phase in 30 wt% Loading	wt % Phase in 40 wt% Loading
Pyrochlore			
(CaU$_x$Zr$_{1-x}$Ti$_2$O$_7$)	50	61	52
Hollandite (BaAl$_2$Ti$_6$O$_{16}$)	22	19	17
Rutile			
(TiO$_2$)	20	13	3
Perovskite (CaTiO$_3$)	5.4	–	–
Brannerite (UTi$_2$O$_6$)	–	4.8	26.7

    Leaching data 0-7 days (see Table IX) on these samples yielded results comparable with those obtained on reference grade Synroc-C (4), although the Cs results were a factor of 2-4 higher.

Table IX: Leach data ($g/m^2/day$) at 90°C on Synroc-B containing 20-40wt% simulated waste.

Sample	20 wt%	30 wt%	40 wt%
Days	0-7	0-7	0-7
Element			
U	0.00015	0.00001	0.00022
Ba	0.002	0.001	0.007
Ca	Below blank	0.013	0.009
Cs	0.261	0.117	0.024
Sr	0.007	0.007	0.059

CONCLUSIONS AND FINAL REMARKS

Titanate ceramics of the Synroc type are appropriate for immobilization of U-rich wastes from radioisotope production. The samples have excellent aqueous durability, comparable to that of synroc-C. Considerable chemical flexibility is available in terms of waste/precursor ratios. Hot isostatic pressing, with its relatively small footprint and absence of secondary wastes in the high temperature step of the process, should be close to ideal for consolidation.

REFERENCES

(1) S.E. Kesson and A.E. Ringwood, "Safe Disposal of Spent Nuclear Fuel", *Radioactive Waste and the Nuclear Fuel Cycle*, **4**, 259-74 (1983).

(2) A.E. Ringwood, S.E. Kesson, K.D. Reeve, D.M. Levins and E.J. Ramm, "Synroc"; pp. 233-334 in *Radioactive Waste Forms for the Future*. Edited by R.C. Ewing and W. Lutze. Elsevier. North-Holland, 1988.

(3) E. Sizgek, J.R. Bartlett, J.L. Woolfrey and E.R. Vance, "Production of Synroc Ceramics from Titanate Gel Microspheres"; pp. 305-12 in *Scientific Basis for Nuclear Waste Management XVII*. Edited by R. Van Konynenburg and A.A. Barkatt. Materials Research Society. PA, 1994.

(4) E.R. Vance, G.R. Lumpkin, M.L. Carter, D.J. Cassidy, C.J. Ball, R.A. Day and B.D. Begg, "The Incorporation of U in Zirconolite ($CaZrTi_2O_7$)", *Journal of the American Ceramic Society*, **85**, 1853-9 (2002).

WASTE FORM DEVELOPMENT FOR THE SOLIDIFICATION OF PDCF/MOX LIQUID WASTE STREAMS

A.D. Cozzi and C.A. Langton
Westinghouse Savannah River Company
773-43A
Aiken, SC 29808

ABSTRACT

At the Savannah River Site, part of the Department of Energy's nuclear materials complex located in South Carolina, cementation is under consideration for the solidification method for high-alpha and low-activity aqueous waste streams generated in the planned plutonium disposition facilities. The potential to construct a Waste Solidification Building (WSB) that would be used to treat and solidify three radioactive liquid waste streams generated by the Pit Disassembly and Conversion Facility and the Mixed Oxide Fuel Fabrication Facility is being investigated. The WSB would treat a transuranic (TRU) waste stream composed primarily of americium and two low–level waste (LLW) streams. The acidic wastes would be concentrated in the WSB evaporator and neutralized prior to solidification.

A series of TRU mixes were prepared to produce waste forms exhibiting a range of processing and cured properties. The LLW mixes were prepared using the premix from the preferred TRU waste form. All of the waste forms tested passed the Toxicity Characteristic Leaching Procedure (TCLP). After processing in the WSB, current plans are to dispose of the solidified TRU waste at the Waste Isolation Pilot Plant in New Mexico and the solidified LLW waste at an approved low-level waste disposal facility.

INTRODUCTION

The Savannah River Site is considering the construction of a Waste Solidification Building (WSB) for the treatment and solidification of the three radioactive liquid waste streams generated by the Pit Disassembly and Conversion Facility (PDCF) and the Mixed Oxide Fuel Fabrication Facility (MFFF). The WSB would treat a high alpha (TRU) waste (HAW) stream and two low alpha waste (LLW) streams. The acidic wastes would be concentrated in an evaporator and then neutralized prior to solidification. Cementation has been identified as a potential solidification method for both the high-activity and low-activity waste streams generated in the PDCF and MFFF.

The HAW waste is an acidic aqueous stream generated in the following processes: the MOX acid recovery and recycle processes; the alkaline treatment process; and the aqueous purification process. The two LLW streams consist of the stripped uranium stream (SUS), an acidic depleted uranium nitrate solution resulting from the uranium stripping process in the MFFF and, the PDCF lab liquids stream (PDCF-LL). The PDCF waste is generated from the laboratory analysis of plutonium oxide, highly enriched uranium (HEU) oxide, process samples, and waste samples. This waste also includes rinse water from equipment and drain flushes and is expected to have a pH <1.

Table I is the composition of each of the waste streams investigated. The surrogates for the radioactive components (Am, Pu, and U) were chosen based on work by Villarreal and Spall[1]. None of the waste streams will contain any listed components in the Resource Conservation and

Recovery Act (RCRA). However, the HAW stream will contain significant quantities of silver that when concentrated in an evaporator can be considered an underlying constituent. The other RCRA metals have been spiked into the waste at a multiple of the expected concentrations to 1) provide quantities that can be accurately measured in the TCLP extract and 2) provide assurance that exceeding the current proposed values would not increase the risk of the cement based waste form.

Table I. Composition of the Expected Waste Streams.

Constituent	Surrogate	HAW Average (g/L)	SUS Maximum (g/L)	PDCF-LL Maximum (g/L)
Am	Eu	1.21E+00	N/A	4.26E-05
Pu	Ce	8.83E-03	5.50E-03	9.34E-04
U	Ce	2.49E-01	4.00E+02[†]	7.95E-04
Acid (M)	--	5.95E+00	2.00E+00	6.12E+00
Ga	--	2.08E+00	N/A	5.73E-05
Na	--	7.20E+00	N/A	N/A
Ag	--	1.50E+01	2.50E-01	6.10E-04
TBP	N/A	5.74E-04	1.10E-03	2.19E-03
Ba	--	8.64E-01	2.50E-01	1.47E-02
Ca	--	2.07E+01	N/A	N/A
Cd	--	1.73E-01	2.50E-01	2.50E-01*
K	--	3.80E+01	N/A	N/A
Mg	--	1.21E+01	N/A	N/A
Pb	--	3.46E-02	2.50E-01	1.13E-06
Hg	--	1.00E-02*	2.50E-01*	1.13E-02
Tl	N/A	5.00E-03*	2.50E-01*	1.13E-06
Cr	--	5.00E-02*	2.50E-01*	1.33E-02
Be	N/A	5.00E-02*	2.50E-01*	6.67E-06
SO₃	--	N/A	N/A	8.00E-02
Cl	--	N/A	N/A	2.00E+00
F	--	N/A	N/A	2.67E-01
Acetone	N/A	N/A	N/A	1.73E-06

[†]Depleted uranium used for uranium in SUS surrogate.
*Not expected to be routinely in the waste, but included by WSB to simulate potential process upsets.

To facilitate materials handling in the proposed facility, each of the waste forms must be prepared using the same premix. The premix is the combination of cement and other materials that, when mixed with the neutralized waste, produce the solidified waste form. The water to premix ratios (w/c) for the mixes are calculated using the weight percent water in the neutralized waste solution.

OBJECTIVE

The objectives of this work are: 1) to develop a waste form that will allow the HAW stream from the MFFF to be shipped to the Waste Isolation Pilot Plant (WIPP) (To accomplish this objective, the waste form must be processable, contain less than one volume percent bleed water, and show evidence of gas generation rates acceptable for shipping.) and 2) the waste form must also be useable for the two LLW streams. The LLW waste forms must be treated so as to be designated as non-hazardous waste for disposal in an approved low-level waste facility. Specific requirements for each of the waste forms are shown in Table II. Other requirements/constraints such as the use of a single premix composition for all of the waste forms are listed in Table III. The table also includes desirable properties that are not specific requirements.

Table II. Requirements for the Three WSB Waste Forms.

	HAW	SUS-LLW	PDCF-LL
Proposed Disposal Site	WIPP	LLW disposal	LLW disposal
Requirements			
Fresh/Cured Properties			
No free liquid	X	X	X
Compressive strength	N/A	N/A	N/A
Set Time	N/A	N/A	N/A
Gel Time	N/A	N/A	N/A
Flowability	N/A	N/A	N/A
Regulatory			
Acceptable gas generation rate	X	N/A	N/A
Pass TCLP	N/A*	X	X

*Non-hazardous designation is desirable but not required.

Table III. Additional Requirements Considered During Waste Form Development.

- One premix for all three waste forms
  - The WSB design specifies one dry materials silo to feed both the HAW and the LLW processes.
- Texture
  - The mixed waste form may be required to be either "crumbly" or self-leveling.
- Minimal or no organic processing admixtures.
- Minimize specific gravity of waste form
  - Reduce weight of waste package.
- Maximum temperature during curing <95°C.
  - Eliminate potential of steam formation during curing.
- Maximize waste loading for LLW waste forms.

DISCUSSION

Surrogate Preparation

Surrogate waste solutions of the evaporator bottoms were prepared for each of the three waste streams. The surrogates were neutralized to 1-M excess hydroxide and the weight percent solids (dissolved and undissolved) were measured. Table IV is a summary of the measured properties of the neutralized surrogates.

Table IV. Summary of Properties of Neutralized Surrogate Waste Streams.

Property	HAW	SUS-LLW	PDCF-LL
Density (g/mL)	1.37	1.7	1.25
Wt. % Water	51.79	56.05	67.01
Dissolved solids (%)	43.38	24.86	32.99
Undissolved solids (%)	4.82	19.09	0.00
Viscosity (cP)	NM	16.8@ 25°C 20.3@ 66°C	NM

NM - Property not measured

HAW: To prepare the surrogate HAW waste, the composition in Table I was prepared from the nitrate salts, nitric acid and water. The surrogate solution omitted thallium and beryllium. Both elements are highly toxic and are underlying constituents for a hazardous designation. Therefore, if the waste form passes the Toxic Characteristic Leaching Procedure (TCLP)[*] for the eight metals, the thallium and beryllium levels are not used for regulatory classification. The organic component, tributyl phosphate (TBP) was not added as it will not be retained by the waste form and is expected to be fully leached during the TCLP. The solution is then neutralized to 1-M hydroxide with 50 wt% sodium hydroxide solution. The amount of sodium hydroxide required is determined by calculating the amount of hydroxide necessary to neutralize the free acid, precipitate the metals as hydroxides, and attain the 1 M free hydroxide. The weight percent water was measured to be 51.79% (48.21% solids; 43.38% dissolved solids and 4.82% undissolved solids).

SUS-LLW: The surrogate for the stripped uranium stream was prepared by dissolving depleted uranium oxide in concentrated nitric acid and water to achieve a 400 g/L uranyl nitrate solution. The minor nitrate salts were then added to the solution to attain the composition for the SUS-LLW in **Table I**. For this surrogate, again the thallium, beryllium and TBP were omitted. The plutonium was also omitted from this composition as it is present in small quantities and does not contribute to the value of the surrogate. The required sodium hydroxide addition was determined in the same manner as the HAW surrogate. The weight percent water of the neutralized solution was measured to be 56.05%% (43.95% solids; 24.86% dissolved solids and 19.09% undissolved solids). The density of the neutralized surrogate was measured to be 1.7 g/mL. The viscosity of the neutralized solution was measured at 25 and 66°C. The maximum viscosity at 25°C was 16.8 cP and 20.3 at 66°C.

PDCF-LL: The PDCF lab liquids waste surrogate was prepared in the same manner as the HAW surrogate. The sodium salts were used to introduce the sulfate, chloride and fluoride. The density and the weight percent solids of the acidic solution was measured (ρ = 1.20 g/mL; wt% solids 0.35%). The surrogate was neutralized and the weight percent water was measured to be 67.01% (32.99% solids, all dissolved).

Waste Form Development

Given that the HAW waste stream has the most restrictive requirements (see Table II and Table III), and that the same premix must be used for all of the waste forms, the HAW waste form will be developed first. The remaining waste forms will be adjusted to succeed with the

---

[*] EPA Manual SW-846 Procedure 1311

premix formulation developed for the HAW waste form. A Hobart N-50 mixer was used to prepare all of the waste forms for this task.

HAW: A sample test matrix was prepared to evaluate the effects of the water to cement (premix) ratio (by mass) w/c, the addition of pozzolans[*], and the use of admixtures on the processability and TCLP response of the HAW waste form. A low w/c ratio is used to minimize the free water available for radiolysis (hydrogen generation). Higher w/c ratios improve flowability of the mix and decrease the mass of the waste form (Am content is fixed in the waste, therefore less premix is required to solidify the waste form). In the range of w/c ratios evaluated in this task, the mixes prepared solely with cement as the premix material adhered to the mixing equipment. To produce a "drier" mix at the low w/c ratio, perlite, a high surface area pozzolan, was added. Perlite is an inorganic silicate mineral with a high water demand. The effect of perlite on the hydrogen generation rate is not known. For the higher w/c ratio mixes, a high range water reducer admixture was used to increase the flowability of the mix. Perlite is an inorganic silicate mineral. The effect of perlite on the gas generation rate is unknown. As the admixture is an organic material, there is a potential for the admixture to adversely effect the gas generation rate. However, due to the more complete mixing attained with admixture, there is also the potential for more complete reaction of the cementitious materials; thus reducing the free water available for radiolysis. The admixture used in this task is Daracem 19[†].

Given that most of the waste constituents form insoluble hydroxides during the neutralization process, cement alone may be sufficient to stabilize the waste. However, the waste contains significant quantities of mercury that may not be satisfactorily retained by a cement waste form. In the SRS Saltstone process, slag is used to chemically reduce the waste and stabilize/precipitate mercury. Other materials identified that can assist in the stabilization of mercury are sodium sulfide and sodium thiosulfate. In this task, slag and sodium thiosulfate were evaluated for mercury stabilization. Table V is a summary of the HAW mixes prepared. The neutralized waste solution used for these mixes was 52.15 wt.% water. Figure 1 is the mix #3 waste form and mixer.

Table V. HAW Mixes Prepared.

Mix	w/c	Premix (%)	Cement (%)	Slag (%)	Perlite (%)	Waste (%)	$Na_2S_2O_3$ (%)	Admixture (%)	TCLP
1	0.2	70.5	64.9	0	5.6	29.5	0	0	Y
2	0.2	70.5	32.4	32.4	5.6	29.5	0	0	Y
3	0.2	70.5	32.4	32.4	5.6	29.5	0.013	0	Y
4	0.3	61.5	30.7	30.7	0	38.5	0	0.56	Y
5	0.3	61.5	30.7	30.7	0	38.5	0.017	0.56	Y
6	0.4	54.5	27.2	27.2	0	45.5	0.02	0	N

---

[*] A siliceous material that, in a finely divided form and in the presence of moisture, chemically react with calcium hydroxide to form compounds having cementitious properties.

[†] Daracem 19, W.R. Grace

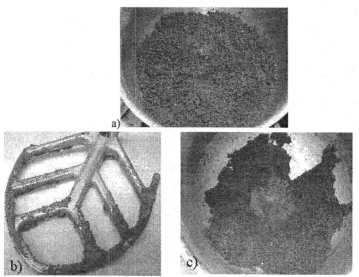

Figure 1. Mix #3. a) Mixed waste form b) uncleaned mixer blade, and c) heel remaining in mixer.

Mixes one through five were analyzed using a modified TCLP[*]. The mixes were cured for 28 days, crushed, and passed through a 20-mesh sieve (841μm – 0.03 in)[†]. A TCLP was performed on the crushed waste forms. The resulting leachate was analyzed by Inductively Coupled Plasma-Atomic Emission Spectrometry (ICP-AES) for barium, cadmium, chromium, lead, and silver and by Cold Vapor Atomic Absorption Spectroscopy (CV-AAS) for mercury. Table VI is the concentration of these elements in the waste solution, neutralized waste solution, and the waste form for each of the mixes tested.

Table VII is a summary of the TCLP results. The EPA Land Disposal Restrictions (LDR) values for the RCRA elements are included for reference. The LDR are treatment standards that ensure hazardous waste is properly treated to immobilize hazardous chemical components before it is land disposed. The results indicate that this waste form would be acceptable as a treatment for hazardous waste. The final determination would require further analysis to obtain results with lower detection limits for silver and mercury. This would be a routine procedure, however it was not included in the original scope of work.

To address a concern that the silver concentration may exceed the 15 grams per liter tested, a mix with a w/c = 0.2 was prepared using a surrogate with a silver concentration of 25 grams per liter. The silver release from the TCLP was <0.3 mg/L, similar to the results for the 15 g/L silver concentration waste surrogate.

---

[*] The test was performed with 10 grams of sample rather than prescribed 100 grams. The leachant to sample ratio remained 20.
[†] The TCLP requires the sample to pass through a 3/8-inch sieve. EPA Manual SW-846, Procedure 1311.

Table VI. Concentration of Elements of Concern in Waste Solutions and Waste Forms.

RCRA Element	Concentration in Waste Solution (mg/L)	Concentration in Neutralized Waste Solution (mg/L; mg/kg)	Concentration in Waste Form (mg/kg)				
			Mix 1 w/c=0.2	Mix 2 w/c=0.2	Mix 3 w/c=0.2	Mix 4 w/c=0.3	Mix 5 w/c=0.3
Ba	864	571; 417	123	123	123	160	160
Cd	173	114; 83	25	25	25	32	32
Cr	50	33; 24	7.1	7.1	7.1	9.3	9.3
Pb	30	23; 17	4.9	4.9	4.9	6.4	6.4
Ag	15000	9911; 7235	2134	2134	2134	2785	2785
Hg	10	7; 5	1.4	1.4	1.4	1.9	1.9

Table VII. TCLP Results from the HAW Mixes.

RCRA Element	TCLP Hazardous Limit (ppm)	LDR[†] Treatment Limit (ppm)	Leachate concentration (mg/L)				
			Mix 1	Mix 2	Mix 3	Mix 4	Mix 5
Ba	100	21	2.15	1.23	1.30	1.26	1.26
Cd	1	0.11	<0.014	<0.014	<0.014	<0.014	<0.014
Cr	5	0.6	0.14	<0.05	<0.05	<0.05	<0.05
Pb	5	0.75	<0.69	<0.69	<0.69	<0.69	<0.69
Ag	5	0.14	<0.3	<0.3	<0.3	<0.3	<0.3
Hg	0.2	0.025	<0.11	<0.11	<0.11	<0.11	<0.11

[†]Land Disposal Restriction limit used for treated waste already declared hazardous.

SUS-LLW: Given that mixes were prepared from the HAW surrogate that met either the crumbly or flowable condition and passed TCLP, both of the premix formulations used to produce the HAW mixes were tested with SUS-LLW surrogate. The constraints applied to this set of tests were as follows: the same premix formulation as the HAW must be used; the waste form must be non-hazardous (pass TCLP); and there must not be any free water after set.

To provide the customer with the option of choosing either a LLW waste form with similar processing properties to the HAW form or a processable waste form with increased waste loading, the premix formulation from Mix #3 in Table V was tested for both of the LLW waste forms. The goal of the first mix was to obtain a "crumbly" waste form. The neutralized SUS-LLW surrogate in Table IV was mixed with the premix in a mixer using an initial w/c ratio of 0.5. The initial consistency was clayey and agglomerated into large masses. Additional premix was mixed in until the desired consistency was attained. The final calculated w/c ratio of the mix was 0.24. The second mix used the same starting materials and w/c ratio, 0.5. With continued mixing, the agglomerations developed into a homogeneous mass (on a visual scale). As the mass was repeatedly broken down and reassembled during the mixing process, the agglomerated masses began to "clean" the bowl. Figure 2 is the waste form after mixing. Table VIII is the composition of the two mixes prepared with the SUS-LLW surrogate. Table IX is the TCLP results for the mix prepared with a w/c ratio of 0.5.

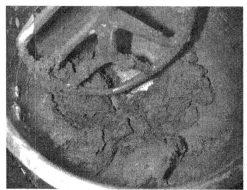

Figure 2. SUS-LLW mix with w/c = 0.5.

Table VIII. Mixes Prepared Using the SUS-LLW Neutralized Surrogate.

Mix	w/c	Premix (%)	Cement (%)	Slag (%)	Perlite (%)	Waste (%)	Na₂S₂O₃ (%)	Admixture (%)	TCLP
SUS-1	0.24	70.1	31.6	31.6	6.9	29.9	0.026	0	N
SUS-2	0.5	52.8	23.8	23.8	2.6	47.2	0.023	0	Y

Table IX. TCLP Results for SUS-LLW Mix SUS-2 with w/c Ratio = 0.5.

RCRA Element	TCLP Hazardous Limit (mg/L)	LDR† Treatment Limit (mg/L)	Concentration			
			Waste Solution (mg/L)	Neutralized Waste Solution (mg/L; mg/kg)	Waste Form (mg/kg)	Leachate (mg/L)
Ba	100	21	250	173; 102	48	0.808
Cd	1	0.11	250	173; 102	48	<0.200
Cr	5	0.6	250	173; 102	48	<0.170
Pb	5	0.75	250	173; 102	48	<2.46
Ag	5	0.14	250	173; 102	48	<0.150
Hg	0.2	0.025	250	173; 102	48	<0.11

†Land Disposal Restriction limit used for treated waste already declared hazardous.

PDCF-LL: Given that mixes were prepared from the HAW surrogate that met either the crumbly or flowable condition and passed TCLP, both of the premix formulations used to produce the HAW mixes were tested with PDCF-LL surrogate.

The constraints applied to this set of tests are as follows: The same premix formulation as the HAW must be used; the waste form must be non-hazardous (pass TCLP); and there must not be any free water after set.

All of the solids in the PDCF-LL neutralized surrogate were dissolved solids. To obtain a "crumbly" mix, the w/c ratio (and therefore the waste loading) would most likely be impracticably low. Therefore, the goal of this test was to produce a fluid waste form. Either TCLP results or the presence of bleed water in the waste form will limit the w/c ratio (waste loading). An initial w/c ratio of 0.4 was used with the same premix formulation from HAW mix

#3. The mix was uniform and poured easily from the mixer. Table X is the formulation of the mix tested.

Table XI is the concentrations of the elements of concern in the waste and waste forms as well as the TCLP results for the waste form.

Table X. Formulation to Prepare a Waste Form with the PDCF-LL Neutralized Surrogate.

Mix	w/c	Premix (%)	Cement (%)	Slag (%)	Perlite (%)	Waste (%)	$Na_2S_2O_3$ (%)	Admixture (%)	TCLP
PDCF	0.4	62.6	28.8	28.8	5	37.4	0.016	0	Y

Table XI. TCLP Results for PDCF-LL Mix with w/c Ratio = 0.4.

RCRA Element	TCLP Hazardous Limit (mg/L)	LDR[†] Treatment Limit (mg/L)	Waste Solution (mg/L)	Concentration Neutralized Waste Solution (mg/L; mg/kg)	Waste Form (mg/kg)	Leachate (mg/L)
Ba	100	21	14.7	11; 9	3.3	0.90
Cd	1	0.11	250	187; 149	56	<0.014
Cr	5	0.6	13.3	10; 8	3	<0.05
Pb	5	0.75	0.0011	0.0008; 0.0007	0.0003	<0.69
Ag	5	0.14	0.28	0.21; 0.17	0.06	<0.3
Hg	0.2	0.025	11.3	8.4; 6.7	2.5	<0.11

[†]Land Disposal Restriction limit used for treated waste already declared hazardous.

CONCLUSIONS

Waste forms have been prepared from surrogates of the three WSB waste streams (HAW, SUS-LLW and PDCF-LL). A mix was prepared for all three waste streams using the same premix formulation; 46% cement, 46% slag, and 8% perlite (by mass). There was no bleed water remaining in any of the waste forms after the mix had set. All of the waste forms tested have passed the Toxicity Characteristic Leaching Procedure (TCLP). The processability of the mixes will require validation after the WSB has selected a mixing system.

REFERENCES
[1] R. Villarreal and D. Spall, "Selection of Actinide Chemical Analogues for WIPP Tests," LA-13500-MS, Los Alamos National Laboratory, Los Alamos, NM August 1998.

SOLIDIFICATION OF SODIUM BEARING WASTE USING HYDROCERAMIC AND PORTLAND CEMENT BINDERS.

Yun Bao and Michael W. Grutzeck
Materials Research Institute
Materials Research Laboratory Building
The Pennsylvania State University
University Park, PA 16802

ABSTRACT

A hydroceramic is a monolithic solid produced by a hydration driven reaction that occurs after mixing metakaolin with concentrated NaOH solution. The chemistry of the starting materials is optimized to encourage the formation of tectosilicates (zeolites, feldspathoids) during curing. Furthermore, the process is forgiving enough so that a variety of caustic waste streams can be accommodated without sacrificing performance. This characteristic makes a hydroceramic perfectly suited to solidify low activity sodium bearing waste (SBW) now in storage at Hanford, Savannah River and INEEL. SBW normally contains three main ingredients, NaOH, $NaNO_3$ and $NaNO_2$. Depending upon the level of nitrate and nitrite salts in the SBW the SBW can be solidified by direct mixing and curing (low concentrations) or alternately must first be heat treated in some manner to decompose its nitrate/nitrite inventory. Two procedures, both using approximately the same blend of aluminosilicate, carbonaceous materials and SBW are being evaluated by the DOE. The first is a calcine formed by heating the blended materials in air at 525°C (this work). The second is a granular product produced during steam reforming in a Studsvik reactor at 725°C. In both cases, nitrate and nitrite are reduced and the nacent Na ions that form react with the metakaolinite forming X-ray amorphous precursors and crystalline zeolites (zeolite A, hydroxysodalite) and feldspathoids (cancrinite) rather than Na_2CO_3. Although the products have low leachabilities in their own right, they can be further solidified and made into a hydroceramic using the approach outlined above for liquid SBW. Performance data for hydroceramics made from a simulated low nitrate/nitrite Savannah River SBW solidified by direct solidification and also via intermediate calcination and then solidification suggest that leachabilities are nearly the same. Thus direct solidification is recommended for this type of SBW. Performance data for a heavily nitrated and nitrited Hanford SBW that was calcined at 525°C with metakaolin and sucrose and then solidified using a hydroceramic binder and a Portland cement binder are compared. The hydroceramic exhibited superior performance vis à vis a more conventional grout formulation.

INTRODUCTION

Radioactive waste produced during DOE's active weapons production program has been stored in tanks at the Hanford, Savannah River and INEEL sites for more than 50 years. In all cases nitric acid was used to dissolve nuclear fuel rods during reprocessing. After extracting fissile materials and other elements of interest the remaining waste liquid was placed into storage. Of the three sites only INEEL decided to calcine its acid waste directly using a fluidized bed calciner. Their calcines are now in storage on site in bins (bin sets)[1]. Savannah River and Hanford chose to neutralize their waste streams with an excess of NaOH to facilitate their storage in underground steel tanks. Once the pH was changed, a significant amount of insoluble precipitates and soluble salts formed in the tanks. These sank to the bottom of the tank and formed a sizeable amount of sludge and hard salt cake. The remaining liquid fraction (supernate) is commonly referred to as sodium bearing waste (SBW). It is a highly alkaline solution containing large amounts of sodium hydroxide, sodium nitrate and sodium nitrite as well as smaller amounts of soluble salts containing other elements. Radioactive decay of most short lived elements has occurred over the years so that most SBW is now considered to be low activity (LA) waste. However, cesium and strontium are still present and as such are the two remaining isotopes of concern to the DOE. Savannah River is planning to separate the cesium and strontium from their SBW further reducing its activity and then plans to stabilize the remaining caustic nitrate and nitrite salt solution with cementitious grout (Saltstone). The bulk of the waste at Hanford is cesium free and is still in storage and plans are not yet finalized.

What we offer here is a process that can be used to solidify SBW. The process is inexpensive and can be implemented using shear mixers commonly used by the ceramic industry to mix its clay. Instead of clay and water however, metakaolin is mixed with just enough 4-12M NaOH SBW to make a thick paste that can then be precured at 40°C and then cured at 90°C to form a monolithic solid (a.k.a. hydroceramic) that is both robust and insoluble enough to qualify as an acceptable waste form. Unfortunately, not all SBW is the same. Some can be solidified as is, some need to be concentrated, and others require pretreatment to reduce nitrate and nitrite content via heat treatment of some sort.

The description that follows provides general guidelines outlining the chemical limits on the SBW that can be solidified using a hydroceramic binder. There are three types of SBW. The first is essentially an impure NaOH solution that may be the result of sludge washing activities. If it is relatively dilute it can be concentrated to 4-12 M NaOH before using it to make the hydroceramic. This type of SBW is more of a resource than a liability. Due to its low activity it should be reserved and used to make monolithic hydroceramics from calcined Type II and III waste described below. The second type of SBW contains a relatively small amount of sodium nitrate and sodium nitrite salts. The mole fraction of $NO_3+NO_2 \div Na$ should not exceed ~25%. Hydroceramics made with this type of SBW usually contain both zeolites (A and hydroxysodalite) plus small amounts of nitrate/nitrite containing cancrinite. However, the ability to accommodate these salts in zeolite and/or feldspathoid matrices is limited and consequently as the concentration of these salts increase in the SBW the corresponding leachability of the hydroceramic degrades rapidly. The third type of SBW is also the most prevalent. It has a molar concentration of Na ranging from 8-12 of which an exceeding large amount consists of sodium nitrate and sodium nitrite (far in excess of the 25% direct solidification can accommodate). These wastes must be pretreated in order to lower their nitrate/nitrite content to a low enough point to allow their solidification using hydroceramic processing. Calcination and pyrolysis (a.k.a. steam reforming) are two means of doing this[2-10]. In both instances the SBW is mixed with an aluminosilicate source (metakaolin or kaolin) and a reducing agent (sucrose, charcoal) prior to heat treatment. The solid that forms is typically granular and has low leach rates for sodium and other elements. These materials may be considered suitable waste forms, but if they are to be shipped, they would have to be solidified. A hydroceramic binder is shown to be superior to a Portland cement grout for this purpose.

DEFINITIONS
Type I SBW
It has been shown that a mixture of concentrated NaOH solution mixed with metakaolinite can be precured in a chamber at 40°C and 100% humidity to develop green strength and then cured at 90°C to 190°C to form a monolithic solid that is both robust and extremely insoluble[2-9].

The chemistry of the reaction is similar to that used by industry to synthesize zeolites, however the processing is very different. To wit, if one is trying to synthesize zeolites one usually carries out reactions in an enclosed vessel filled with boiling caustic and metakaolinite. To make a hydroceramic one mixes only enough concentrated NaOH with the metakaolinite to make a thick paste. The paste is molded, precured to give it enough strength to demold, and then it is cured at 90°C-190°C. It has been shown that the end result is much the same. Zeolite phases such as zeolite A form more slowly at 90°C than at 190°C, but after a month at 90°C leachabilities begin to approach each other[9]. Curing at 90°C seems the best choice because pressure vessels or autoclaves are not required to cure the samples.

Type II SBW

Hydroceramics can be made from SBW without any processing if the mole fraction of nitrate and nitrite ions in the waste does not exceed ~25% of the sodium ions in the waste. The liquid waste is simply mixed with metakaolinite and cured. This is demonstrated later in the paper where we take a simulated waste based on the composition of an actual SBW derived from Savannah River's Tank 44 and solidify it directly and also after first calcining it at 525°C and then solidifying it. The simulated waste is representative of the radioactive waste that Savannah River is currently using to evaluate equivalent but radioactive hydroceramic samples. Unlike the original tank waste this waste has had its Cs and Sr removed to make it safe to handle in a glove box/hood rather than remotely. Data are presented that suggest direct solidification works.

Type III SBW

This SBW has a great deal of sodium bearing salts other than NaOH dissolved in it. Since hydroceramics are zeolite/feldspathoid based and even pure cancrinite can not host all of the nitrate and nitrite. As a consequence, soluble salts remain behind as a second phase dispersed in the zeolite matrix. These are soluble and easily leached, making the waste form unacceptable. Due to this fact, this type of SBW must first be pretreated in some fashion to reduce the nitrate/nitrite concentration. INEEL used fluidized bed calcination to calcine their acid waste. Calcination in a rotary or vertical kiln could be used instead. Laboratory experiments described later show that the calcination process can be run in a crucible at 525°C overnight forming a tectosilicate-like precursor that has very low leachability. The DOE is currently testing a Studsvik steam reformer using basically the same starting materials injected into a fluidized bed containing alumina, charcoal, and iron oxide heated by superheated steam and oxygen at 725C[2]. In both processes, nitrate and nitrite are destroyed and the product that forms consists of tectosilicates (amorphous at ~525°C and nepheline and sodalite at ~725°C). The resulting calcines (we are including steam reformed material in the calcine category) have varying degrees of granularity ranging from fluffy powders to flakes to granules. It is demonstrated later in the paper that a simulated waste based upon an average Hanford Tank SBW could be calcined as a function of temperature and then solidified using a hydroceramic consisting of additional metakaolinite and 4M NaOH solution. It is further demonstrated that the hydroceramic performed much better than a similarly solidified Portland cement based monolith. The data suggest that hydroceramic monoliths containing calcine or steam reformed material can be solidified forming monoliths that have strength, low leachability and phase composition that is compatible with Yucca Mountain strata should the need arise to ship these materials off site.

EXPERIMENTAL METHODS

Two sets of experiments were carried out. In the first set it is demonstrated that a border-line SBW that is close to the cusp between Type I and II SBW can be solidified in two ways: directly as a hydroceramic, and, after first calcining it and then solidifying it by mixing it with a hydroceramic binder. A partial SBW analysis supplied by Savannah River for the composition of a Cs and Sr free Tank 44 SBW they are using to perform parallel experiments is given in Table I. Its composition reflects two things: the SBW has been processed to remove Cs and Sr in order to facilitate its use in a hood rather than in a remote facility, and during processing to remove Cs and Sr the volume of SBW increased having the effect of reducing its sodium ion concentration by more than one half that of the original Tank 44 SBW. Our intent is to compare our results with theirs (simulated versus radioactive) when data become available. The formula for the simulant made at Penn State is given in Table II. These experiments were initiated to illustrate the differences in outcome when a single SBW was solidified using two different but related procedures. In short, a directly solidified monolith performed well enough, eliminating the need to calcine Type II waste prior to solidifying it.

Table I. Composition of Tank 44 SBW.

Ions present	Concentration (M)
Na^+	5.1
NO_3^-	0.5
NO_2^-	0.5
AlO_2^-	0.2
OH^-	4.5

Table II. Composition of Savannah River SBW simulant.

Composition	MW	M	g/l
$CsNO_3$	194.909	0.5	97.4545
KNO_2	85.107	0.1	8.5107
$NaNO_2$	69.000	0.4	27.6000
$NaAlO_2$	81.979	0.2	16.3958
$NaOH$	39.998	4.5	179.9910

The second set of experiments demonstrate that a given calcine made with metakaolin and sucrose heated in air at 525°C overnight could be solidified using either a metakaolin + NaOH or Portland cement + water binder, but by no means was performance the same. The simulated SBW used in the experiments was based upon a recipe originally published by Brough et al.[11] representing an average of Hanford's SBW supernate (Table III). The calcine was made by preblending the SBW simulant with metakaolinite and sucrose in the proper proportions needed to reduce nitrate and nitrite ions and form precursor tectosilicate phases[9]. The resulting calcine was a powder, but much like a steam reformed product[10] it has reasonably low leachability in its own right as determined by measuring sodium ion leachability after 1 day PCT tests. Nevertheless, in order to ship the calcines off site calcines might have to be solidified unless current NRC regulations are changed. Currently, waste in transit must be monolithic and have a minimum compressive strength. Anticipating this possibility, experiments were carried out to evaluate the performance of a hydroceramic binder vis à vis a conventional grout binder to solidify the Hanford calcine described above. The results indicate that a

Table III. Composition of Hanford SBW.

Compound	(g added to a L of H_2O)
NaOH	82.68
$Al(NO_3)_3 \cdot 9H_2O$	133.50
$NaNO_2$	36.91
Na_2CO_3	36.21
$NaNO_3$	8.58
Na_2HPO_4	27.79
KCl	1.83
NaCl	2.51
$Na_2B_4O_7 \cdot 10H_2O$	0.11
Na_2SO_4	3.90
$Ni(NO_3)_2 \cdot 6H_2O$	0.31
$Ca(NO_3)_2 \cdot 4H_2O$	0.44
$Mg(NO_3)_2 \cdot 6H_2O$	0.03

hydroceramic binder performs much better than a cement binder. Although strength was adequate, the leachabilities of the Portland cement solidified samples were higher than those of companion samples made with metakaolin.

Savannah River Simulant

The moles of nitrate+nitrite divided by the total moles of Na in the SBW is ~20%. For this reason two experiments were run and the results compared. In one case the hydroceramic was made using the solution itself without any pretreatment. Metakaolinite was mixed with varying amounts of SBW directly, cured at 90°C and 190°C in a steam saturated atmosphere (Teflon-lined Parr bombs) for 24 hours. Mix proportions, crystalline phases observed in the hydroceramic, Na leachability as calculated using a 1 day *unwashed* sample PCT test, and a calculated value for the % Na leached versus total Na content of the hydroceramic are given in Table IV (90°C cured) and Table V (190°C cured).

Table IV. Metakaolin mixed with different amounts of supernate, cured at 90°C for 24 hrs.

Sample Number	#1	#2	#3	#4	#5	#6
Metakaolin	1 g	1 g	1 g	1 g	1 g	1 g
SBW Simulant	0.5 ml	1.0 ml	1.5 ml	2.0 ml	3.0 ml	3.5 ml
Crystalline Phases	Mk	A	A	HS	HS	HS
Conductivity (mS/cm)	--	7.70	8.90	9.10	--	--
% Na leached	--	10.6	8.9	7.4	--	--

Mk=metakaolin, A=zeolite A, HS=hydroxysodalite, -- not analyzed

Table V. Metakaolin mixed with different amounts of supernate, cured at 190°C for 24 hrs.

Sample Number	#1	#2	#3	#4	#5
Metakaolin	1 g	1 g	1 g	1 g	1 g
SBW Simulant	0.5 ml	1.0 ml	1.5 ml	2.0 ml	3.0 ml
Crystalline Phases	Mk+Q	Mk+Q+HS	MK+Q+HS	Q+HS+A	Q+HS+A
Conductivity (mS/cm)	--	4.80	5.30	9.20	--
% Na leached	--	6.6	5.3	7.5	--

Mk= metakaolin, Q=quartz, HS=hydroxysodalite, A=zeolite A, -- not analyzed

PCT procedures were modified as follows: The sample was ground to a powder in an agate mortar and sieved using piggy-backed 100 mesh and 200 mesh screens (75-149 micron size). One gram of the *unwashed* sized sample was placed in a Teflon lined Parr bomb with 10 cc deionized water at 90°C for 1 day. The solution was filtered and its electrical conductivity was determined using a conductivity probe and a Quickcheck Model 118 Conductivity-2 meter (Orion). Conductivity rather than a chemical analysis was used to screen the samples. It has been demonstrated that the sodium ion dominates the leachate in these tests. It is so abundant in the SBW and in the zeolite and feldspathoid phases present in the hydroceramic that it simply overwhelms simply the solution with sodium ions. A plot of NaOH versus conductivity is a reasonably accurate means of correlating conductivity to concentration of Na in solution. Even better however is the plot reproduced here in Figure 1. These numbers were obtained using PCT chemical analyses and comparing them with conductivity tests throughout the tenure of these

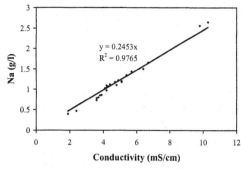

Figure 1. Correlation of conductivity versus g/l determined analytically

experiments. For sake of discussion, a 7 day PCT of EA glass tested out at 6.35 mS/cm which is equivalent to ~11.9 % total Na leached[1].

In the second case the SBW was first calcined with sucrose, metakaolinite at 525°C overnight. Sugar was chosen as a reducing agent and 3.8 g sugar was added to 100 mL SBW simulant at the ratio of 38 g/mol NO_3+NO_2. In order to reduce the formation of sodium carbonate during calcination 79.3 and 113.3 g metakaolin

were added to the SBW at the mole ratio of metakaolin : Na = 0.7 : 1 and 1 : 1. These have been found to be the ratios that most favor the production of tectosilicate phases in the calcine. After slurrying the dry ingredients with the SBW, the slurry was dried at 90°C and then fired at 525°C for 10 hours. The resulting powders are referred to as calcines. Weight lost during firing, crystalline phase formation, conductivity measurements performed using a Modified PCT leaching test and % Na leached for the two calcines studied are given in Table VI.

Table VI. Leachability of the Savannah River SBW calcine

Sample Number	Calcine #1 Mk:Na=0.7:1	Calcine #2 Mk:Na=1.0:1
Weight lost during firing	11.6%	9.2%
Crystalline phase	Amorphous + Quartz	Amorphous + Quartz
Conductivity (mS/cm)	7.10	5.10
% Na leached	9.6	9.0

Monolith hydroceramics were prepared by mixing 12 g calcine and 8 g metakaolin with 18 mL 4 M NaOH to form a paste. The pastes were molded in 2 by 1 inch diameter cylinder molds and precured at 40°C and 100% humidity overnight. They were then demolded, and cured at 90° or 190°C in a steam saturated atmosphere (Teflon-lined Parr bombs) for 24 hours. The samples were characterized and tested as before. See Tables VII and VIII.

Table VII. Hydroceramic waste forms cured at 90°C for 24 hours

Sample	Monolith #1 Mk:Na=0.7:1	Monolith #2 Mk:Na=1.0:1
Crystalline phase	Zeolite A	Zeolite A
density	1.06	1.07
Compressive strength (mPa)	3.21	3.07
Conductivity (mS/cm)	3.20	3.00
% Na leached	3.9	4.0

---

[1] EA Glass contains 16.9 % Na_2O. A conductivity of 6.35 is approximately equivalent to 1.5 g/l Na. Doing the math and converting g Na to moles Na and then to moles Na_2O and then to grams in 10 cc and then dividing that number by 0.169 g Na_2O/g EA Glass one gets ~11.9% leached.

Table VIII. Hydroceramic waste forms cured at 190°C for 24 hours

Sample	Monolith #1 Mk:Na=0.7:1	Monolith #2 Mk:Na=1.0:1
Crystalline phase	Analcime + HS	Zeolite A + HS
density	1.05	1.06
Compressive strength (mPa)	3.51	3.42
Conductivity (mS/cm)	2.10	1.10
% Na leached	2.6	1.5

The percentages of Na leached from the monolithic solids are the lowest of the tested samples, lower than the directly solidified waste. Strengths are also adequate to meet NRC shipping requirements (>500 psi). The data suggest that a two step process will produce a much better waste form, but the fact is that the single step process may in fact be sufficient to do the job in its own right. Waste specific testing will have to be carried out in order to show that this is true. If it is not, a two step process could be used instead.

Hanford Simulant

In this set of experiments an optimized Hanford calcine was solidified using both a hydroceramic binder and a conventional Portland cement grout. The calcine was prepared from the Table 3 simulant by mixing it with sucrose and metakaolinite at the weight ratio 20:2.5:11.9 which produced a slurry with the consistency of mud. After drying at 90°C for 24 hours, the dried material was calcined at 525°C for 18 hours. The calcine was granular and the X-ray diffraction pattern suggested that it was mainly X-ray amorphous with an amorphous hump centered at a location suggesting that it contains network structured phases. A chemical analysis of this type of calcine suggests that its nitrate and nitrite content are quite low (< 1 wt%) but carbonate still weighs in at ~8 wt%[4]. As an aside, it is of interest to mention at this point that the calcine in question contains almost the identical raw materials used by Westinghouse in its SBW reforming trials run at 725°C in a Studsvik steam reformer[10]. Other differences related to temperature are improved crystallinity and lower leachability but in principle the end product is the same as the clay calcined SBW we have been studying for 6 years[2-9]. It seems that the DOE has recognized the merits of making hydroceramic waste forms from SBW in part due to the current experimental program based at Penn State and in part based upon past demonstrations of steam reforming technology both off site[10] and currently onsite at INEEL and Savannah River.

Should either technology be adopted, the end product will eventually have to be solidified and transported to a repository of some sort-be it Yucca Mountain or some other site. In order to make the waste "road ready" it will have to be solidified. Here we report the results of an experiment in which the Hanford calcine described above was solidified using metakaolinite and a variety of solutions ranging from water to 10M NaOH solutions.

Table IX gives the mechanical and phase properties of a set of hydroceramics made from the Hanford calcine. Different amounts of various molar NaOH solutions were used to make the pastes that were then cured at 90° and 190°C. It is notable that strength development is adequate for all of the NaOH containing samples and phases that developed include zeolite A and hydroxysodalite. Table 10 provides data on leachability. The sample made with 4M NaOH had the lowest Na loss as percentage Na in the sample.

Table IX. Crystalline phases present and compressive strengths of hydroceramics made from Hanford calcine, metakaolin and various solutions that were cured at 90° and 190°C for 24 hrs.

5g Solids Mixed with 4.5 mL of Solution below	Total Mol Si/Na	Total Mol Al/Na	90°C Cured			190°C Cured		
			Crystalline Phases Present	Green Strength (MPa)	Cured Strength (MPa)	Crystalline Phases Present	Green Strength (MPa)	Cured Strength (MPa)
H_2O	2.689	2.103	Mk	0.21	0.21	Mk	0.21	0.21
2M NaOH	1.689	1.321	A	0.27	2.37	A	0.72	4.20
4M NaOH	1.231	0.963	A	2.13	3.55	A	2.32	4.00
6M NaOH	0.969	0.758	A+HS	5.27	6.34	A+HS	4.80	4.56
8M NaOH	0.799	0.624	HS	2.54	6.13	HS	5.45	4.48
10MNaOH	0.679	0.531	HS	1.45	3.96	HS	4.56	4.20

Mk= metakaolin, A=zeolite A, HS=hydroxysodalite

Table X. Continuation of Table IX-Leachability of Hydroceramics made with metakaolin binder.

Total Mol of Na in PCT Samples	90°C Cured				190°C Cured			
	Leachability (mS/cm)	Na Leached (mol $\cdot$ 10^4)	% Na Leached	pH of Leachate	Leachability (mS/cm)	Na Leached (mol $\cdot$ 10^4)	% Na Leached	pH of Leachate
0.00304	3.40	1.70	5.5	10.6	3.60	1.80	6.0	10.8
0.00484	4.80	2.40	5.0	10.2	3.20	1.60	3.5	10.4
0.00664	4.80	2.40	3.5	10.7	2.00	1.00	1.5	10.7
0.00844	7.00	3.50	4.0	10.6	4.90	2.45	3.0	10.8
0.01024	11.40	5.70	5.5	10.7	15.30	7.65	7.5	12.6
0.01204	>100	--	--	13.0	>100	--	--	13.0

-- = not determined

Tables XI and XII contain data for the same Hanford calcine but this time it was solidified with a Portland cement binder. Strengths are similar but phase formation and leachability are dramatically different. The percentage Na leached is at best roughly 10 times as high as the hydroceramic.

Table XI. Crystalline phases present and compressive strength of waste forms made from Hanford calcine, Portland cement and various solutions that were cured at 90° and 190°C for 24 hrs.

5g Solids Mixed with 4.5 mL of Solution below	Mol Si/Na	Mol Al/Na	90°C cured			190°C cured		
			Crystalline Phases Present	Green Strength (MPa)	Cured Strength (MPa)	Crystalline phases Present	Green Strength (MPa)	Cured Strength (MPa)
H_2O	1.877	1.291	C_3S^*	0.29	0.03	C_3S	0.34	-
2M NaOH	1.180	0.812	A+C_3S	3.82	4.09	HS+ C_3S	3.94	2.58
4M NaOH	0.861	0.592	HS+C_3S	3.05	3.52	HS+C_3S	2.98	1.61
6M NaOH	0.677	0.466	HS+ C_3S	1.61	1.00	HS+ C_3S	2.43	1.38
8M NaOH	0.558	0.384	HS+C_3S	2.14	1.25	HS+C_3S	2.05	1.00
10MNaOH	0.475	0.327	HS+C_3S	1.47	0.89			

*C_3S is an abbreviation for the anhydrous calcium silicate phase that comprises ~60 wt% of Portland cement (Ca_3SiO_5), A=zeolite A, HS=hydroxysodalite.

Table XII. Continuation of Table XI-Leachability of Waste Forms made with Portland Cement Binder.

Total mol of Na in PCT samples	90°C cured				190°C cured			
	Leachability (mS/cm)	Na Leached (mol • 10⁴)	% Na leached	pH of Leachate	Leachability (mS/cm)	Na Leached (mol • 10⁴)	% Na Leached	pH of Leachate
0.00305	8.40	4.20	14.0	12.3	6.6	3.30	11.0	12.2
0.00485	15.50	7.75	16.0	12.8	11.0	5.50	11.5	12.8
0.00665	>100	--	--	12.9	>100	--	--	12.8
0.00845	>100	--	--	13.2	>100	--	--	13.1
0.01025	>100	--	--	13.2	>100	--	--	13.2
0.01205	>100	--	--	13.2	>100	--	--	--

-- = not determined

DISCUSSION

The leachability of the directly solidified Savannah River Type II SBW and its calcined equivalent had approximately the same leachability (Tables IV-VI). The amount of sodium lost by each sample in terms of the total amount of sodium present in the samples suggests that approximately 5-10% of the sodium present is leached in 1 day. However, it is also concluded that the calcine made from the SBW mixed with sucrose and metakaolinite and fired at 525°C overnight did not perform as well. However, once the calcine was solidified with additional metakaolinite and concentrated NaOH as described above leachability improved significantly (Tables VII and VII). However, the reduction in sodium leachability comes with a huge amount of extra cost associated with the thermal processing necessary to make the calcine or steam reformed product. Thus a Type II SBW is better accommodated by direct solidification.

Portland cement is commonly used to solidify low level wastes of all kinds. Savannah River uses OPC blended with blast furnace slag and fly ash to solidify its Cs- and Sr-free nitrate/nitrite bearing SBW to produce a product they call Saltstone. British Nuclear Fuel Limited (BNFL) has used OPC based grouts to solidify Great Britain's "historic" reprocessing waste. Because OPC is widely used for solidification, conceivably OPC might be considered at some future time as a binder for pretreated and calcined/reformed Hanford SBW. Given this possibility, the current study was undertaken in order to determine how a sample of OPC solidified Hanford calcine would perform relative to an equivalent hydroceramic sample. The results show that OPC can be used, but its leachability could be many times higher than a similarly solidified metakaolin/NaOH sample. The zeolitic phases in the calcine acted like pozzolans and reacted with the Ca(OH)$_2$ in the Portland cement binder forming additional calcium silicate hydrate (C-S-H). Typically C-S-H is unable to host large amounts of sodium ions in its structure, thus a majority of the sodium present in the zeolites became concentrated in the pore solution present in the Portland cement binder and readily entered the leachant during PCT testing. In this instance metakaolin mixed with NaOH proved to be a superior binder for solidification purposes.

CONCLUSIONS

The "sodalite" rule of thumb has been suggested as a guide to formulating a hydroceramic waste form[2-9]. The minimum in conductivity exhibited by the hydroceramic sample made with 4M NaOH mixing solution had an overall 1:1:0.8 Na:Al:Si molar ratio and thus came closest to sodalite's 1:1:1 molar ratio, apparently confirming the validity of the rule. Based on performance, hydroceramics made using metakaolin additions to SBW prior to calcination and water as a binder phase are worthy of consideration as a contingency waste form for vitrification of low activity SBW in storage at Hanford, Savannah River and INEEL. All of the values for % Na lost by the samples fell below the 11.9 % value for DOE's EA Glass making all of them at least in principle adequate waste forms. Based on selectivity of zeolite A for Cs and Sr, it seems possible that hydroceramic waste forms can be prepared from SBW without removing Cs and Sr.

REFERENCES

[1]T.B. Edwards, D.K. Peeler, I.A. Reamer, G.F. Piepel, J.D. Vienna, H. Li, Phase 2B Experimental Design for the INEEL Glass Variation Study," WSRC-TR-99-00224, Westinghouse Savannah River Company, Aiken, South Carolina 1999.

[2] M.W. Grutzeck, D.D. Siemer, "Zeolites Synthesized from Class F Fly Ash and Sodium Aluminate Slurry", Journal American Ceramic Society 80, 2449-2453 (1997).

[3]D.D. Siemer, M.W. Grutzeck, D.M. Roy, B.E. Scheetz, "Zeolite Waste Forms Synthesized from Sodium Bearing Waste and Metakaolinite", WM98, Tucson AZ, March 1-5, 1998.

[4]D.D Siemer, M.W. Grutzeck, B.E. Scheetz, "Comparison of Materials for Making Hydroceramic Waste Forms," pp. 161-167 in Environmental Issues and Waste Management Technologies in the Ceramic and Nuclear Industries V, Edited by G.T. Chandler and X. Feng, American Ceramic Society, Westerville, 2000.

[5]N. Krishnamurthy, M.W. Grutzeck, S. Kwan and D.D. Siemer, "Hydroceramics for Savannah River Laboratory's Sodium Bearing Waste," pp. 337-344 in Environmental Issues and Waste Management Technologies in the Ceramic and Nuclear Industries VI, Edited by D.R. Spearing, G.L. Smith and R.L. Putnam, American Ceramic Society, Westerville, 2001.

[6]D.D. Siemer, J. Olanrewaju, B.E. Scheetz, N. Krishnamurthy and M.W. Grutzeck, "Development of Hydroceramic Waste Forms," pp. 383-390 in Environmental Issues and Waste Management Technologies in the Ceramic and Nuclear Industries VI, Edited by D.R. Spearing, G.L. Smith and R.L. Putnam, American Ceramic Society, Westerville, 2001.

[7]D.D. Siemer, J. Olanrewaju, B.E. Scheetz, and M.W. Grutzeck, "Development of Hydroceramic Waste Forms for INEEL Calcined Waste," pp. 391-398 in Environmental Issues and Waste Management Technologies in the Ceramic and Nuclear Industries VI, Edited by D.R. Spearing, G.L. Smith and R.L. Putnam, American Ceramic Society, Westerville, 2001.

[8]D.D. Siemer, Performance of Hydroceramic Concretes on Radwaste Leach Tests," pp. 369-397 in Environmental Issues and Waste Management Technologies in the Ceramic and Nuclear Industries VII, Edited by D.R. Spearing, G.L. Smith and S.K. Sundaram, American Ceramic Society, Westerville, 2001.

[9]Y. Bao, S. Kwan, D.D. Siemer and M.W. Grutzeck, "Binders for Radioactive Waste Forms made from Pretreated Calcined Sodium Bearing Waste", Journal of Materials Science, 39 481-488 (2004).

[10]C.M. Jantzen, "Engineering Study of the Hanford Low Activity Waste (LAW) Steam Reforming Process," WSRC-TR-2002-00317 and SRT-RPP-2002-00163, Westinghouse Savannah River Company, Aiken, South Carolina 2002

[11]A.R. Brough, A. Katz, T. Bakharev, G-K Sun, R.J. Kirkpatric, L.J. Struble and J.F. Young, "Microstructural Aspects of Zeolite Formation in Alkali Activated Cements Containing High Levels of Fly Ash"; pp. 199-208 in Microstructure of Cement-Based Systems/Bonding and Interfaces in Cementitious Materials, Edited by S. Diamond, S. Mindess, F.P. Glasser, L.W. Roberts, J.P. Skalny, L.D. Wakeley, Materials Research Society, Pittsburgh, 1995

ACKNOWLEDGEMENT:

The support of DOE EMSP Grant DE-FG07-98ER45726 is gratefully acknowledged.

GROUT FORMULATIONS FOR CLOSING HANFORD HIGH-LEVEL WASTE TANKS – BENCH-SCALE STUDY

T.H. Lorier, D.H. Miller, J.R. Harbour, and C.A. Langton
Westinghouse Savannah River Company
Savannah River National Laboratory
Aiken, SC 29808

W.L. Mhyre
Washington Quality Inspection Company
Savannah River National Laboratory
Aiken, SC 29808

ABSTRACT

Hanford has 149 single-shell HLW tanks that were constructed during the period from 1943 to 1964. Many of these tanks have leaked and so a major effort is proceeding to transfer the liquid portion of the waste to 28 newer, double-shell tanks and "close" the single-shell tanks. To initiate and accelerate this closure process, the Accelerated Tank Closure Demonstration (ATCD) Project was implemented. This Project focuses only on the retrieval and subsequent filling with grout of HLW Tank 241-C-106 in the 200 East Area of the Hanford Site. The Savannah River National Laboratory (SRNL) was tasked with developing the grout formulations for the three layers that will be placed in the tanks. In addition to designing fill materials for Tank C-106, the task included identifying grouts for the next six tanks scheduled to be closed: C-201, C-202, C-203, C-204, S-102, and S-112. Phase I fill is a high-strength grout that will stabilize the residual waste in the tanks. Mix HRG4 was selected for this layer, referred to as the "stabilization layer," based on its 28-day compressive strength of ~2200 psi. The Phase II fill (>80% of the grout fill in the tank) is a grout that provides structural stability to the tank system and prevents subsidence. Mix HRG2 was selected for this "structural layer" based on its relatively low compressive strength of ~1150 psi after 28 days of curing. A final Phase III fill ("capping layer") is a grout of very high strength to provide protection against intrusion. Mix HRG9, which is an interpolated composition chosen to provide a projected 28-day compressive strength of 2400 psi, was selected for this capping layer.

INTRODUCTION

The Accelerated Tank Closure Demonstration (ATCD) project is an interim action that was initiated to accelerate the closure of single-shell High-Level Waste (HLW) tanks at Hanford [1]. Recommendations from the Environmental Top-to-Bottom Review [2] and from ideas developed by the Hanford Cleanup, Constraints, and Challenges Team (C3T) [3] were used to develop the ATCD strategy. The remaining single-shell tanks will be closed following the Record of Decision (ROD) on the Environmental Impact Statement (EIS) for the Retrieval, Treatment, and Disposal of Tank Waste and Closure of Single-Shell Tanks at the Hanford Site [4].

The ATCD project is focused on only closing Tank 241- C-106, which is located in the 200 East Area of the Hanford Site. As a first step, the project will "identify the technical and regulatory framework under which tank closures are conducted" [3]. As currently defined, the tasks accomplished under the ATCD will be component closures of Tank C106 and will not constitute final closure of the landfill.

The closure process includes waste retrieval, heel stabilization, and subsidence abatement by filling the empty tanks with physically stable inert material. Portland cement-based grouts were selected for filling these HLW tanks and for stabilizing residual radionuclide contamination (Tc-99) in the tanks after waste retrieval.

The strategy for filling the tanks with grout involves placement of three separate grout fill layers in the tank after the waste is retrieved and DOE and Washington Department of Ecology approval/concurrence is obtained. The grout fill materials are described as follows in the Environmental Assessment (EA) for the ATCD Project [3]:

- Phase I fill shall be a high-strength grout that will stabilize the residual waste in the tank. This is referred to as the stabilization layer.
- The Phase II layer will be a grout layer that provides structural stability to the tank system and prevents subsidence. This layer is referred to as the structural layer.
- A final Phase III layer will consist of a very high strength grout as protection against intrusion. This layer is referred to as the capping layer.

CH2M HILL tasked SRNL to develop the three grout layers for the ATCD project for Tank C-106 [5]. The Scope of Work (SOW) [5] expanded the grout development to include Tanks 241-C-201, 241-C-202, 241-C-203, 241-C-204, 241-S-102, and 241-S-112. These additional 6 tanks will be the next six tanks to be closed following the Record of Decision for the Tank Closure EIS.

The work included in this report was performed according to the Bench-Scale Task Technical and Quality Assurance Plan [6], which was approved by CH2M HILL. These results will be used for identifying grout mix designs that will be used in the scale-up testing and ultimately recommended for Tank C-106 closure.

Objectives

The objectives of the bench-scale testing were to leverage Savannah River Site experience in closing HLW tanks to:

- Compile and review the requirements for the Hanford tank closure fill(s) with respect to material properties.
- Identify test methods for evaluating grout properties.
- Develop formulations for the Phase 1 Stabilization Grout, the Phase 2 Structural Grout, and the Phase 3 Capping Grout that meet the placement (fresh property) and cured material requirements identified for the Hanford fill materials.

Approach

A directive of the SOW was to leverage the knowledge gained through grout development and closure of the Savannah River Site Tanks 17-F and 20-F in 1997 and 1998 [7]. In addition, the SOW required that to the extent practical, the Hanford tank closure grouts be developed with materials available at the Hanford site.

EXPERIMENTAL METHODS

The Hanford grout trial mixes were prepared with materials obtained from suppliers in the Hanford area except for the admixtures, which are distributed nationwide, and slag. The slag is manufactured by Holcim, Inc., Birmingham, AL and supplied to the Augusta, GA area by a distributor in Duluth, GA. SRS process water was used as the mixing water in the laboratory samples. The ingredients were concrete sand, portland cement (Type II), fly ash (Class F), slag (Grade 100), ADVA™ Flow, Kelco-Crete®, and Methocel™. Sand is the major component in the Hanford tank closure grouts. Consequently, concrete sand obtained from the Hanford area and delivered to SRS was used in all mixtures. The absorption (amount of moisture associated with the sand) was determined in accordance with ASTM C128. The particle-size distribution was determined according to ASTM C136.

Test Methods

The bench-scale testing was performed in the SRS Civil Engineering Test Laboratory, which is operated by Washington Quality Inspection Company. Samples were prepared according to ASTM C192 and cured in a constant temperature (73°F) curing room at 100 % relative humidity. A one cubic foot Hobart mixer equipped with a planetary paddle mixer was used to prepare all test mixes (see Figure 1). The batch size was approximately 0.5 cubic feet.

The laboratory samples were prepared by placing all of the sand in the mixing bowl. The mixer was turned on at low speed and a portion of the mixing water, followed by the all of the fly ash and slag were added. If deemed necessary, more water was added followed by the cement. Water was added as needed up to the predetermined amount. The admixture(s) were added last. The final grout was mixed for about 3 minutes at low speed.

Figure 1. Hobart mixer used to mix Hanford tank closure grouts. Binder ingredients for a 0.5 cubic feet batch are shown on the left. The mixer paddle is illustrated in one of the self-leveling mixes.

A description of each parameter measured is provided below:

Flow: Ability of the grout to spread evenly without vibration (self-level).
Set time: Time after mixing at which the grout responds as a solid.
Bleed Water: Water that separates from the grout as the result of solids settling.
Segregation: Separation of sand from binder as the result of impact and separation of water from

grout as the result of gravity settling of the solids from the grout slurry.

Air content: Amount of air incorporated into the grout as the result of mixing and placement.

Unit weight: Weight of a unit volume, typically one cubic foot.

Yield: Actual volume produced relative to the volume calculated from the ingredients.

Compressive strength: Force per unit area required to break an unconfined grout sample.

Young's modulus: Measurement of the grout elasticity.

Poisson's ratio: Poisson's ratio is the ratio of lateral to axial strain.

Hydraulic conductivity: Velocity of water flow through saturated grout.

Porosity: Bulk porosity.

Shrinkage: Percent length change of grout samples cured at 73°F as a function of curing time in saturated and drying environments (50 ± 5 % relative humidity).

MIX PROPORTIONS

Grout and concrete mixes are typically proportioned on a volumetric basis with the objective of incorporating as much aggregate as possible per unit volume while still achieving fresh properties, i.e., placement properties and cured properties, typically compressive strength. Mixes with too much aggregate (insufficient binder or paste) are referred to as "harsh". Harsh mixes are difficult to mix and pump and will not flow. Increasing the water content results in segregation and bleed water rather than improving the processing and placement properties.

Mixes with too much binder are referred to as "fat." The extra paste in this type of mix makes the fresh grout "sticky" and therefore placement is impaired unless extra processing admixtures are used. In addition high-paste mixes are subject to excessive shrinkage and heat generation. The hydrated paste is subject to both chemical and drying shrinkage. The more paste, the more shrinkage. The hydration reactions of the binder/paste ingredients in water are exothermic. Consequently, the more paste, the less suitable a grout is for mass placements such as those planned for the Hanford tank closure.

The ratio of paste/binder to sand (by volume) was approximately the same as that identified as suitable for self-leveling grout in earlier studies at SRS. For the purpose of obtaining reproducible results in the laboratory, volumetric proportions were converted to weights, which have a higher degree of reproducibility on a small scale.

All of the trial mixes tested for tank fill applications contained processing admixtures. Mixes typically contained both a high range water reducer and a thickener/viscosifier. A Kelco-Crete®/ADVA[TM] Flow mixture (used in the SRS tank closure grouts) and combinations of Methocel[TM] and high-range water reducers (HRWRs) were evaluated.

The grout mix designs tested are listed in Table I with the corresponding results for fresh and cured properties.

Table I. Trial mixes tested – mix designs and results.

	HG1	HG2	HG3	HG4	HG5	HG6	HG7	HG8	HG9
Ingredients									
Portland cement (lbs/yd^3)	50	75	125	225	450	75	75	50	280
Slag (lbs/yd^3)	210	210	210	210	210	210	210	120	210
Fly ash (lbs/yd^3)	400	375	325	225	---	375	375	490	170
Sand (lbs/yd^3)	2530	2530	2530	2530	2530	2530	2530	2530	2530
Water (lbs/yd^3)	392	406	416	442	506	336	460	355	506
Water/binder ratio	0.59	0.62	0.63	0.67	0.77	0.51	0.70	0.54	0.77
Kelco-Crete® (grams/yd^3)	275	275	275	275	275	---	---	275	275
Methocel (grams/yd^3)	---	---	---	---	---	---	378	---	---
ADVA™ Flow (fl.oz/yd^3)	90	90	90	90	90	90	---	90	90
Fresh Properties									
Yield (wt/ft^3)	1	1.01	0.99	1.02	1	0.94	1.07	1.03	1.01
Unit weight	132	132.1	134	133.1	137.1	139.1	128.1	132.9	134
Flow (inches)	13	13.25	13.5	13.25	13	14.5	12.5	12.75	13.5
Bleed water	Min.	Min.	Min.	Min.	Min.	12 ml	7 ml	Min.	Min.
Set time (hours)	<24	<24	<24	<24	<24	<24	<24	<24	<24
Casting temp. (°F)	73	73	73	73	73	73	73	75	75
Cohesiveness	Good	Good	Good	Good	Good	Good	Good	Good	Good
Comments	N/A	N/A	N/A	N/A	N/A	No bulking	Extra bulking	N/A	N/A
Cured Properties									
Strength:									
14-day	490	740	770	770	1760	1460	430	230	770
14-day	490	750	770	770	1740	1430	450	230	790
28-day	710	1130	1560	2170	3130	1680	610	290	N/A
28-day	720	1170	1640	2200	3010	1880	650	310	N/A
90-day	1130	1650	2630	3510	5270	2740	970	430	N/A
90-day	1120	1650	2700	3690	5130	2550	940	440	N/A
Poisson's ratio (28-day)	N/A	N/A	N/A	0.5	0.41	0.42	N/A	N/A	N/A

RESULTS – DESIGN MIX SELECTION

Three design mixes, a Stabilizing grout, a Structural grout, and a Capping grout, were selected from the trial mix data generated for the HRG 1 to 5 series. All three design mixes contain slag to stabilize Tc-99, since in addition to sludge on the bottom of the tanks, the tank walls, dome roof, ancillary equipment, and associated debris may be slightly contaminated with Tc. All of the mixes contain the Kelco-Crete®/ADVA™ Flow admixture system since combination of admixtures gave the best processing properties.

The design mixes were based on the compressive strength data for the trial mixes. 28-day compressive strength data were used to select the design mixes even though the strengths continued to increase at least through 90 days. Based on SRS experience, confirmation of acceptable results as early as practical is desirable. In addition, field mixing and placement may not be as reliable as laboratory batching.

Stabilizing Grout

Mix HRG4 was selected as the design mix for the Stabilizing grout. This mix is a strong grout and contains 225 pounds (less than 2 ½ bags) of cement and 210 pounds of slag per cubic yard. Compressive strengths of 770, 2185 and 3600 psi were measured for this mix after curing for 14, 28, and 90 days, respectively. The Grade 100 slag not only provides stabilization for the Tc-99, it also significantly contributes to the strength gain as a function of curing time. The slag hydration products are also expected to contribute to relatively low hydraulic conductivity for this mix and thereby reduce water infiltration into the stabilized/solidified heel. The hydraulic conductivity will be measured in the next phase of this program in which properties that impact the Performance Assessment are measured for the design mixes. The ingredients and fresh and cured properties for Mix HRG4 are listed in Table II.

Structural Grout

Mix HRG2 was selected as the design mix for the Structural grout. This mix contains 75 pounds (less than one bag) of cement and 210 pounds of slag per cubic yard. Compressive strengths of 745, 1150, and 1650 psi were measured for this mix after 14, 28 and 90 days curing, respectively. Slag in the mix contributes significantly to the strength gain over this time period.

This mix was selected for the Structural Grout even though the compressive strength is much higher than the minimum required for support of the overburden. Although a hydraulic conductivity was not specified, lower values will result in less water infiltration than higher values. Maintaining a reducing environment in the tank (210 lbs. of slag per cubic yard) and reducing the hydraulic conductivity of the Structural grout (bulk of the tank) were factors in selecting a mix design that significantly exceeded the strength requirement. Hydraulic conductivity as well as heat generation as the result of hydration reactions will be measured in the next stage of this program. The ingredients and fresh and cured properties for Mix HRG2 are listed in Table II.

Capping Grout

The capping grout is intended to deter an inadvertent intruder and was derived from the HRG trial mix test data. The 28-day compressive strength data for the HRG trial mix series versus cement contents were plotted, and HRG, the mix corresponding to a cement content of 2400 psi, was selected as the Capping grout design mix. The design requirement was a compressive strength at 28 days of greater than 2000 psi.

Compressive strength data for the HRG series of trial mixes is plotted in Figure 2 as a function of cement content. Mix HRG9 contains 280 pounds (about three bags) of cement and 210 pounds of slag per cubic yard. This mix is projected to have a compressive strength of about 2400 after curing 28 days. This mix also contains 210 pounds of slag per cubic yard and will provide stabilization for any Tc-99 in residue on the tank dome, riser piping, and ancillary equipment. Hydration of the slag will contribute to the strength gain of this mix and the 90-day compressive strength is expected to approach 4000 psi.

Table II. Hanford tank closure grout design mix proportions and properties.

	HRG2 (Structural)	HRG4 (Stabilizing)	HRG9 (Capping)
Ingredients			
Portland cement (lbs/yd^3)	75	225	280
Slag (lbs/yd^3)	210	210	210
Fly ash (lbs/yd^3)	375	225	170
Sand (lbs/yd^3)	2530	2530	2530
Water (gal/yd^3)	48.7	53.0	60.6
Kelco-Crete® (g/yd^3)	275	275	275
ADVA™ Flow (fl.oz/yd^3)	90	90	90
Fresh Properties			
Yield	1.01	1.02	1.01
Unit weight (lbs/ft^3)	132.06	133.07	134.7
Flow (inches)	13.25	13.25	13.5
Bleed water	Trace/None	Trace/None	Trace/None
Set time (hours)	<24	<24	<24
Cohesiveness	Good	Good	Good
Cured Properties			
Compressive Strength (psi @ 14 days)	745	770	1000
Compressive Strength (psi @ 28 days)	1150	2185	2400 (projected)
Compressive Strength (psi @ 90 days)	1650	3600	4000 (projected)

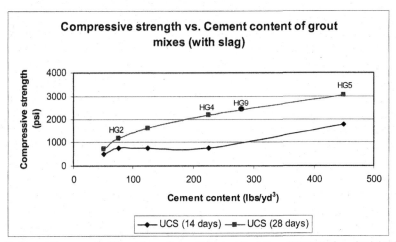

Figure 2. Plot of Compressive Strength vs. Cement Content for Mixes HRG1 Series of Trial Mixes, including HRG9.

CONCLUSIONS

Design mixes (grout formulations) have been selected for the stabilization, structural and capping layers as part of the overall strategy for closure of Hanford's single-shell HLW tanks. These design mixes were selected based on a series of trial mixes that were produced and tested during the bench-scale-testing phase of this work. The selection process was based on the following:

1. Blast furnace slag shall be present in each layer in order to increase the binding of Tc-99 within the grout waste form,

2. Kelco-Crete®/ADVA^{TM} Flow is the preferred processing admixture as revealed by experimental results, and

3. Compressive strengths of the cured grout layers were based on design requirements.

These design mixes (see Table II for ingredients and proportions for each mix) are:

- Stabilization layer. Mix HRG4 was selected for the Stabilization layer based on its 28-day compressive strength of ~220 psi. The selection of HRG4 is consistent with the ES for ATCD that calls for a high compressive strength stabilization layer.

- Structural layer. Mix HRG2 was selected for the Structural layer based on its relatively low compressive strength of ~1150 psi after 28 days of curing. This lower compressive strength grout also has a lower heat of hydration that is suited for more rapid pouring of this bulk layer (>80% of the grout fill in the tank).

- Capping layer. Mix HRG9 was selected for the Capping layer based on the requirement of a very high compressive strength for this layer. An interpolated composition was chosen to provide a projected 28-day compressive strength of 2400 psi. This 2400 psi target corresponds to 280 lbs/yd^3 of portland cement.

REFERENCES

[1] DOE/RL-2002-47, Rev. D, Performance Management Plan for the Accelerated Cleanup of the Hanford Site, U.S. Department of Energy, Richland Operations Office, Richland, Washington, 2002.

[2] A Review of the Environmental Management Program, Presented to the Assistant Secretary for Environmental Management by the Top-to-Bottom Review Team, U.S. Department of Energy, Washington, D.C., February 4, 2002.

[3] DOE/EA-1462, Rev. 0, *Environmental Assessment for The Accelerated Tank Closure Demonstration Project*, U.S. Department of Energy, Richland Operations Office, Richland, Washington, June 2003.

[4] "Notice of Intent to Prepare an Environmental Impact Statement for Retrieval, Treatment, and Disposal of Tank Waste and Closure of Single-Shell Tanks at the Hanford Site, Richland, WA", *68 FR 03318*, Federal Register, January 3, 2003.

[5] Interoffice Work Order (IWO) No. MOSRLE81 from Richland Operations Office, February 24, 2003.

[6] C. A. Langton, "ATCD Project Grout Design and Testing – Bench-Scale Task Technical and QA Plan," WSRC-RP-2003-00341, US DOE Report, Westinghouse Savannah River Company, Aiken, South Carolina, 2003.

[7] C. A. Langton, R. D. Spence, and J. Barton, "State of the Art Report on High-Level Waste Tank Closure", WSRC-TR-2001-00359, US DOE Report, Westinghouse Savannah River Company, Aiken, South Carolina, July, 2001.

CHEMICAL SOLUTION DEPOSITION OF $CaCu_3Ti_4O_{12}$ THIN FILMS

Mark D. Losego and Jon-Paul Maria
North Carolina State University
1001 Capability Drive, Raleigh, NC 27695

ABSTRACT

A chemical solution deposition method is used to prepare calcium copper titanate thin films on platinized silicon substrates. The impact of annealing temperature and stoichiometry on phase formation and dielectric properties is investigated. Through x-ray diffraction analysis, an intermediary phase, identified here as Cu_3TiO_4, is shown to emerge before the crystallization of $CaCu_3Ti_4O_{12}$ at 725°C. The temperature at which conversion occurred was mildly dependent on copper stoichiometry. Permittivities between 200 and 400 were observed for all cases; these values are significantly smaller than others reported recently. Field dependent measurements show a voltage variable permittivity which is linked to mobile charges in the CCT crystals and Schottky barriers at both electrode interfaces.

INTRODUCTION

For more than twenty-five years, calcium copper titanate (CCT) and its related isomorphic $(AB_3M_4O_{12})$ compounds have been known to exhibit a perovskite-like structure. However, because the MO_6 octahedra in this structure are tilted[1], the unit cell becomes a centrosymmetric supercell of 8 perovskite-like units. Unlike common ferroelectric perovskites like barium titanate and lead zirconate titanate, that are well known for their high dielectric permittivities, materials with the CCT structure are crystallographically forbidden from exhibiting ferroelectricity, and, until recently, dielectric measurements were not even reported.

In 2000, Subramanian et al. reported for the first time on the "giant" dielectric constant observable in CCT.[2] Since that discovery, dielectric constants of 80,000 for bulk single crystals[3] and 10,000 for bulk polycrystalline samples[2,3] have been reported at test frequencies of 10 kHz. Epitaxial CCT films grown by pulsed-laser deposition (PLD) have been reported to display relative permittivities of 10,000 at 1 MHz.[4] Further work revealed that although other $AB_3M_4O_{12}$ isomorphs do not show permittivities as high as CCT, they do exhibit dielectric constants that are 10 to 50 times higher than predicted by the Clausius-Mosotti equation.[5]

A complete explanation for this behavior has not yet been identified. Most current models suggest a large extrinsic contribution. For instance, an internal barrier layer mechanism was proposed by Sinclair, based on impedance spectroscopy analysis of polycrystalline bulk CCT pellets.[6,7] However, classical internal barrier layer capacitors rely on the dielectric contributions of non-conducting grain boundaries and as such are inconsistent with the results reported for single-crystal CCT. One proposed explanation is that twinning, which appears to be densely present in bulk samples, may act as the internal barrier.[5,8]

This paper will discuss a chemical solution deposition (CSD) method for preparing CCT thin films. To date, all CCT films reported in the open literature have been deposited using PLD and most of these have exhibitied epitaxial morphologies. Synthesizing films by CSD however, offers an important advantage for investigating the origin of giant permittivity in CCT: The stoichiometry can be controlled with the precision of bulk synthesis methods, while the thin layer geometry affords easy measurement of field dependent dielectric properties and electrical transport. Finally, CSD deposition represents a reasonable method to cost-effectively mass-produce commercial CCT thin films.

EXPERIMENTAL DETAILS

Methanol-based, chelated solutions were prepared using calcium acetate hydrate (Aldrich, 99.99% pure, <6 mol% H_2O), copper (II) acetate (Aldrich, 98% pure), and titanium (IV) isopropoxide (Alfa-Aesar, >98% pure) as cation sources. Solutions containing each metallic species were prepared separately and then combined by mass to maintain stoichiometric accuracy. During preparation solutions were maintained at 85°C and continually stirred at 300 rpm. Final CCT solutions had concentrations of about 0.055 M and appear to be stable for between 3 and 6 weeks at room temperature. To evaluate the effects of composition, off stoichiometric solutions were also prepared, including $Ca_{0.26}Cu_{0.74}TiO_4$ (Ca-rich); $Ca_{0.24}Cu_{0.76}TiO_4$ (Cu-rich); $Ca_{0.25}Cu_{0.75}Ti_{0.99}O_4$ (Ti-deficient); $Ca_{0.25}Cu_{0.75}Ti_{1.01}O_4$ (Ti-rich). Difficulties in ascertaining the exact amount of hydration in the calcium source may have led to solutions containing less calcium than expected. However, this error is believed to be less than 1 mol%, thereby maintaining the validity of compositional property trends.

Solutions were deposited through a 0.2 μm filter onto platinized silicon substrates. Films were spun at 3000 rpm for 30 s and then dried on a hotplate at 250°C for 5 min. Pyrolysis was carried out in an open air furnace at 400°C for 15 min. These steps were repeated to reach the desired thickness. Crystallization anneals were performed between 500°C and 800°C. Samples prepared for electrical characterization contained four deposition layers and were crystallized at 725°C and 750°C. Platinum top electrodes were dc magnetron sputtered on these crystallized samples at 30 mTorr Ar, 250 W for 30 s.

Crystallization behavior and phase development in these films was studied with a Bruker AXS D-5000 x-ray diffractometer equipped with a HighStar® area detector. Grain size analysis was performed using the lineal intercept procedure on images taken with a Digital Instruments Dimension 3100 atomic force microscope (AFM). AFM scans were taken over a 1 μm² area in tapping mode. Film thickness was ascertained from cross-sectional images taken with a Hitachi S-3200 scanning electron microscope. Capacitance and dielectric loss data was collected with an HP 4192A impedance analyzer. Leakage current information was obtained using a Keithley 617 Electrometer.

CRYSTALLIZATION BEHAVIOR

X-ray diffraction (XRD) scans demonstrating the phase behavior during crystallization of stoichiometric CCT films are shown in figure 1. The intensities of the CCT reflections shown in these scans are indicative of randomly oriented polycrystalline CCT. Furthermore, these scans clearly indicate the presence of a low temperature intermediary phase prior to the crystallization of CCT. The location of this peak matches well with the *101* reflection for Cu_3TiO_4 (PDF Card #27-199). Figure 2 summarizes the phase behavior data collected from XRD scans performed on non-stoichiometric CCT films. This data reveals that the intermediary phase has the shortest lifetime in the calcium rich sample. Since the calcium rich sample would lessen the amount of

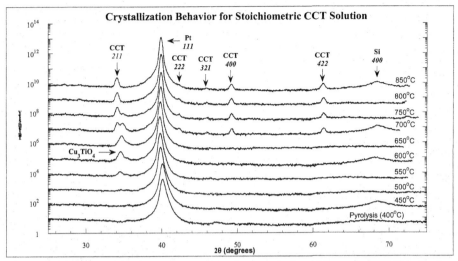

Figure 1: X-ray diffraction data collected for CSD deposited stoichiometric CCT films fired between 400°C and 850°C

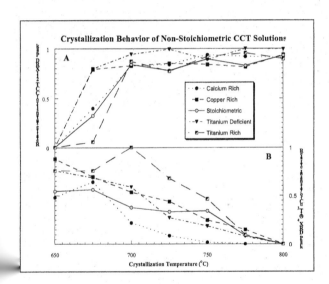

Figure 2: Relative XRD peak intensities for (A) the 211 CCT reflection and (B) the Cu_3TiO_4 in CCT films of varying composition crystallized over a range of temperatures

copper available for forming Cu_3TiO_4, this observation appears to be consistent with the proposed intermediary phase.

The electrical characterization data presented later in this paper was collected from samples crystallized at 725°C. At this temperature, all samples still contain some remaining intermediary phase, but this phase appears to have the highest volume fraction in the titanium rich and copper rich samples.

MICROSTRUCTURAL ANALYSIS

Scanning electron microscopy performed on cross-sections of CCT films (figure 3) reveals an isometric grain morphology. Grain size is determined with atomic force microscopy (see figure 4). The 0.3 μm thick (4 layer) sample has an average grain size of 67 nm and an RMS roughness of 4.4 nm while the 0.5 μm thick (7 layers) sample had an average grain size of 76 nm and an RMS roughness of 5.2 nm. These grain sizes are believed to be the smallest reported for CCT. As such, given the importance of microstructural features to determine the dielectric response of barrier layer capacitors, these small sizes should be pertinent in future interpretation.

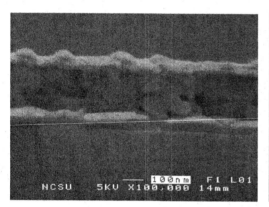

Figure 3: Cross-sectional image of CCT film taken with a scanning electron microscope.

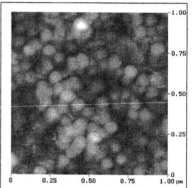

Figure 4: AFM image taken of the 0.3 μm thick CCT film surface.

ELECTRICAL RESULTS

The electrical data shown in Figures 5 and 6 is collected from CCT films crystallized at 725°C. Most of the samples processed in parallel with these films and then annealed at 750°C were extremely conductive with an impedance phase angle of approximately zero at 10 kHz. Similar results were observed in samples that were less than 4 layers thick (0.3 μm). Only the titanium-rich and titanium-deficient samples exhibited dielectric behavior within the examined frequency range. The loss tangent was consistently lower for these 750°C samples. The difference was approximately 25% at 10 kHz. An explanation for why most of these samples were conductive is ongoing and may indicate the necessity for a post-anneal oxidation treatment, which has been shown to lower conductivity in PLD deposited CCT films.[9]

As illustrated in figures 5A and 6A, the permittivity of these CSD derived CCT films is 4 to 10 times lower than values recently reported in the literature for PLD deposited films.[10,11,12] This could be due to several factors including the residual intermediary phase or the fine grained microstructure.

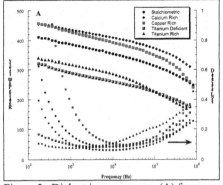
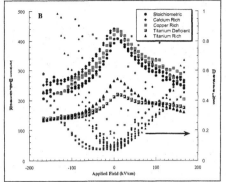

Figure 5: Dielectric response vs. (A) frequency and (B) applied electric field for CCT films with varying composition. Electric field dependent data was collected at 10 kHz.

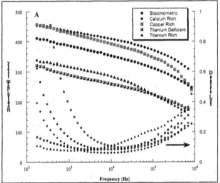
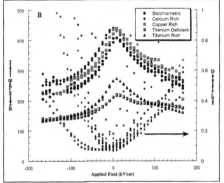

Figure 6: Dielectric response vs. (A) frequency and (B) applied electric field for CCT films of varying thickness. Electric field dependent data was collected at 10 kHz.

XRD data indicates that the 7 layer sample contains less residual intermediary phase than the 8 layer sample, thereby supporting the necessity of complete transformation for optimal dielectric response. Figure 5A also appears to indicate that maintaining the proper divalent to tetravalent cation ratio is necessary for achieving higher permittivities.

However, these films do exhibit tunability with applied bias as shown in figures 5B and 6B. Such a response has not been reported previously for CCT films. Compared to ferroelectric materials, where dielectric loss peaks near zero bias due to domain switching, the tan δ in these CCT films increases with increasing bias. This increase in conductivity with bias may result from the redistribution of charge carriers to a film/electrode interface, and a subsequent reduction in the Schottky barrier. This effect is also likely to be responsible for the apparent dielectric tuning.

Leakage current data is plotted in figure 7. Thicker films exhibit lower conductivity which is consistent with the hypothesis that most of the dielectric loss is due to mobile charge carriers.

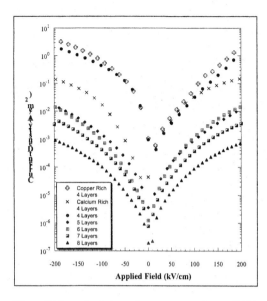

Figure 7: Leakage current data for CCT films of varying thickness and composition.

The addition of calcium also appears to lower DC losses; this result may reflect the fact that the calcium rich sample contained less of the intermediary phase.

CONCLUSIONS

Chemical solution deposition has proven to be a possible means for preparing calcium copper titanate thin films. Currently, however, the dielectric constant of these films is significantly lower than the ultra-high values reported by several authors. Since the ultra-high permittivities are believed to result from internal barriers, like crystallographic twins, we conclude that such boundaries are not present in the current CSD films. The reason for this absence is not understood, however, possibilities include very fine grain size and low process temperatures (as compared to bulk ceramic methods). We show that dielectric loss and leakage current can be improved with increased film thickness and possibly the addition of excess calcium, however, these changes likely correspond to the intrinsic permittivity of the CCT lattice and not the ultra-high permittivity associated with the barrier-layer mechanism.

This material is based upon work supported under a National Science Foundation Graduate Research Fellowship

REFERENCES

[1]B. Bochu et al. "Synthèse et Caractèrisation d'une Sèrie de Titanates Pèowskites Isotypes de [CaCu$_3$](Mn$_4$)O$_{12}$," J. Solid State Chem. **29** 291-298 (1979).
[2]M. A. Subramanian et al. "High Dielectric Constant in ACu$_3$Ti$_4$O$_{12}$ and ACu$_3$Ti$_3$FeO$_{12}$ Phases," J. Solid State Chem. **151** 323-325 (2000).

[3]C. C. Homes et al. "Optical Response of High-Dielectric-Constant Perovskite-Related Oxide," *Science* **293** 673-676 (2001).

[4]Y. Lin et al. "Epitaxial Growth of Dielectric $CaCu_3Ti_4O_{12}$ Thin Films on (001) $LaAlO_3$ by Pulsed Laser Deposition," *Appl. Phys. Let.* **81** [4] 631-633 (2002)

[5]M A. Subramanian and A. W. Sleight, "$ACu_3Ti_4O_{12}$ and $ACu_3Ru_4O_{12}$ Perovskites: High Dielectric Constants and Valence Degeneracy," *Solid State Sciences* **4** 347-351 (2002)

[6]D. C. Sinclair et al. "$CaCu_3Ti_4O_{12}$: One-Step Internal Barrier Layer Capacitor," *Appl. Phys. Let.* **80** [12] 2153-2155 (2002).

[7]T. B. Adams, D. C. Sinclair, and A. R. West, "Giant Barrier Layer Capacitance Effects in $CaCu_3Ti_4O_{12}$ Ceramics," *Adv. Mater.* **14** [18] 1321-1323 (2002).

[8]M. H. Cohen et al. "Extrinsic Models for the Dielectric Response of $CaCu_3Ti_4O_{12}$," *J. Appl. Phys.* **94** [5] 3299-3306 (2003).

[9]Tselev et al. "Intrinsic Dielectric Response of the Colossal Dielectric Constant Material $CaCu_3Ti_4O_{12}$," *Condensed Matter,* under review.

[10]L. Fang and M. Shen, "Deposition and Dielectric Properties of $CaCu_3Ti_4O_{12}$ Thin Films on $Pt/Ti/SiO_2/Si$ Substrates using Pulsed-Laser Deposition," *Thin Solid Films*, **440** 60-65 (2003).

[11]W. Si et al. "Epitaxial Thin Films on the Giant-Dielectric-Constant Material $CaCu_3Ti_4O_{12}$ Grown by Pulsed-Laser Deposition," *Appl. Phys. Let.* **81** [11] 2056-2058 (2002).

[12]Y. L. Zhao et al. "High Dielectric Constant in $CaCu_3Ti_4O_{12}$ Thin Films Prepared by Pulsed Laser Deposition," *Thin Solid Films* **445** [1] 7-13 (2003).

Author Index

Keyword Index

Polystyrene butadiene
   rubber emulsions, 99
Pu burning, 91
Pu-238 doped ceramic
   waste form, 191

Radioactive glass sample
   analysis, 153
Redox activity, 163
Rhenium, 163
River Protection Project
   Waste Treatment
   Management Plant,
   21

Salt supernate waste, 69
Savannah River Defense
   Waste Processing
   Facility, 21, 31, 51,
   69, 153, 233, 243

Silicate glasses, 163
Slurry-fed melt rate
   furnace, 51
Soda-lime-silicate glass
   melt, 163
Sodalite-based ceramic
   waste forms, 191
Sodium-bearing waste,
   81, 91, 243
Solidification, 233, 243
Spinel-forming
   constituents, 121
Steam reformation, 81
Sulfate solubility, 141
Synroc-B, 225
Synroc-C, 91, 225
Synroc-F, 225

Thermal treatment
   conditions, 109

Thermochemistry of the
   liquid/glass phase,
   121
Titanate ceramics, 91, 225
Titanate-bearing waste
   form products, 91
T_L models, 133
Transuranic waste stream,
   233
Trivalent rare earth ions,
   199

UO_2 fissile kernels, 109
U-rich liquid wastes, 225

Vitrification, 3, 21, 69

Waste loading, 31
WTP melters, 3

9 781574 98189